Pinch Analysis and Process Integration

To Dad and Sue

Pinch Analysis and Process Integration

A User Guide on Process Integration for the Efficient Use of Energy

Second edition

Ian C Kemp

The authors of the First Edition were: B. Linnhoff, D.W. Townsend, D. Boland, G.F. Hewitt, B.E.A. Thomas, A.R. Guy and R.H. Marsland
The IChemE Working Party was chaired by B.E.A. Thomas.

AMSTERDAM • BOSTON • HEIDELBERG • LONDON • NEW YORK • OXFORD
PARIS • SAN DIEGO • SAN FRANCISCO • SINGAPORE • SYDNEY • TOKYO
Butterworth-Heinemann is an imprint of Elsevier

Butterworth-Heinemann is an imprint of Elsevier
Linacre House, Jordan Hill, Oxford OX2 8DP, UK
30 Corporate Drive, Suite 400, Burlington, MA 01803, USA

First edition 1982
Revised First edition 1994
Second edition 2007

British Library Cataloguing in Publication Data
A catalogue record for this book is available from the British Library

Library of Congress Cataloging in Publication Data
A catalogue record for this book is available from the Library of Congress

ISBN 13: 978 0 75068 260 2

Transferred to Digital Printing in 2010

Contents

Foreword

The original User Guide was published more than 20 years ago and it is probably a case of . ... "from small acorns big oak trees grow".

Innovation is fascinating. John Lennon once said: "Reasonable people adapt to the world. Unreasonable people want the world to adapt to them. It follows that all innovation is due to unreasonable people."

I never thought of Ian Kemp as unreasonable but as a young engineer he did join up with those of us who innovated a (then) novel and unorthodox approach to energy management in process design. He became one of the most committed practitioners I remember meeting. It's fitting that it is Ian who showed the staying power to produce, 20 years on, this real labour of love, the second edition, with more than double the number of pages.

Detail, complexity and sheer volume are often a sign of maturity. As a technology develops, the books get longer. It's a common trend and often a thankless task. On behalf of many process design professionals I thank Ian for tackling this task.

Bodo Linnhoff
Berlin, 30th October 2006

Foreword to the first edition

Every now and then there emerges an approach to technology which is brilliant – in concept and in execution. Of course it turns out to be both simple and practical. Because of all these things it is a major contribution to the science and art of a profession and discipline.

Bodo Linnhoff and the other members of this team have made a major contribution to chemical engineering through their work. It is already recognised worldwide and I have personal experience of the acclaim the techniques embodied in this guide have received in the USA.

There is no need to underline the necessity for more efficient use of energy: the chemical industry is a very large consumer, as a fuel and as a feedstock. What is equally important is that conceptual thinking of a high order is necessary to our industry to keep advancing our technologies in order to reduce both capital and operating costs. The guide provides new tools to do this, which forces the sort of imaginative thinking that leads to major advances.

It is also important to note that the emphasis in the guide is on stimulating new concepts in process design which are easily and simply implemented with the aid of no more than a pocket calculator. In these days, when the teaching and practice of many applied sciences tend heavily toward mathematical theory and the need for sophisticated computer programs, a highly effective, simple tool which attains process design excellence is very timely.

R. Malpas
President and Chief Executive Officer
Halcon International Inc.

Preface

When the first edition of the User Guide on Process Integration appeared in 1982, it was instantly recognised as a classic for the elegance and simplicity of its concepts, and the clarity with which they were expressed. Instead of reams of equations or complex computer models, here were straightforward techniques giving fundamental new insights into the energy use of processes. Rigorous thermodynamically based targets enabled engineers to see clearly where and why their processes were wasting energy, and how to put them right. A key insight was the existence of a "pinch" temperature, which led to the term "pinch analysis" to describe the new methodology.

Since then pinch analysis has evolved and deepened in many ways, and can now be regarded as a mature technology. Much research has been performed, and many new techniques have been developed, but the original core concepts still largely hold good. The aim of this book is to follow in the footsteps of the original User Guide and to bring it up-to-date with the main advances made since then, allowing the techniques to be applied in almost any energy-consuming situation. It does not attempt to duplicate or replace the detailed research papers and texts on the subject that have appeared in the last 25 years, but makes reference to them as appropriate.

Chapter 1 sets the scene and Chapter 2 describes the key concepts – energy targeting, graphical representation through the composite and grand composite curves, and the idea of the pinch, showing how this is central to finding a heat exchanger network that will meet the targets. Hopefully, this will whet the reader's appetite for the more detailed discussion of targeting for energy, area and cost (Chapter 3) and network design and optimisation (Chapter 4). Chapter 5 describes the interaction with heat and power systems, including CHP, heat pumps and refrigeration, and the analysis of total sites. Beneficial changes to operating conditions can also be identified, as described in Chapter 6, especially for distillation, evaporation and other separation processes; while Chapter 7 describes application to batch processes, start-up and shutdown, and other time-dependent situations. Chapter 8 takes a closer look at applying the methodology in real industrial practice, including the vital but often neglected subject of stream data extraction.

Two case studies run like constant threads through the book, being used as appropriate to illustrate the various techniques in action. Five further complete case studies are covered in Chapter 9, and others are mentioned in the text.

It is a myth that pinch analysis is only applicable to large complex processes, such as oil refineries and bulk chemicals plants. Even where complex heat exchanger networks are unnecessary and inappropriate, pinch analysis techniques provide the key to understanding energy flows and ensuring the best possible design and operation. Thus, as will be seen in the text and in particular the case studies, it is relevant to smaller-scale chemicals processes, food and drink, consumer products, batch processing and even non-process situations such as buildings. Often, small and simple plants still reveal worthwhile savings, because nobody has really systematically looked for opportunities in the past. Reducing energy usage benefits the company (every pound, dollar or euro saved reduces direct costs and goes straight on the bottom line as increased profit) and the environment (both from reduced fossil fuel usage and lower emissions). And even if no major capital projects result, the engineer gains substantially in his understanding and "feel" for his plant. In several cases, a pinch study has led to improved operational methods giving a substantial saving – at zero cost.

One barrier to the more widespread adoption of pinch analysis has been a lack of affordable software. To remedy this, the Institution of Chemical Engineers ran a competition for young members to produce a spreadsheet for pinch analysis. The entrants showed a great deal of ingenuity and demonstrated conclusively that the key targeting calculations and graphs could be generated in this way, even without widespread use of programming techniques such as macros. Special congratulations are due to Gabriel Norwood, who produced the winning entry which is available free of charge with this book.

Nowadays, therefore, there is no reason why every plant should not have a pinch analysis as well as a heat and mass balance, a process flowsheet and a piping and instrumentation diagram. (That being said, it is salutary to see how many companies do not have an up-to-date, verified heat and mass balance; this is often one of the most valuable by-products of a pinch study!)

My hope is that this revision will prove to be a worthy successor to the original User Guide, and that it will inspire a new generation of engineers, scientists and technologists to apply the concepts in processes and situations far beyond the areas where it was originally used.

Ian C Kemp
Abingdon, Oxfordshire

Supporting material for this book is available online. To access this material please go to http://books.elsevier.com/companions/0750682604 and then follow the instructions on screen.

Acknowledgements

Much of the material in the User Guide has stood the test of time, and it is pointless to reinvent the wheel. A significant proportion of the text and figures in Chapters 1–5 and 9 of this book have been reproduced from the first edition, often verbatim. I am grateful to the IChemE and Professor Bodo Linnhoff for permission to use this material, which made the writing of this book a manageable task rather than an impossible one, and I am only too happy to acknowledge my debt to the original team of authors: B. Linnhoff, D.W. Townsend, D. Boland, G.F. Hewitt, B.E.A. Thomas, A.R. Guy and R.H. Marsland, plus the additional contributors J.R. Flower, J.C. Hill, J.A. Turner and D.A. Reay.

Many other people have had an influence on this book. I was fortunate enough to attend one of Bodo Linnhoff's early courses at UMIST and to be trained by several members of the pioneering ICI research and applications teams, particularly Jim Hill, Ajit Patel and Eric Hindmarsh. I am profoundly grateful to them, and also to my colleagues at Harwell, particularly Ewan Macdonald who gave me much valuable guidance as a young engineer. Of the many others who have influenced me over the years, I would particularly like to mention John Flower and Peter Heggs.

Robin Smith, Geoff Hewitt, Graham Polley and Alan Deakin worked with me when the idea of a second edition of the User Guide was first being mooted, and made significant contributions. I am also grateful to Audra Morgan and Caroline Smith at the IChemE, and to Jonathan Simpson and his colleagues at Elsevier, for their practical help in bringing this book to fruition after a long gestation period.

Last but not least, my thanks go to my wife Sue for her support and patience, especially when the adage that applies to many books and software projects was proven true again; the first 90% of the work takes 90% of the time: and the last 10% takes 90% of the time …

Figure acknowledgements

The author acknowledges with thanks the assistance given by the following companies and publishers in permitting the reproduction of illustrations from their publications:

Elsevier Ltd for Figure 3.20 from Linnhoff, B. and Ahmad, S. (1990). *Computers and Chemical Engineering*, vol. 7, p. 729 and Figure 5.19 from Klemes, J. *et al.* (1997), *Applied Thermal Engineering*, vol. 17, p. 993.
John Wiley and Sons for Figures 5.20, 5.21, 5.22, 5.24 and 6.16 from Smith, R. (2005). *Chemical Process Design and Integration.*
Johnson Hunt Ltd for Figure 4.4 and Table 4.2.
The Institution of Chemical Engineers (IChemE) for Figure 6.15, from Smith, R. and Linnhoff, B. (1988), TransIChemE Part A, vol. 66, p. 195.

And special thanks to the IChemE and Professor Bodo Linnhoff for permission to use many of the figures from the first edition.

1 Introduction

1.1 What is pinch analysis?

Figure 1.1(a) shows an outline flowsheet representing a traditional design for the front end of a specialty chemicals process. Six heat transfer "units" (i.e. heaters, coolers and exchangers) are used and the energy requirements are 1,722 kW for heating and 654 kW for cooling. Figure 1.1(b) shows an alternative design which was generated by Linnhoff *et al.* (1979) using pinch analysis techniques (then newly developed) for energy targeting and network integration. The alternative flowsheet uses only four heat transfer "units" and the utility heating load is reduced by about 40%

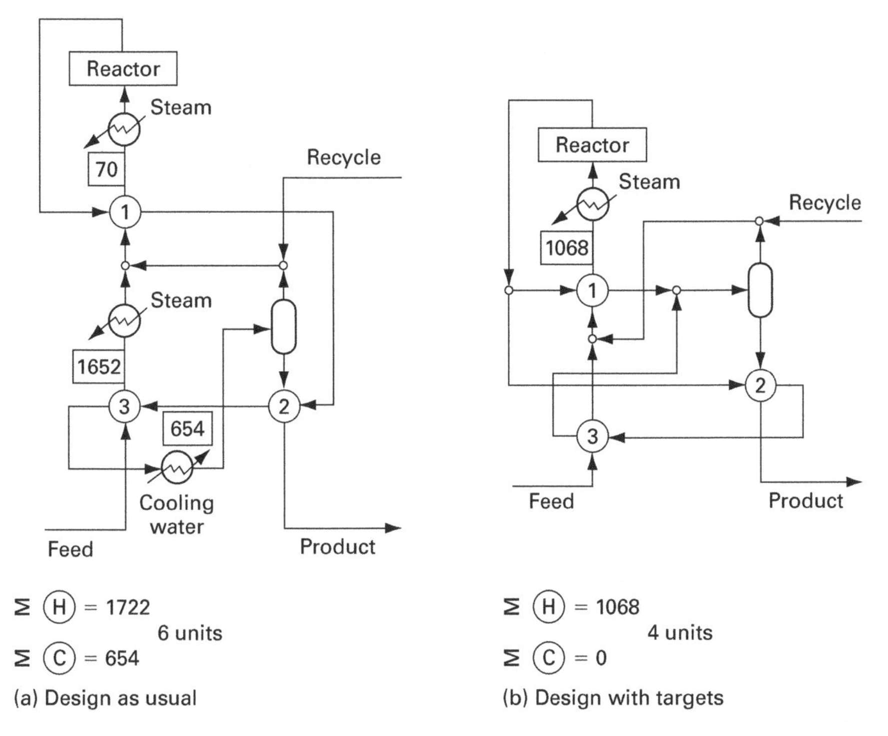

Figure 1.1 Outline flowsheets for the front end of a specialty chemicals process

with cooling no longer required. The design is as safe and as operable as the traditional one. It is simply better.

Results like this made pinch analysis a "hot topic" soon after it was introduced. Benefits were found from improving the integration of processes, often developing simpler, more elegant heat recovery networks, without requiring advanced unit operation technology.

There are two engineering design problems in chemical processes. The first is the problem of unit operation design and the second is the problem of designing total systems. This book addresses the system problem, in particular design of the process flowsheet to minimise energy consumption.

The first key concept of pinch analysis is **setting energy targets**. "Targets" for energy reduction have been a key part of energy monitoring schemes for many years. Typically, a reduction in plant energy consumption of 10% per year is demanded. However, like "productivity targets" in industry and management, this is an arbitrary figure. A 10% reduction may be very easy on a badly designed and operated plant where there are many opportunities for energy saving, and a much higher target would be appropriate. However, on a "good" plant, where continuous improvement has taken place over the years, a further 10% may be impossible to achieve. Ironically, however, it is the manager of the efficient plant rather than the inefficient one who could face censure for not meeting improvement targets!

Targets obtained by pinch analysis are different. They are absolute thermodynamic targets, showing what the process is inherently capable of achieving if the heat recovery, heating and cooling systems are correctly designed. In the case of the flowsheet in Figure 1.1, the targeting process shows that only 1,068 kW of external heating should be needed, and no external cooling at all. This gives the incentive to find a heat exchanger network which achieves these targets.

1.2 History and industrial experience

The next question is, are these targets achievable in real industrial practice, or are they confined to paper theoretical studies?

Pinch analysis techniques for integrated network design presented in this guide were originally developed from the 1970s onwards at the ETH Zurich and Leeds University (Linnhoff and Flower 1978; Linnhoff 1979). ICI plc took note of these promising techniques and set up research and applications teams to explore and develop them.

At the time, ICI faced a challenge on the crude distillation unit of an oil refinery. An expansion of 20% was required, but this gave a corresponding increase in energy demand. An extra heating furnace seemed the only answer, but not only was this very costly, there was no room for it on the plant. It would have to be sited on the other side of a busy main road and linked by pipe runs – an obvious operability problem and safety hazard. Literally at the 11th hour, the process integration teams were called in to see if they could provide an improved solution.

Within a short time, the team had calculated targets showing that the process could use much less energy – even with the expansion, the targets were lower than the current energy use! Moreover, they quickly produced practical designs for a heat exchanger network which would achieve this. As a result, a saving of over a million pounds per year was achieved on energy, and the capital cost of the new furnace with its associated problems was avoided. Although new heat exchangers were required, the capital expenditure was actually lower than for the original design, so that both capital and operating costs had been slashed! Full details of the project are given as the first of the case studies in Chapter 9 (Section 9.2).

It is hardly surprising that after this, ICI expanded the use of pinch analysis throughout the company, identifying many new projects on a wide variety of processes, from large-scale bulk chemical plants to modestly sized specialty units. Energy savings averaging 30% were identified on processes previously thought to be optimised (Linnhoff and Turner 1981). The close co-operation between research and application teams led to rapid development; new research findings were quickly tried out in practice, while new challenges encountered on real plant required novel analysis methods to be developed. Within a few years, further seminal papers describing many of the key techniques had been published (Linnhoff and Hindmarsh 1983; Linnhoff *et al.* 1983; Townsend and Linnhoff 1983). From this sprang further research, notably the establishment of first a Centre and then the world's first dedicated Department of Process Integration at UMIST, Manchester (now part of the new School of Chemical Engineering and Analytical Science at Manchester University).

The techniques were disseminated through various publications, including the first edition of this user guide (Linnhoff *et al.* 1982) and three ESDU Data Items (1987–1990), and through training courses at UMIST. Applications in industry also forged ahead; Union Carbide, USA, reported even better results than ICI, mainly due to progress in the understanding of how to effect process changes (Linnhoff and Vredeveld 1984). BASF, Germany, reported completing over 150 projects and achieving site-wide energy savings of over 25% in retrofits in their main factory in Ludwigshafen (Korner 1988). They also reported significant environmental improvements. There have been many papers over the years from both operating companies and contractors reporting on the breadth of the technology, on applications, and on results achieved. In all, projects have been reported in over 30 countries. Studies partially funded by the UK Government demonstrated that the techniques could be applied effectively in a wide range of industries on many different types of processes (Brown 1989); these are described further in Chapter 8. Pinch-type analysis has also been extended to situations beyond energy usage, notably to wastewater minimisation (Wang and Smith 1994, 1995; Smith 2005) and the "hydrogen pinch" (Alves 1999; Hallale and Liu 2001); these are extensive subjects in their own right and are not covered in this book.

Pinch analysis was somewhat controversial in its early years. Its use of simple concepts rather than complex mathematical methods, and the energy savings and design improvements reported from early studies, caused some incredulity. Moreover, pinch analysis was commercialised early in its development when there was little know-how from practical application, leading to several commercial failures. Divided opinions resulted; Morgan (1992) reported that pinch analysis significantly improves both the "process design and the design process", whereas Steinmeyer (1992) was concerned that pinch analysis might miss out on major opportunities for improvement. Nevertheless, the techniques have now been generally accepted (though more widely adopted in some countries than others), with widespread inclusion in undergraduate lecture courses, extensive academic research and practical application in industry. Pinch analysis has become a mature technology.

1.3 Why does pinch analysis work?

The sceptic may well ask; why should these methods have shown a step change over the many years of careful design and learning by generations of highly competent engineers? The reason is that, to achieve optimality in most cases, particular insights are needed which are neither intuitively obvious nor provided by common sense.

Let us simplify the question initially to producing a heat recovery arrangement which recovers as much heat as possible and minimises external heating and cooling (utilities). At first sight, in a problem comprising only four process streams, this may seem an easy task. The reader might therefore like to try solving a simplified example problem comprising four process streams (two hot and two cold) similar to the process example of Figure 1.1, the data for which are given in Table 1.1. Interchangers may not have a temperature difference between the hot and cold streams (ΔT_{min}) of less than 10°C. Steam which is sufficiently hot and cooling water which is sufficiently cold for any required heating and cooling duty is available. After trying this example, the reader will probably agree that it is not a trivial task. Admittedly it is relatively easy to produce some form of basic heat recovery system, but how do you know whether it is even remotely optimal? Do you continue looking for better solutions, and if so, how? However, if you know *before starting* what the energy targets are for this problem, and the expected minimum number of heat exchangers required, this provides a big stimulus to improving on first attempts. If you are then given key information on the most constrained point in the network, where you must start the design, this shows you how to achieve these targets. We will be returning to this example dataset in Chapter 2 and throughout the guide. The value of the pinch-based approach is shown by the fact that a plausible "common-sense" heat recovery system, developed in Chapter 2, falls more than 10% short of the feasible heat exchange and uses no less than two-and-a-half times the calculated hot utility target!

How does this relate to practical real-life situations? Imagine a large and complex process plant. Over the years, new ideas are thought of for ways to reduce energy.

Table 1.1 Data for four-stream example

Process stream number and type	*Heat capacity flowrate (kW/°C)*	*Initial (supply) temperature (°C)*	*Final (target) temperature (°C)*	*Stream heat load (kW) (positive for heat release)*
(1) cold	2.0	20	135	$2.0 \times (20 - 135) = -230$
(2) hot	3.0	170	60	$3.0 \times (170 - 60) = 330$
(3) cold	4.0	80	140	$4.0 \times (80 - 140) = -240$
(4) hot	1.5	150	30	$1.5 \times (150 - 30) = 180$

$\Delta T_{min} = 10°C$.

However, as "retrofitting" – changes to an existing plant – is more difficult and expensive than altering the design of a new plant; many of these ideas have to wait for implementation until a "second generation" plant is designed. Further experience then leads to further ideas, and over many years or decades, the successive designs are (hopefully!) each more energy-efficient than the last.

Boland and Linnhoff (1979) gave an example of this from one of the earliest pinch studies. Figure 1.2 shows the improvement in energy consumption which was achieved by successive designs for a given product. The successive designs lie on a "learning curve". However, calculation of energy targets as described later revealed suddenly that the ultimate performance, given correct integration, would lie quite a bit further down the "learning curve". This information acted as an enormous stimulus to the design team. Within a short period they produced a flowsheet virtually "hitting" the ultimate practical target.

Obviously, if a completely new process is being designed, pinch analysis allows one to hit the target with the first-generation plant, avoiding the learning curve completely.

Although improvement targets can be stated based on learning curves (e.g. aim for a 10% reduction in the next generation plant), we see that these are merely based on an extrapolation of the past, while pinch analysis sets targets based on an objective analysis.

1.4 The concept of process synthesis

"But pinch analysis is just about heat exchanger networks, isn't it?" That's a common response from people who've heard about the techniques in the past. Implicit in this is the question; isn't it only applicable to oil refineries and large bulk chemical plants, and maybe not to my process?

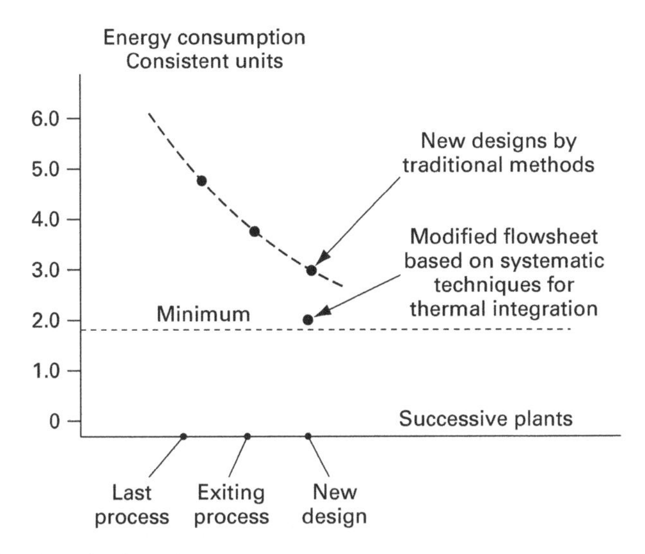

Figure 1.2 Beating the learning curve

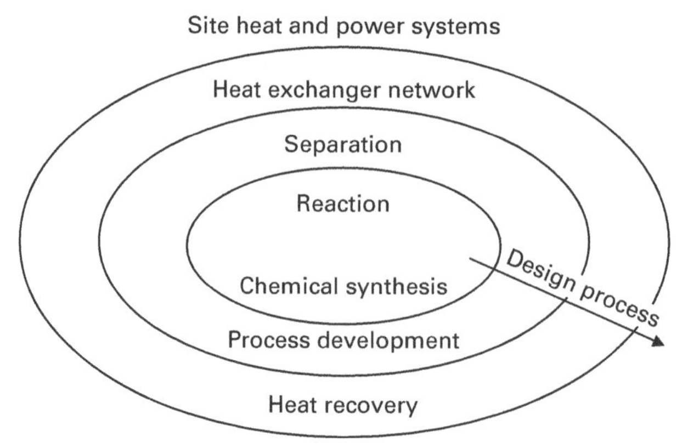

Figure 1.3 The onion diagram for process synthesis

In fact, experience has showed that pinch analysis can bring benefits in a huge range of plants and processes, large and small, both within and outside the "traditional" process industries. This is borne out by the applications and case studies described in Chapters 8 and 9. Improvements come not only from heat recovery projects, but also from changing process conditions, improved operability and more effective interfacing with utility systems, all underpinned by better process understanding. Pinch analysis has broadened a long way beyond the original studies. It is now an integral part of the overall strategy for process development and design, often known as **process synthesis**, and the optimisation of existing plants.

The overall design process is effectively represented by the **onion diagram**, Figure 1.3. Process synthesis is hierarchical in nature (Douglas 1988). The core of the process is the chemical reaction step, and the reactor product composition and feed requirements dictate the separation tasks (including recycles). Then, and only then, can the designer determine the various heating and cooling duties for the streams, the heat exchanger network and the requirements for heating and cooling. The design basically proceeds from the inside to the outside of the "onion".

Figure 1.4 shows a more detailed flowsheet for the front end of the specialty chemicals process which was shown in Figure 1.1. The four tasks in the layers of the onion are all being performed, namely reaction, separation, heat exchange and external heating/cooling.

The design of the reactor is dictated by yield and conversion considerations, and that of the separator by the need to flash off as much unreacted feed as possible. If the operating conditions of these units are accepted, then the design problem that remains is to get the optimum economic performance out of the *system* of heat exchangers, heaters and coolers. The design of the heat exchange system or "network" as it stands in Figure 1.4 may not be the best and so it is necessary to go back to the underlying data that define the problem.

The basic elements of the heat recovery problem are shown in Figure 1.5. All the exchangers, heaters and coolers have been stripped out of the flowsheet and what remains therefore is the definition of the various heating and cooling tasks. Thus

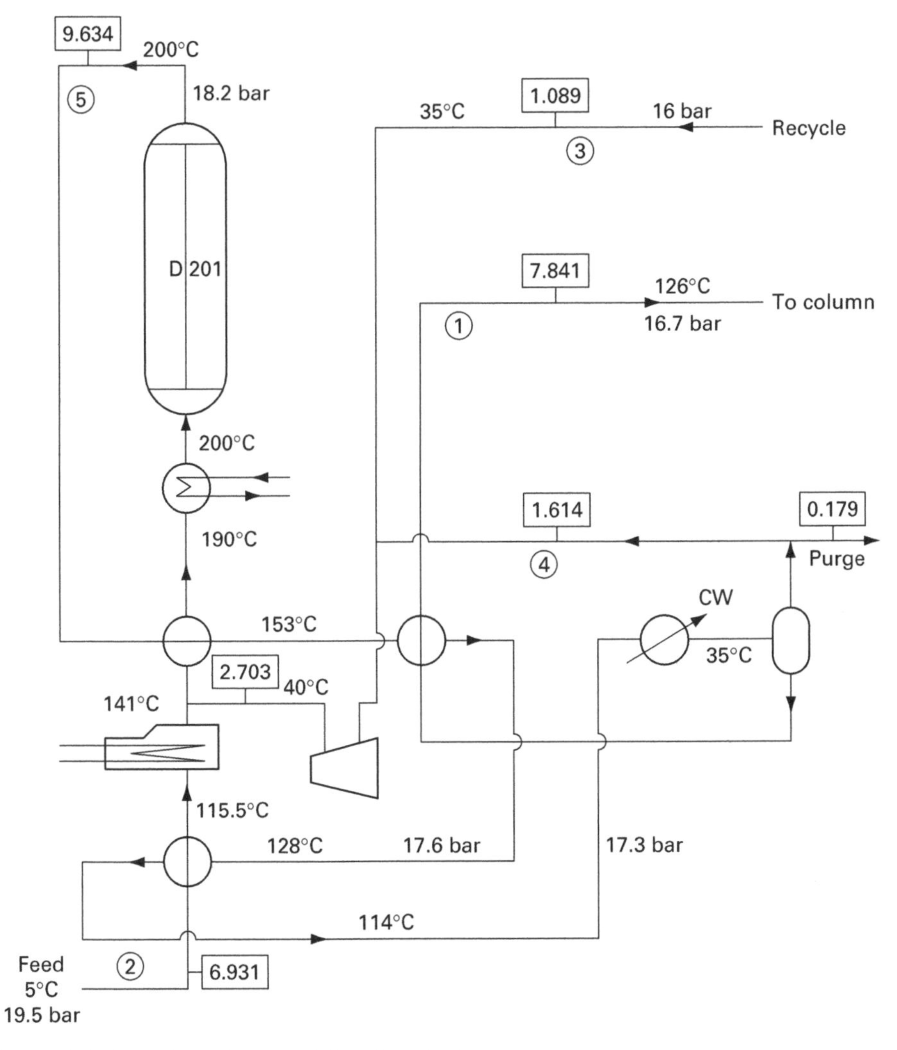

Figure 1.4 Flowsheet for specialty chemicals process

one stream, the reactor product, requires cooling from reactor exit temperature to separator temperature. Three streams require heating, these being reactor feed (from fresh feed storage temperature to reactor inlet temperature), recycle (from recycle temperature to reactor inlet temperature) and the "front end" product (from separator temperature to the temperature needed for downstream processing). Therefore the problem data comprise a set of four streams, one requiring cooling and three requiring heating, whose endpoint temperatures are known and whose total enthalpy changes are known (from the flowsheet mass balance and physical properties). The design task is to find the best network of exchangers, heaters and coolers, that handles these four streams at minimum operating and annualised capital cost, consistent with other design objectives such as operability. This was the

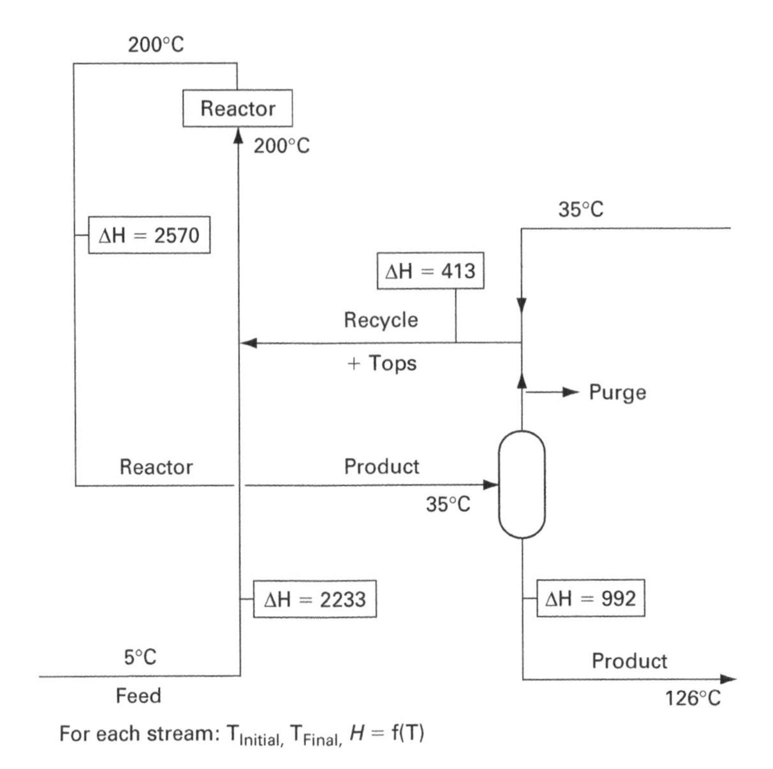

Figure 1.5 Heat exchange duties for specialty chemicals process

scope of pinch analysis in its first applications, exemplified by the network design techniques in Chapter 4.

However, the process can be optimised by going beyond the "one-way street" described above. For example, the configuration and operating conditions of the separation system (and, more rarely, the reactor) can be altered to fit better with the rest of the heating and cooling tasks in the process, as explained in Chapter 6. The pressures, temperatures and phase equilibria in the process determine the need for pumps, compressors and expanders, but this is also affected by the network configuration, especially pressure drops through exchangers and long pipe runs. The overall heat and power needs of the site are evaluated, and a combined heat and power (CHP) system can be considered to fulfil these (Chapter 5). This may alter the relative costs of different utility levels, and thus change the incentive for heat recovery. Total site analysis becomes important, and a wider range of targeting techniques (Chapter 3) helps us to understand the complex interactions. Batch processes require refinements to the analysis, and these can also be applied to other time-dependent situations, such as start-up and shutdown, as described in Chapter 7. Thus, pinch analysis and process integration have grown from a methodology for the heat recovery problem alone to a holistic analysis of the total process. The practical outworking of this is described in Chapter 8 and in the range of case studies in Chapter 9.

1.5 The role of thermodynamics in process design

1.5.1 How can we apply thermodynamics practically?

Most of us involved in engineering design have somewhat unhappy memories thinking back to thermodynamics in college days. Either we did not understand, gave up hope that we ever would, and remember with dread the horror that struck on examination day. Alternatively, we were amongst the chosen few whose photographic memory would allow us to reiterate the definitions of entropy, Gibbs free energy and all those differential equations faultlessly, but without real understanding. Afterwards, we could never help asking ourselves: what is it all for? What do I do with it? In the best of cases, thermodynamics seemed to be a fascinating science without a real application.

Pinch analysis is based on straightforward thermodynamics, and uses it in a practical way. However, the approach is largely non-mathematical. Although (classical) thermodynamics itself may be a thoroughly developed subject, we need to apply it the context of practical design and operation. This is the aim of the following chapters. We distinguish between "inevitable" and "avoidable" thermodynamic losses, and "practical" or "ideal" performance targets, to achieve both energy savings and other process benefits.

1.5.2 Capital and energy costs

Sometimes, it is believed that energy recovery is only important if energy costs are high and capital costs are low. Consider, for example, Figure 1.6, which shows a heat exchanger network that would seem appropriate to most when energy is cheap and capital expensive. There is no process heat recovery – only utility usage. Conversely, Figure 1.7 shows a network which might seem appropriate when energy is expensive. There is as much process heat recovery as is possible in preference to utility usage. The implicit assumption is that heat recovery (instead of utility use) saves energy but costs capital.

Consider now Figure 1.8. This shows a simpler network which still achieves maximum energy recovery. Based on a uniform heat transfer coefficient and sensible steam and cooling water temperatures, the total surface area for both designs has been evaluated. To our surprise, the "network for minimum capital cost" turns out to have the higher total surface area, and is more expensive in capital cost as well as operating (energy) cost!

From this example we realise that in networks there are two basic thermodynamic effects influencing capital costs. One is the effect of driving forces and the other is the effect of heat loads. Evidently, as we go to tighter designs (i.e. to reduce driving forces) we need less utility and the overall heat load decreases. Capital cost then increases with reduced driving forces (we all know that) but decreases with reduced heat load (we rarely consider this point). The design without process heat recovery in Figure 1.6 handles twice as much heat as is necessary. As a result, capital costs are increased even though the driving forces are large!

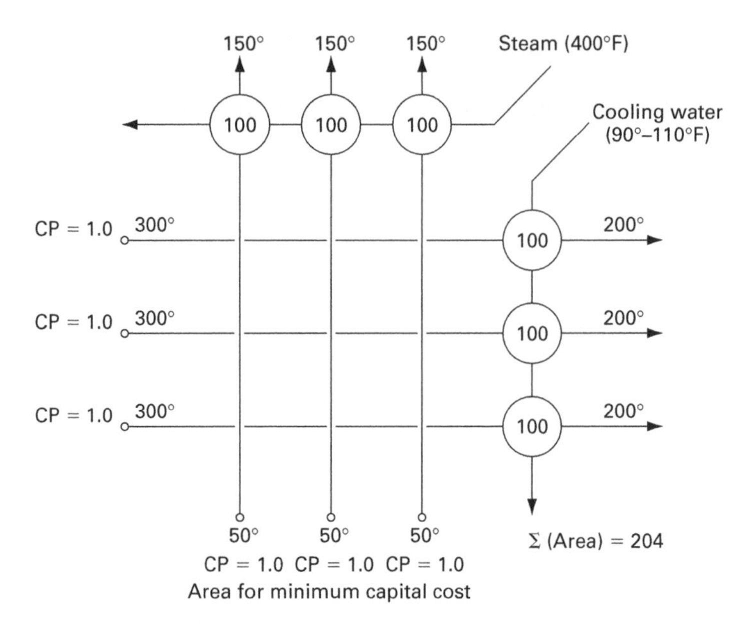

Figure 1.6 Outline heat exchanger network for "minimum capital cost"

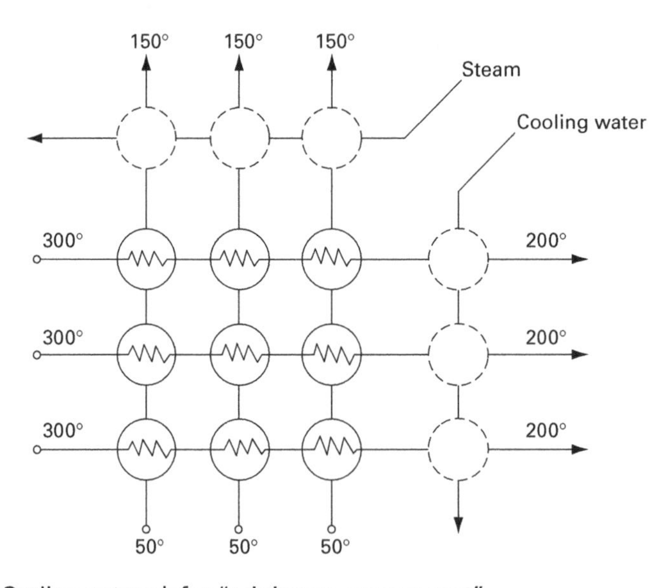

Figure 1.7 Outline network for "minimum energy cost"

Although this is obviously a contrived example, it helps to shows that there is not necessarily a trade-off between energy and capital cost, and helps to explain the frequent capital savings (as well as energy) observed in practical case studies.

Thermodynamics-based techniques can help in many other ways. For example, the analysis of driving forces may be used not to reduce them but to distribute them

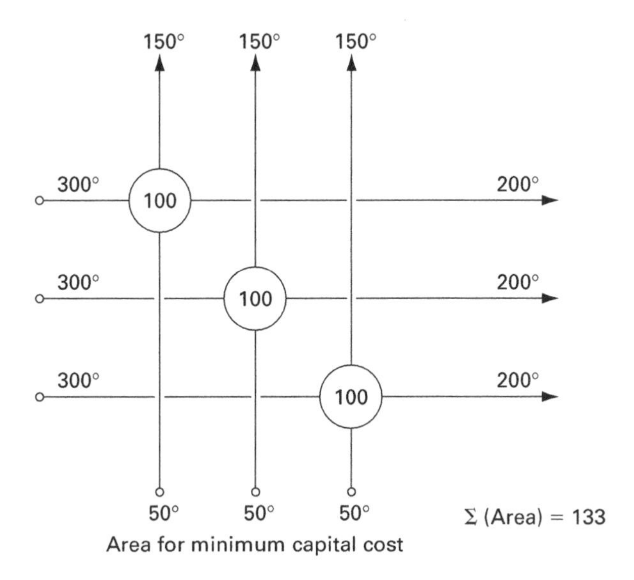

Figure 1.8 Optimal network for minimum energy cost

differently. This can help to clarify options in design, say, for better operability and/ or lower capital costs at a constant level of energy recovery.

1.6 Learning and applying the techniques

This book, like the original User Guide, is intended to be a self-teaching document. Studying Chapters 2–7, solving the example problems and reading the outline case studies should take the user 1–2 weeks of concentrated effort. Thereafter, he should be able to tackle his own problems generating better energy recovery networks.

However, a word of warning seems appropriate. Like most techniques based on concepts rather than rules, the techniques require a good understanding and some creative flexibility on behalf of the user. Without these assets the user will not be able to take full advantage of the generality and the flexibility offered by the techniques. Both systematic and lateral thinking are needed. An inkling of the type of *ad-hoc* arguments necessary when applying the techniques to specific projects can be obtained from Chapters 8 to 9 which describe practical application and case studies.

The book aims to be a summary of the most useful techniques, for practical application by the user, and naturally cannot cover all the refinements and nuances discovered in the last 30 years. Readers wishing to extend their knowledge of the methods are advised to consult the detailed research papers in the list of references in each chapter. Furthermore, short courses (such as those run for many years by the University of Manchester and its predecessor UMIST) are an obvious aid to an in-depth understanding and appreciation of the tricks and subtleties involved in practical applications.

References

Alves, J. (1999). *Design and Analysis of Refinery Hydrogen Distribution Systems*, PhD Thesis, UMIST, Manchester, UK.

Boland, D. and Linnhoff, B. (1979). The preliminary design of networks for heat exchange by systematic methods, *Chem Eng*, 9–15, April.

Brown, K. J. (1989). Process Integration Initiative. A Review of the Process Integration Initiatives Funded under the Energy Efficiency R&D Programme. (Energy Technology Support Unit (ETSU), Harwell Laboratory, Oxfordshire, UK.)

Douglas, J. M. (1988). *Conceptual Design of Chemical Processes.* McGraw-Hill, New York.

ESDU Data Item 87001 (1987). Process Integration. Available by subscription from ESDU International Ltd, London. (Engineering Sciences Data Unit.)

ESDU Data Item 89030 (1989). Application of process integration to utilities, combined heat and power and heat pumps. ESDU International.

ESDU Data Item 90017 (1987–1990). Process integration; process change and batch processes. ESDU International.

Hallale, N. and Liu, P. (2001). Refinery hydrogen management for clean fuel production, *Adv Environ Res*, 6: 81–98.

Korner, H. (1988). Optimal use of energy in the chemical industry, *Chem Ing Tech*, 60(7): 511–518.

Linnhoff, B. (1979). *Thermodynamic Analysis in the Design of Process Networks*, PhD Thesis, University of Leeds.

Linnhoff, B., Dunford, H. and Smith, R. (1983). Heat integration of distillation columns into overall processes, *Chem Eng Sci*, 38(8): 1175–1188.

Linnhoff, B. and Flower, J. R. (1978). Synthesis of heat exchanger networks. Part I: Systematic generation of energy optimal networks, *AIChE J*, 24(4): 633–642. Part II: Evolutionary generation of networks with various criteria of optimality, *AIChE J*, 24(4): 642–654.

Linnhoff, B. and Hindmarsh, E. (1983). The pinch design method of heat exchanger networks, *Chem Eng Sci*, 38(5): 745–763.

Linnhoff, B., Mason, D. R. and Wardle, I. (1979). Understanding heat exchanger networks, *Comp Chem Eng*, 3: 295.

Linnhoff, B., Townsend, D. W., Boland, D., Hewitt, G. F., Thomas, B. E. A., Guy, A. R. and Marsland, R. H. (1982). *User Guide on Process Integration for the Efficient Use of Energy*, 1st edition. IChemE, Rugby, UK. Revised 1st edition 1994.

Linnhoff, B. and Turner, J. A. (1981). Heat-recovery networks: new insights yield big savings, *Chem Eng*, 56–70, November 2.

Linnhoff, B. and Vredeveld, D. R. (1984). Pinch technology has come of age, *Chem Eng Prog*, 33–40, July.

Morgan, S. (1992). Use process integration to improve process designs and the design process, *Chem Eng Prog*, 62–68, September.

Polley, G. T. and Heggs, P. J. (1999). Don't Let the "Pinch" Pinch You, Chemical Engineering Progress, AIChE, Vol. 95, No. 12, pp 27–36, December.

Smith, R. (2005). *Chemical Process Design and Integration.* John Wiley, Chichester and New York.

Steinmeyer, D. (1992). Save energy, without entropy, *HydroCarb Process*, 71: 55–95, October.

Townsend, D. W. and Linnhoff, B. (1983). Heat and power networks in process design. Part 1: Criteria for placement of heat engines and heat pumps in process networks, *AIChE J*, 29(5): 742–748. Part II: Design procedure for equipment selection and process matching, *AIChE J*, 29(5): 748–771.

Wang, Y. P. and Smith, R. (1994). Wastewater minimisation, *Chem Eng Sci*, 49: 981.

Wang, Y. P. and Smith, R. (1995). Wastewater minimisation with flowrate constraints, *Trans I ChemE Part A*, 79: 889–904.

Linnhoff, B., Tainsh, R. and Wasilewski, M. (1999). Hydrogen network management. A systems approach from using paper presented at: *The European Refining Technology Conference*, Paris, November.

2 Key concepts of pinch analysis

In this section, we will present the key concepts of pinch analysis, showing how it is possible to set energy targets and achieve them with a network of heat exchangers. These concepts will then be expanded for a wide variety of practical situations in the following chapters.

2.1 Heat recovery and heat exchange

2.1.1 Basic concepts of heat exchange

Consider the simple process shown in Figure 2.1. There is a chemical reactor, which will be treated at present as a "black-box". Liquid is supplied to the reactor and needs to be heated from near-ambient temperature to the operating temperature of the reactor. Conversely, a hot liquid product from the separation system needs to be cooled down to a lower temperature. There is also an additional unheated make-up stream to the reactor.

Any flow which requires to be heated or cooled, but does not change in composition, is defined as a **stream**. The feed, which starts cold and needs to be heated up, is known as a **cold stream**. Conversely, the hot product which must be cooled down is called a **hot stream**. Conversely, the reaction process is not a stream, because it involves a change in chemical composition; and the make-up flow is not a stream, because it is not heated or cooled.

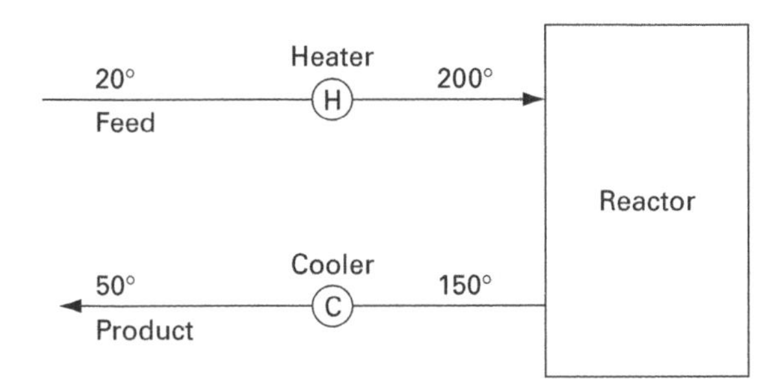

Figure 2.1 Simple process flowsheet

Table 2.1 Data for simple two-stream example

	Mass flowrate W *(kg/s)*	*Specific heat capacity* C_P *(kJ/kgK)*	*Heat capacity flowrate* CP *(kW/K)*	*Initial (supply) temperature* T_S *(°C)*	*Final (target) temperature* T_T *(°C)*	*Heat load* H *(kW)*
Cold stream	0.25	4	1.0	20	200	−180
Hot stream	0.4	4.5	1.8	150	50	+180

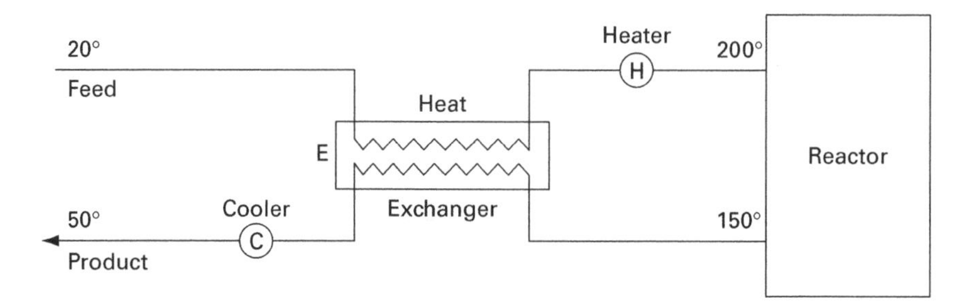

Figure 2.2 Simple process flowsheet with heat exchange

To perform the heating and cooling, a steam heater could be placed on the cold stream, and a water cooler on the hot stream. The flows are as given in Table 2.1. Clearly, we will need to supply 180 kW of steam heating and 180 kW of water cooling to operate the process.

Can we reduce energy consumption? Yes; if we can recover some heat from the hot stream and use it to heat the cold stream in a heat exchanger, we will need less steam and water to satisfy the remaining duties. The flowsheet will then be as in Figure 2.2. Ideally, of course, we would like to recover all 180 kW in the hot stream to heat the cold stream. However, this is not possible because of temperature limitations. By the Second Law of Thermodynamics, we can't use a hot stream at 150°C to heat a cold stream at 200°C! (As in the informal statement of the Second Law, "you can't boil a kettle on ice"). So the question is, how much heat can we actually recover, how big should the exchanger be, and what will be the temperatures around it?

2.1.2 The temperature–enthalpy diagram

A helpful method of visualisation is the temperature–heat content diagram, as illustrated in Figure 2.3. The heat content *H* of a stream (kW) is frequently called its enthalpy; this should not be confused with the thermodynamic term, specific

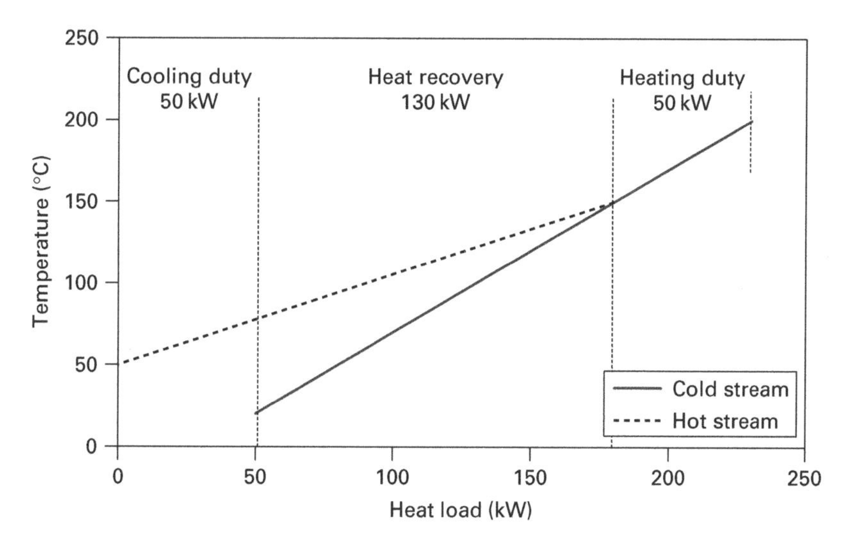

Figure 2.3 Streams plotted on temperature/enthalpy (*T/H*) diagram with $\Delta T_{min} = 0$

enthalpy (kJ/kg). Differential heat flow dQ, when added to a process stream, will increase its enthalpy (H) by $CP\,dT$, where:

CP = "heat capacity flowrate" (kW/K) = mass flow W (kg/s) × specific heat C_P (kJ/kgK)
dT = differential temperature change

Hence, with CP assumed constant, for a stream requiring heating ("cold" stream) from a "supply temperature" (T_S) to a "target temperature" (T_T), the total heat added will be equal to the stream enthalpy change, i.e.

$$Q = \int_{T_S}^{T_T} CP\,dT = CP(T_T - T_S) = \Delta H \tag{2.1}$$

and the slope of the line representing the stream is:

$$\frac{dT}{dQ} = \frac{1}{CP} \tag{2.2}$$

The *T/H* diagram can be used to represent heat exchange, because of a very useful feature. Namely, since we are only interested in enthalpy *changes* of streams, a given stream can be plotted anywhere on the enthalpy axis. Provided it has the same slope and runs between the same supply and target temperatures, then wherever it is drawn on the H-axis, it represents the same stream.

Figure 2.3 shows the hot and cold streams for our example plotted on the *T/H* diagram. Note that the hot stream is represented by the line with the arrowhead pointing to the left, and the cold stream *vice versa*. For feasible heat exchange between the two, the hot stream must at all points be hotter than the cold stream, so it should be plotted above the cold stream. Figure 2.3 represents a limiting case; the hot stream cannot be moved further to the right, to give greater heat recovery, because the temperature difference between hot and cold streams at the cold end of the exchanger is already zero. This means that, in this example, the balance of heat required by the cold stream above 150°C (i.e. 50 kW) has to be made up from steam heating. Conversely, although 130 kW can be used for heat exchange, 50 kW of heat available in the hot stream has to be rejected to cooling water. However, this is not a practically achievable situation, as a zero temperature difference would require an infinitely large heat exchanger.

In Figure 2.4 the cold stream is shown shifted on the H-axis relative to the hot stream so that the minimum temperature difference, ΔT_{min} is no longer zero, but positive and finite (in this case 20°C). The effect of this shift is to increase the utility heating and cooling by equal amounts and reduce the load on the exchanger by the same amount – here 20 kW – so that 70 kW of external heating and cooling is required. This arrangement is now practical because the ΔT_{min} is non-zero. Clearly, further shifting implies larger ΔT_{min} values and larger utility consumptions.

From this analysis, two basic facts emerge. Firstly, there is a correlation between the value of ΔT_{min} in the exchanger and the total utility load on the system. This means that if we choose a value of ΔT_{min}, we have an **energy target** for how much heating and cooling we should be using if we design our heat exchanger correctly.

Secondly, if the hot utility load is increased by any value α, the cold utility is increased by α as well. *More in, more out*! As the stream heat loads are constant, this also means that the heat exchanged falls by α.

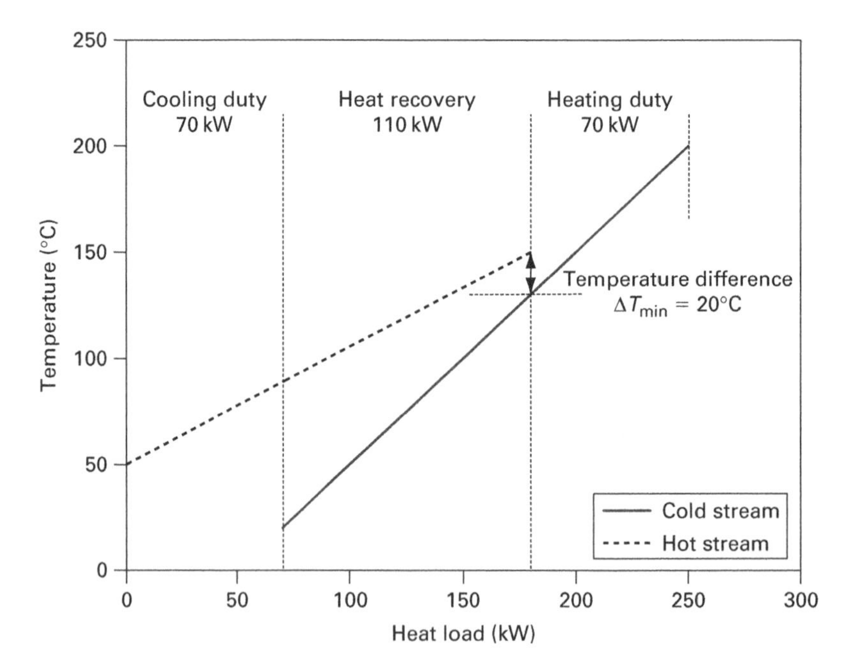

Figure 2.4 *T/H* diagram with $\Delta T_{min} = 20°C$

The reader will rightly point out that a method confined to a single hot and cold stream is of little practical use. What is needed is a methodology to apply this to real multi-stream processes. The composite curves give us a way of doing so.

2.1.3 Composite curves

To handle multiple streams, we add together the heat loads or heat capacity flowrates of all streams existing over any given temperature range. Thus, a single composite of all hot streams and a single composite of all cold streams can be produced in the *T/H* diagram, and handled in just the same way as the two-stream problem.

In Figure 2.5(a) three hot streams are plotted separately, with their supply and target temperatures defining a series of "interval" temperatures T_1–T_5. Between T_1 and T_2, only stream B exists, and so the heat available in this interval is given by

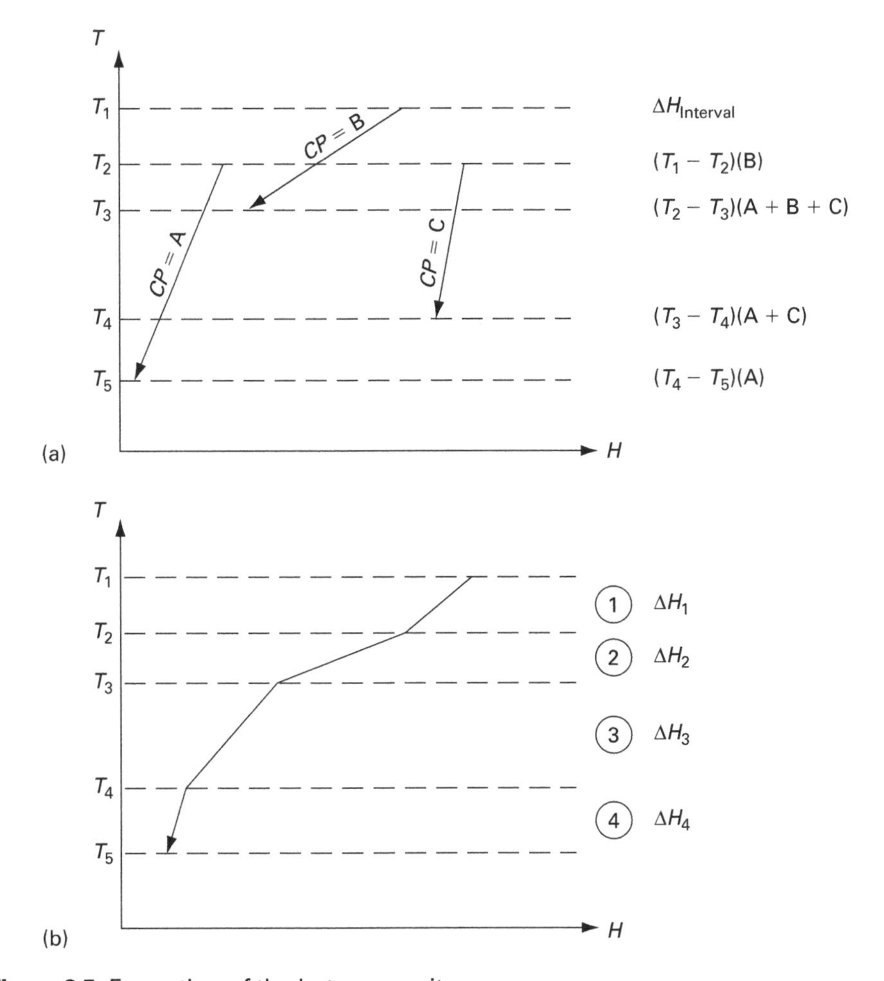

Figure 2.5 Formation of the hot composite curve

$CP_B(T_1 - T_2)$. However between T_2 and T_3 all three streams exist and so the heat available in this interval is $(CP_A + CP_B + CP_C)(T_2 - T_1)$. A series of values of ΔH for each interval can be obtained in this way, and the result re-plotted against the interval temperatures as shown in Figure 2.5(b). The resulting T/H plot is a single curve representing all the hot streams, known as the **hot composite curve**. A similar procedure gives a **cold composite curve** of all the cold streams in a problem. The overlap between the composite curves represents the maximum amount of heat recovery possible within the process. The "overshoot" at the bottom of the hot composite represents the minimum amount of external cooling required and the "overshoot" at the top of the cold composite represents the minimum amount of external heating (Hohmann 1971).

Figure 2.6 shows a typical pair of composite curves – in fact, for the four-stream problem given in Table 1.1 and repeated as Table 2.2. Shifting of the curves leads to

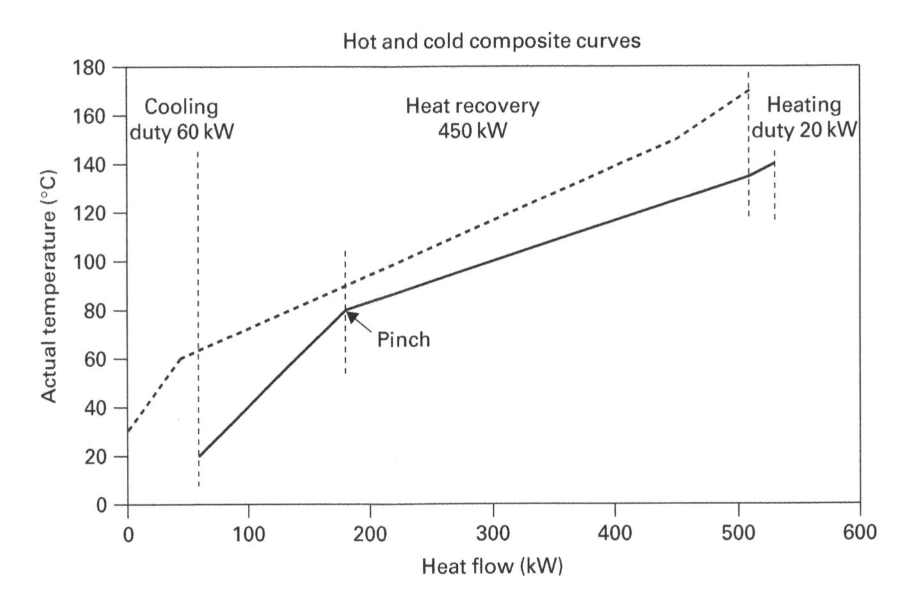

Figure 2.6 Composite curves for four-stream problem

Table 2.2 Data for four-stream example from Chapter 1

Stream number and type	CP (kW/K)	Actual temperatures		Shifted temperatures	
		T_S (°C)	T_T (°C)	S_S (°C)	S_T (°C)
1. Cold	2	20°	135°	25°	140°
2. Hot	3	170°	60°	165°	55°
3. Cold	4	80°	140°	85°	145°
4. Hot	1.5	150°	30°	145°	25°

$\Delta T_{min} = 10°C$.

behaviour similar to that shown by the two-stream problem. Now, though, the "kinked" nature of the composites means that ΔT_{min} can occur anywhere in the interchange region and not just at one end. *For a given value of ΔT_{min}, the utility quantities predicted are the minima required to solve the heat recovery problem.* Note that although there are many streams in the problem, in general ΔT_{min} occurs at only one point of closest approach, which is called the **pinch** (Linnhoff *et al.* 1979). This means that it is possible to design a network which uses the minimum utility requirements, where *only the heat exchangers at the pinch* need to operate at ΔT values down to ΔT_{min}. Producing such a design will be described in Section 2.3. It will be seen later that the pinch temperature is of great practical importance, not just in network design but in all energy-related aspects of process optimisation.

2.1.4 A targeting procedure: the "Problem Table"

In principle, the "composite curves" described in the previous sub-section could be used for obtaining energy targets at given values of ΔT_{min}. However, it would require a "graph paper and scissors" approach (for sliding the graphs relative to one another) which would be messy and imprecise. Instead, we use an algorithm for setting the targets algebraically, the "Problem Table" method (Linnhoff and Flower 1978).

In the description of the construction of composite curves (Figure 2.6), it was shown how enthalpy balance intervals were set up based on stream supply and target temperatures. The same can be done for hot and cold streams together, to allow for the maximum possible amount of heat exchange within each temperature interval. The only modification needed is to ensure that within any interval, hot streams and cold streams are at least ΔT_{min} apart. This is done by using **shifted temperatures**, which are set at $½\Delta T_{min}$ (5°C in this example) *below* hot stream temperatures and $½\Delta T_{min}$ *above* cold stream temperatures. Table 2.2 shows the data for the four-stream problem including shifted temperatures. Figure 2.7 shows the streams in a schematic representation with a vertical temperature scale, with interval boundaries superimposed (as shifted temperatures). So for example in interval number 2, between shifted temperatures 145°C and 140°C, streams 2 and 4 (the hot streams) run from 150°C to 145°C, and stream 3 (the cold stream) from 135°C to 140°C. Setting up the intervals in this way *guarantees* that full heat interchange within any interval is possible. Hence, each interval will have either a net surplus or net deficit of heat as dictated by enthalpy balance, *but never both*. Knowing the stream population in each interval (from Figure 2.7), enthalpy balances can easily be calculated for each according to:

$$\Delta H_i = (S_i - S_{i+1})\,(\Sigma CP_H - \Sigma CP_C)_i \qquad (2.3)$$

for any interval i. The results are shown in Table 2.3, and the last column indicates whether an interval is in heat surplus or heat deficit. It would therefore be possible to produce a feasible network design based on the assumption that all "surplus"

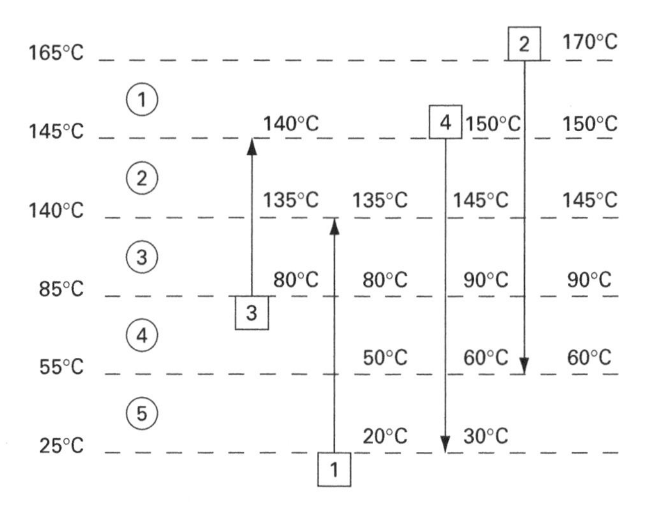

Figure 2.7 Streams and temperature intervals

Table 2.3 Temperature intervals and heat loads for four-stream problem

	Interval number i	$S_i - S_{i+1}$ *(°C)*	$\sum CP_{HOT} - \sum CP_{COLD}$ *(kW/°C)*	ΔH_i *(kW)*	*Surplus or deficit*
S_1 = 165°C					
	1	20	+3.0	+60	Surplus
S_2 = 145°C					
	2	5	+0.5	+2.5	Surplus
S_3 = 140°C					
	3	55	−1.5	−82.5	Deficit
S_4 = 85°C					
	4	30	+2.5	+75	Surplus
S_5 = 55°C					
	5	30	−0.5	−15	Deficit
S_6 = 25°C					

intervals rejected heat to cold utility, and all "deficit" intervals took heat from hot utility. However, this would not be very sensible, because it would involve rejecting and accepting heat at inappropriate temperatures.

We now, however, exploit a key feature of the temperature intervals. Namely, *any heat available in interval* i *is hot enough to supply any duty in interval* i + 1. This is shown in Figure 2.8, where intervals 1 and 2 are used as an illustration. Instead of sending the 60 kW of surplus heat from interval 1 into cold utility, it can be sent down into interval 2. It is therefore possible to set up a heat "cascade" as shown in Figure 2.9(a). Assuming that no heat is supplied to the hottest interval 1 from hot utility, then the surplus of 60 kW from interval 1 is cascaded into interval 2. There it joins the 2.5 kW surplus from interval 2, making 62.5 kW to cascade into

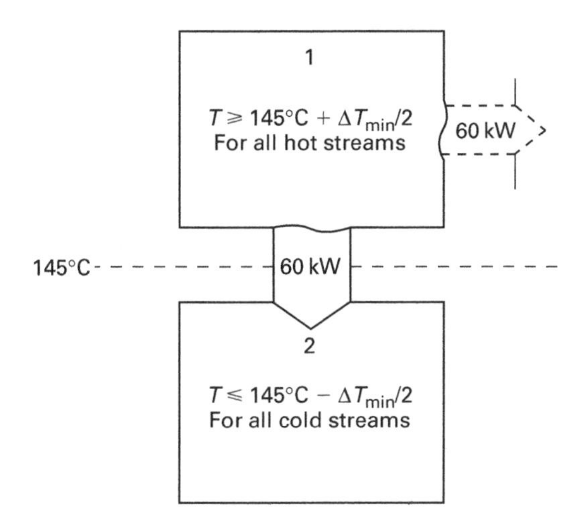

Figure 2.8 Use of a heat surplus from an interval

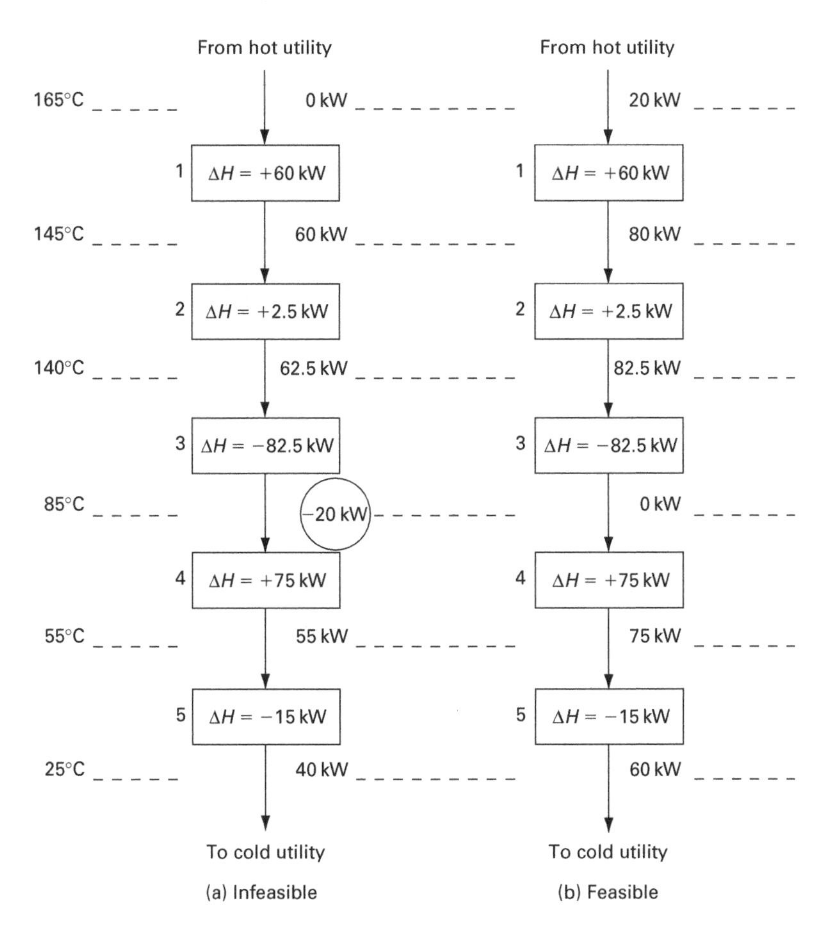

Figure 2.9 Infeasible and feasible heat cascades

interval 3. Interval 3 has a 82.5 kW deficit, hence after accepting the 62.5 kW it can be regarded as passing on a 20 kW deficit to interval 4. Interval 4 has a 75 kW surplus and so passes on a 55 kW surplus to interval 5. Finally, the 15 kW deficit in interval 5 means that 40 kW is the final cascaded energy to cold utility. This in fact is the net enthalpy balance on the whole problem (i.e. cold utility will always exceed hot utility by 40 kW, whatever their individual values). Looking back at the heat flows between intervals in Figure 2.9(a), clearly the negative flow of 20 kW between intervals 3 and 4 is thermodynamically infeasible. To make it just feasible (i.e. equal to zero), 20 kW of heat must be added from hot utility as shown in Figure 2.9(b), and cascaded right through the system. By enthalpy balance this means that all flows are increased by 20 kW. The net result of this operation is that the minimum utilities requirements have been predicted (i.e. 20 kW hot and 60 kW cold). Furthermore, the position of the pinch has been located. This is at the interval boundary with a shifted temperature of 85°C (i.e. hot streams at 90°C and cold at 80°C) where the heat flow is zero.

Compare the results obtained by this approach to the results from the composite curves, as shown in Figure 2.6. The same information is obtained, but the Problem Table provides a simple framework for numerical analysis. For simple problems it can be quickly evaluated by hand. For larger problems, it is easily implemented as a spreadsheet or other computer software. It can also be adapted for the case where the value of ΔT_{min} allowed depends on the streams matched, and is not simply a "global" value (see under Section 3.3.1). Finally it can be adapted to cover other cases where simplifying assumptions (e.g. $CP =$ constant) are invalid (Section 3.1.3).

The total heat recovered by heat exchange is found by adding the heat loads for all the hot streams and all the cold streams – 510 and 470 kWh, respectively. Subtracting the cold and hot utility targets (60 and 20 kWh) from these values gives the total heat recovery, 450 kWh, by two separate routes. The cold utility target minus the hot utility target should equal the bottom line of the infeasible heat cascade, which is 40 kWh. These calculations provide useful cross-checks that the stream data and heat cascades have been evaluated correctly.

With the Problem Table algorithm, the engineer has a powerful targeting technique at his or her fingertips. Data can be quickly extracted from flowsheets and analysed to see whether the process is nearing optimal, or whether significant scope for energy saving exists. The targets are easily obtained and provide enormous stimulus to break away from the "learning curve" (Figure 1.2). A step-by-step algorithm for calculating the Problem Table is given in Section 3.11.

Note 1: In early papers and the first edition of this User Guide, Equation (2.3) was reversed; temperature intervals with a net demand were shown as positive and with a net surplus as negative. However, this is counter-intuitive, and more recent practice has been for a heat surplus to be positive, as shown here.

Note 2: There are in fact three possible ways of moving the hot and cold composite curves closer together by ΔT_{min}, so that they touch at the pinch. This may be achieved in three ways:

1. Express all temperatures in terms of hot stream temperatures and increase all cold stream temperatures by ΔT_{min}.

2. Express all temperatures in terms of cold stream temperatures and reduce all hot stream temperatures by ΔT_{min}.
3. Use the shifted temperatures, which are a mean value; all hot stream temperatures are reduced by $\Delta T_{min}/2$ and all cold stream temperatures are increased by $\Delta T_{min}/2$.

Approach 3 has been the most commonly adopted, although approach 1 has also been used significantly. In this book, shifted temperatures 3 will be used from now on without further comment. Early papers sometimes called them "interval temperatures".

2.1.5 The grand composite curve and shifted composite curves

If the composite curves are re-plotted on axes of shifted temperature, we obtain the **shifted composite curves**, Figure 2.10. The shifted curves just touch at the pinch temperature, and show even more clearly than the composite curves that the pinch divides the process into two.

Now consider what happens at any shifted temperature S. The heat flow of all the hot streams Q_H, relative to that at the pinch Q_{HP} (fixed), is ΔQ_H. Likewise the heat flow of all cold streams relative to that at the pinch is ΔQ_C. There is an imbalance which must be supplied by utilities – external heating and cooling. Above the pinch, $\Delta Q_C > \Delta Q_H$ and the difference must be supplied by hot utility. Likewise, below the pinch $\Delta Q_H > \Delta Q_C$ and the excess heat is removed by cold utility.

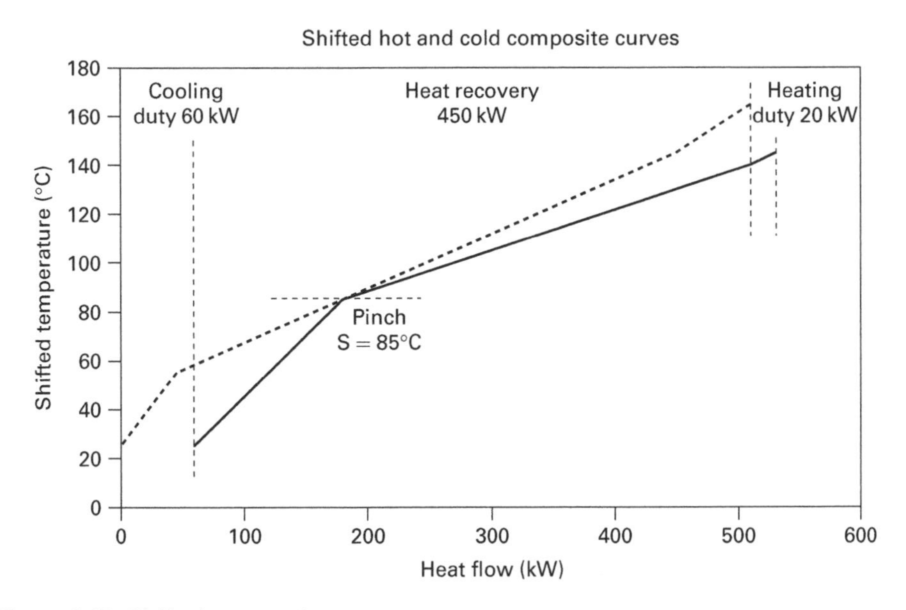

Figure 2.10 Shifted composite curves

Hence, knowing the shifted composite curves, we can find the minimum amount of heating or cooling that needs to be supplied at any given temperature. A graph of net heat flow (utility requirement) against shifted temperature can then easily be plotted. This is known as the **grand composite curve** (hereafter abbreviated to **GCC**). It represents the difference between the heat available from the hot streams and the heat required by the cold streams, relative to the pinch, at a given shifted temperature. Thus, the GCC is a plot of the net heat flow against the shifted (interval) temperature, which is simply a graphical plot of the Problem Table (heat cascade).

The GCC for the four-stream example from Chapter 2 is shown in Figure 2.11. The values of net heat flow at the top and bottom end are the heat supplied to and removed from the cascade, and thus tell us the hot and cold utility targets. But not only does the GCC tell us how much net heating and cooling is required, it also tells us what temperatures it is needed at. There is no need to supply all the utility heating at the highest temperature interval; much of it can, if desired, be supplied at lower temperatures. The pinch is also easily visualised, being the point where net heat flow is 0 and the GCC touches the axis. Moreover, we can see whether the pinch occurs in the middle of the temperature range or at one end (a "threshold" problem), and identify other regions of low net heat flow, or even double or multiple pinches.

Energy targeting can also be used to settle quickly disputes along the lines of "to integrate, or not to integrate?" Processes often fall into distinct sections ("A" and "B") by reason of layout or operability considerations. The question is often "can significant savings be made by cross integration?" The Problem Table algorithm can be applied

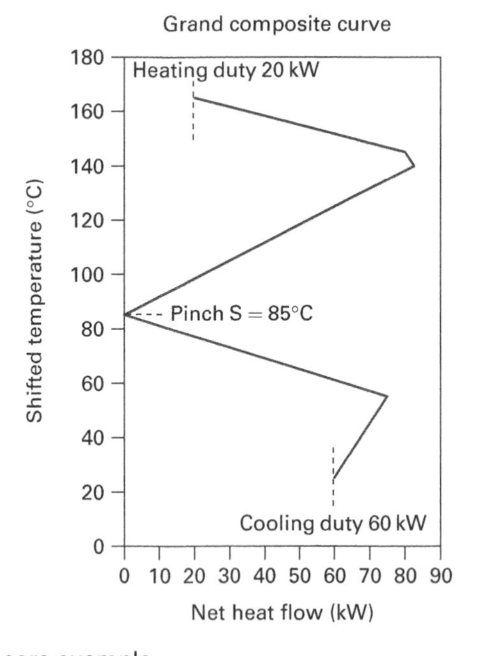

Figure 2.11 GCC for core example

to areas A and B separately, and then to all the streams in A and B together. The results of the analysis will quickly settle the question. For example, if the answer is:

A alone: 10% savings in total fuel bill possible
B alone: 5% savings in total fuel bill possible
A and B together : 30% savings possible

then there is a 15% energy incentive for cross-integrating areas A and B. This is **zonal targeting**, described further in Section 3.5.1.

To summarise this section on energy targeting:

- Composite curves give conceptual understanding of how energy targets can be obtained.
- The Problem Table and its graphical representation, the GCC, give the same results (including the pinch location) more easily.
- Energy targeting is a powerful design and "process integration" aid.

2.2 The pinch and its significance

Figure 2.12(a) shows the composite curves for a multi-stream problem dissected at the pinch. "Above" the pinch (i.e. in the region to the right) the hot composite transfers all its heat into the cold composite, leaving utility *heating only* required. The region above the pinch is therefore a *net heat sink*, with heat flowing into it but no heat flowing out. It involves heat exchange and hot utility, but *no cold utility*. Conversely below the pinch *cooling only* is required and the region is therefore a *net heat source*, requiring heat exchange and cold utility but *no hot utility*. The problem therefore falls into two thermodynamically distinct regions, as indicated by the enthalpy balance envelopes in Figure 2.12(b). Heat Q_{Hmin} flows into the problem above the pinch and Q_{Cmin} out of the problem below, but the heat flow across the pinch is zero. This result was observed in the description of the Problem Table algorithm in Section 2.1.4. The way in which the pinch divides the process into two is even more clearly seen from the shifted composite curves (Figure 2.10).

It follows that any network design that transfers heat α across the pinch must, by overall enthalpy balance, require α more than minimum from hot and cold utilities, as shown in Figure 2.12(c). As a corollary, any utility cooling α above the pinch must incur extra hot utility α, and *vice versa* below the pinch. This gives three **golden rules** for the designer wishing to produce a design achieving minimum utility targets:

- Don't transfer heat across the pinch.
- Don't use cold utilities above the pinch.
- Don't use hot utilities below the pinch.

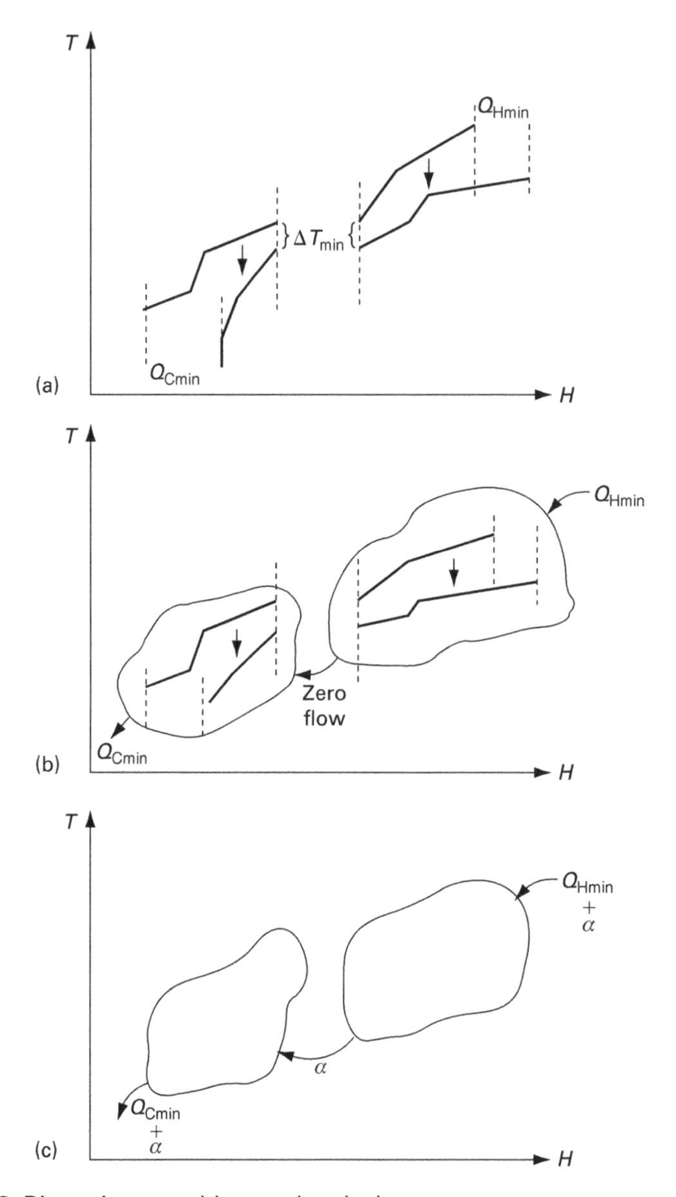

Figure 2.12 Dissecting a problem at the pinch

Conversely, if a process is using more energy than its thermodynamic targets, it must be due to one or more of the golden rules being broken. This may be a deliberate trade-off, as will be seen, but it is important to know that it is happening. The decomposition of the problem at the pinch turns out to be very useful when it comes to network design (Linnhoff and Hindmarsh 1983).

These insights give us five simple and effective concepts:

- **Targets:** Once the composite curves and Problem Table are known, we know exactly how much external heating is unavoidably required. *Near-optimal processes are confirmed as such and non-optimal processes are identified with great speed and confidence.*
- **The Pinch:** Above the pinch the process needs external heating and below the pinch it needs external cooling. This tells us where to place furnaces, steam heaters, coolers, *etc.* It also tells us what site steam services should be used and how we should recover heat from the exhaust of steam and gas turbines.
- **More in, more out:** An off-target process requires more than the minimum external heating and therefore more than the minimum external cooling; see Figure 2.12(c). We coin the catch phrase "more in, more out" and note that for every unit of excess external heat in a process we have to provide heat transfer equipment twice. In some cases this may allow us to improve both energy and capital cost.
- **Freedom of choice:** The "heat sink" and the "heat source" in Figure 2.12(b) are separate. As long as the designer obeys this constraint he can follow his heart's delight in choosing plant-layouts, control arrangements, *etc.* If he has to violate this constraint, he can evaluate the cross-pinch heat flow and therefore predict what overall penalties will be involved.
- **Trade-offs:** A simple relationship exists between the number of streams (process streams plus utilities) in a problem and the minimum number of heat exchange "units" (i.e. heaters, coolers and interchangers); see Section 3.6. A network which achieves the minimum energy targets, with the "heat source" and "heat sink" sections separate, needs more units than if the pinch division had been ignored. This type of trade-off, between energy recovery and number of units, adds to the traditional concept of a trade-off between energy and surface area (Section 2.4).

Thus, we do not necessarily need "black-box" computing power, but we are developing key concepts which the designer can blend with his intuition and experience of the individual process technology. It is this blend which ultimately gives better designs.

2.3 Heat exchanger network design

2.3.1 Network grid representation

For designing a heat exchanger network, the most helpful representation is the "grid diagram" introduced by Linnhoff and Flower (1978) (Figure 2.13). The streams are drawn as horizontal lines, with high temperatures on the left and hot streams at the top; heat exchange matches are represented by two circles joined by a vertical line. The grid is much easier to draw than a flowsheet, especially as heat exchangers can be placed in any order without redrawing the stream system. Also,

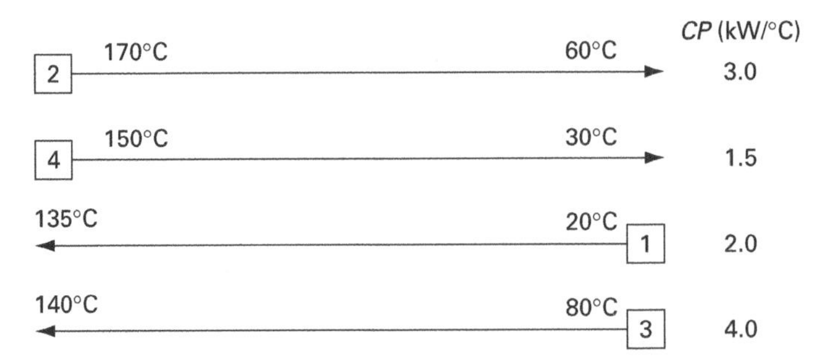

Figure 2.13 Initial grid diagram for four-stream problem

the grid represents the countercurrent nature of the heat exchange, making it easier to check exchanger temperature feasibility. Finally, the pinch is easily represented in the grid (as will be shown in the next sub-section), whereas it cannot be represented on the flowsheet.

2.3.2 A "commonsense" network design

We will now produce a simple heat exchanger network for the four-stream problem and represent it on the grid diagram. Figure 2.13 shows the initial situation. We want to exchange heat between the hot and cold streams, and logically we should start at one end of the temperature range. Matching the hottest hot stream 2 against the hottest cold stream 3 should give the best temperature driving forces and ensure feasibility. If we match the whole of the heat load on stream 4 (240 kW), we can calculate that stream 2 has been brought down to 90°C, which is just acceptable for the given ΔT_{min} of 10°C. Then we match stream 4 against stream 1, and find that we can use the whole of the 180 kW in stream 4 while again achieving the ΔT_{min} = 10°C criterion at the bottom end. This raises stream 1 to 110°C, so it needs an additional 50 kW of hot utility to raise it to the required final temperature of 135°C. Finally, we add a cooler to stream 2 to account for the remaining 90 kW required to bring it down to 60°C. The resulting design is shown as a network grid in Figure 2.14. We have achieved 420 kW of heat exchange, so that only 50 kW external heating is required, with two exchangers, and have every reason to feel pleased with ourselves.

However, when we see the energy targets, we are brought firmly down to earth. We should be able to use just 20 kW of heating and 60 kW of cooling, but our design needs 50 and 90 kW, respectively. What have we done wrong? We could attempt to produce a better network by trial-and-error. However, the reader will find that, even with the knowledge of the targets, it is no easy task to design a network to achieve them by conventional means.

Instead, it is better to use the insights given by the pinch concept to find out why we have missed the target. The pinch is at a shifted temperature of 85°C, corresponding to 90°C for hot streams and 80°C for cold streams. We know there are

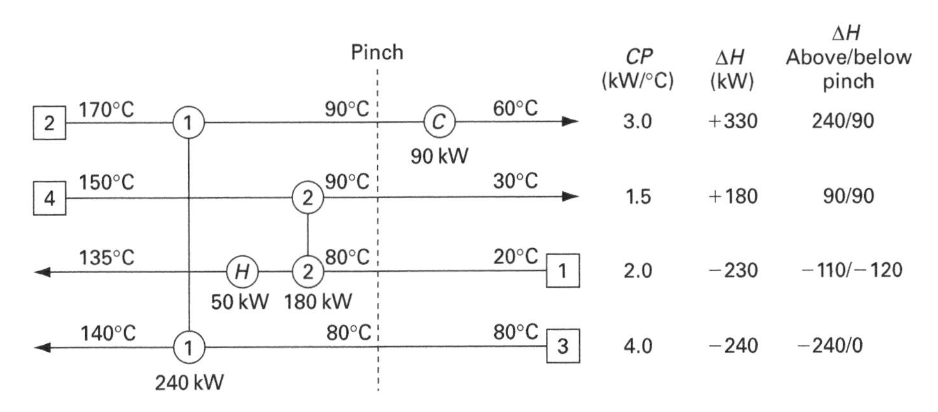

Figure 2.14 "Commonsense" network design for four-stream problem

three, and only three, reasons for a process to be off-target. Looking at the grid diagram (and assisted by having the pinch drawn in) we can see that there are no heaters below the pinch, nor coolers above it. Therefore, we must have heat transfer across the pinch, and the culprit must be heat exchanger 2. Although it spans the pinch on both streams, we can calculate stream 2 is releasing 90 kW above 90°C and 90 kW below it, while stream 3 is receiving only 60 kW above 80°C and 120 kW below it. Therefore, 30 kW is being transferred across the pinch, corresponding precisely to the 30 kW excess of our hot and cold utility over the targets.

How can we guarantee meeting the targets without violating the three golden rules? Answer: only by starting the design at the most constrained point – the pinch itself – and working outwards. This in itself explains why "traditional" heat exchanger network designs are almost invariably off-target; the designer has to start at an intermediate point in the temperature range, and how can he know what it is, without the insights of the pinch?

2.3.3 Design for maximum energy recovery

Let us return to our grid diagram and start to construct a new network. Notice that stream number 3 starts at the pinch. In fact in problems where the streams all have constant *CP*s, the pinch is always caused by the entry of a stream, either hot or cold.

We know that above the pinch, no utility cooling should be used. This means that above the pinch, *all hot streams must be brought to pinch temperature by interchange against cold streams*. We must therefore start the design *at the pinch*, finding matches that fulfil this condition. In this example, above the pinch there are two hot streams at pinch temperature, therefore requiring two "pinch matches". In Figure 2.15(a) a match between streams 2 and 1 is shown, with a *T*/*H* plot of the match shown in inset. (Note that the stream directions have been reversed so as to mirror

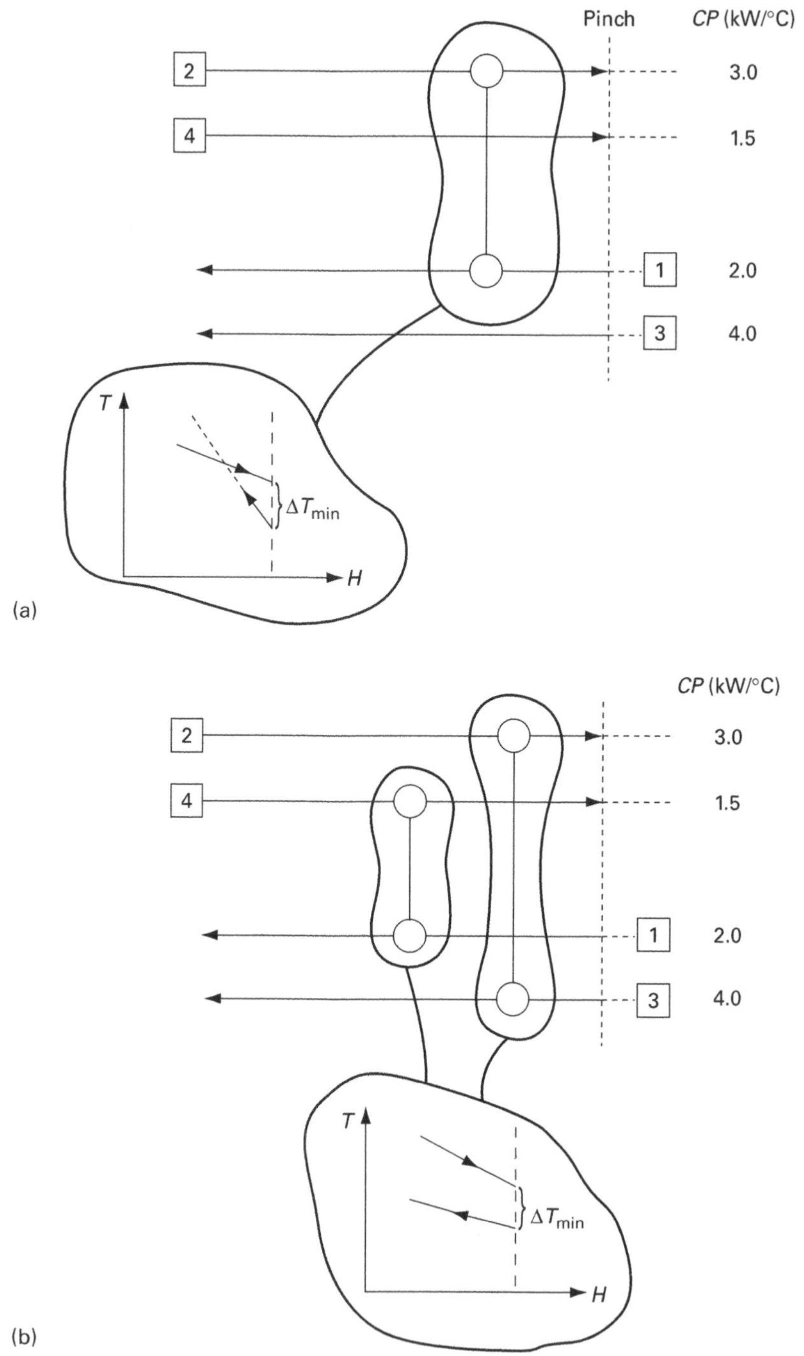

Pinch CP (kW/°C)

2 3.0

4 1.5

1 2.0

3 4.0

T

ΔT_{min}

H

(a)

CP (kW/°C)

2 3.0

4 1.5

1 2.0

3 4.0

T

ΔT_{min}

H

(b)

Figure 2.15 Consequences of matching streams of different *CP*s

the directions in the grid representation.) Because the CP of stream 2 is greater than that of stream 1, as soon as any load is placed on the match, the ΔT in the exchanger becomes less than ΔT_{min} at its hot end. The exchanger is clearly infeasible and therefore we must look for another match. In Figure 2.15(b), streams 2 and 3 are matched, and now the relative gradients of the T/H plots mean that putting load on the exchanger opens up the ΔT.

This match is therefore acceptable. If it is put in as a firm design decision, then stream 4 must be brought to pinch temperature by matching against stream 1 (i.e. this is the only option remaining for stream 4). Looking at the relative sizes of the CPs for streams 4 and 1, the match is feasible ($CP_4 < CP_1$). There are no more streams requiring cooling to pinch temperature and so we have found a feasible design at the pinch. It is the only feasible pinch design because only two pinch matches are required. Summarising, in design immediately above the pinch, we must meet the criterion:

$$CP_{HOT} \leqslant CP_{COLD}$$

Having found a feasible pinch design it is necessary to decide on the match heat loads. The recommendation is "maximise the heat load so as to completely satisfy one of the streams". This ensures the minimum number of heat exchange units is employed. So, since stream 2 above the pinch requires 240 kW of cooling and stream 3 above the pinch requires 240 kW of heating, co-incidentally the 2/3 match is capable of satisfying both streams. However, the 4/1 match can only satisfy stream 4, having a load of 90 kW and therefore heating up stream 1 only as far as 125°C. Since both hot streams have now been completely exhausted by these two design steps, stream 1 must be heated from 125°C to its target temperature of 135°C by external hot utility as shown in Figure 2.16. This amounts to 20 kW, as predicted by the Problem Table analysis. This is no coincidence! The design has been put together obeying the constraint of not transferring heat across the pinch (the "above the pinch" section has been designed completely independently of the "below the pinch" section) and not using utility cooling above the pinch.

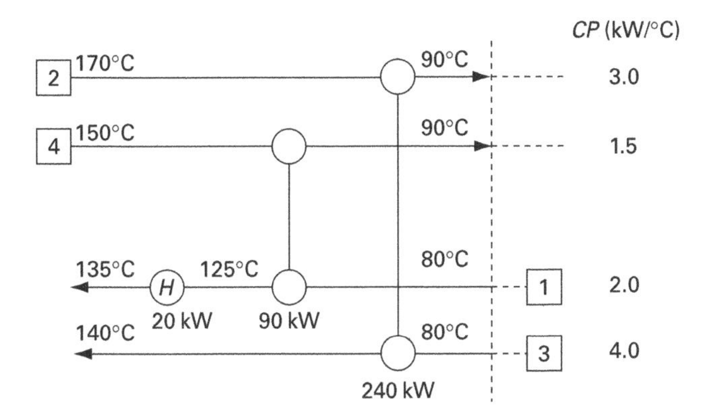

Figure 2.16 Above-pinch network design for four-stream problem

Below the pinch, the design steps follow the same philosophy, only with design criteria that mirror those for the "above the pinch" design. Now, it is required to bring cold streams to pinch temperature by interchange with hot streams, since we do not want to use utility heating below the pinch. In this example, only one cold stream 1 exists below the pinch, which must be matched against one of the two available hot streams 2 and 4. The match between streams 1 and 2 is feasible because the *CP* of the hot stream is greater than that of the cold, and the temperature difference increases as we move away from the pinch to lower temperatures. The other possible match (stream 1 with stream 4) is not feasible. Immediately below the pinch, the necessary criterion is:

$$CP_{\mathrm{HOT}} \geqslant CP_{\mathrm{COLD}}$$

which is the reverse of the criterion for design immediately above the pinch.

Maximising the load on this match satisfies stream 2, the load being 90 kW. The heating required by stream 1 is 120 kW and therefore 30 kW of residual heating, to take stream 1 from its supply temperature of 20–35°C, is required. Again this must come from interchange with a hot stream (not hot utility), the only one now available being stream 4. Although the *CP* inequality does not hold for this match, the match is feasible because *it is away from the pinch*. That is to say, it is not a match that has to bring the cold stream up to pinch temperature. So the match does not become infeasible (though a temperature check should be done to ensure this). Putting a load of 30 kW on this match leaves residual cooling of 60 kW on stream 4 which must be taken up by cold utility. Again, this is as predicted by the Problem Table analysis. The below-pinch design including the *CP* criteria is shown in Figure 2.17.

Putting the "hot end" and "cold end" designs together gives the completed design shown in Figure 2.18. It achieves best possible energy performance for a ΔT_{min} of 10°C incorporating four exchangers, one heater and one cooler. In other words, six units of heat transfer equipment in all. It is known as an **MER** network (because it achieves the **minimum energy requirement**, and **maximum energy recovery**).

Summarising, this design was produced by:

- Dividing the problem at the pinch, and designing each part separately.
- Starting the design at the pinch and moving away.
- Immediately adjacent to the pinch, obeying the constraints:
  $CP_{\mathrm{HOT}} \leqslant CP_{\mathrm{COLD}}$ (above) for all hot streams
  $CP_{\mathrm{HOT}} \geqslant CP_{\mathrm{COLD}}$ (below) for all cold streams
- Maximising exchanger loads.
- Supplying external heating only above the pinch, and external cooling only below the pinch.

These are the basic elements of the "pinch design method" of Linnhoff and Hindmarsh (1983). Network design is not always so easy – for example, streams may have to be split to meet the *CP* criteria at the pinch. Usually, trade-offs are made, known as "relaxing" the network to reduce the number of exchangers, at the expense of some increase in utility loads. All of this will be further elaborated in Chapter 4.

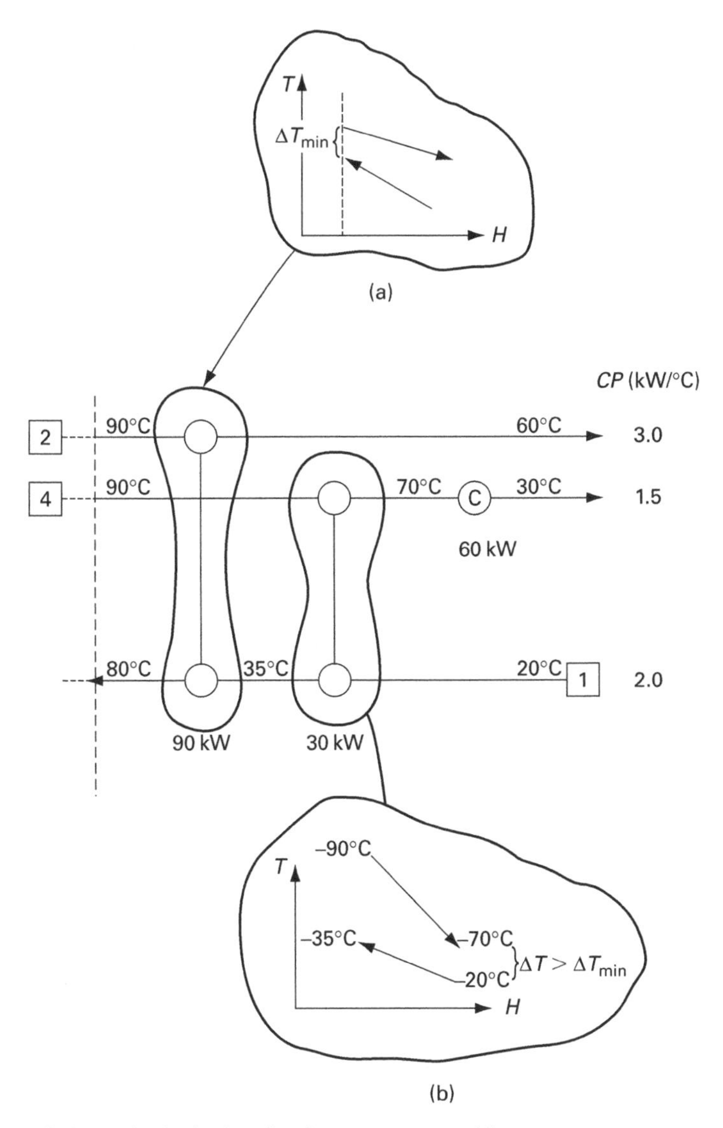

Figure 2.17 Below-pinch design for four-stream problem

2.3.4 A word about design strategy

The method just described does not follow the traditional intuitive method for heat exchanger network design. Left to his own devices, the engineer normally starts to design from the hot end, working his way towards the cold. However, the "pinch design method" *starts the design where the problem is most constrained*. That is, at the pinch. The thermodynamic constraint of the pinch is "used" by the designer to help him identify matches that must be made in order to produce efficient designs. Where it is possible to identify options at the pinch (and this will be discussed later), the

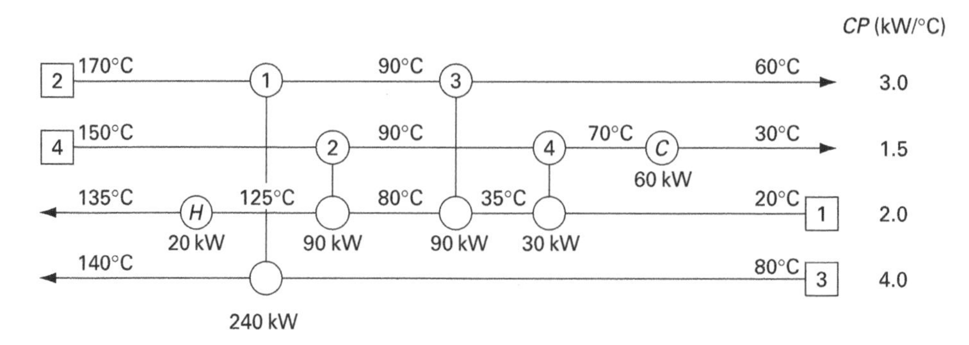

Figure 2.18 Network design achieving energy targets

designer may choose the one he likes for control, layout, safety or other reasons, and still be sure that an energy-efficient design will result. Likewise, away from the pinch, the design is less constrained and the designer knows that he has more leeway.

2.4 Choosing ΔT_{min}: supertargeting

2.4.1 Further implications of the choice of ΔT_{min}

So far, we have seen that higher values of ΔT_{min} give us higher hot and cold utility requirements, and it therefore seems that we want a ΔT_{min} as low as possible, to give maximum energy efficiency. However, there is a drawback; lower ΔT_{min} values give larger and more costly heat exchangers. In a heat transfer device, the surface area A required for heat exchange is given by:

$$A = \frac{Q}{U \Delta T_{LM}} \tag{2.4}$$

A is in m^2, Q is the heat transferred in the exchanger (kW), U is the overall heat transfer coefficient (kW/m^{2}K) and ΔT_{LM} is the log mean temperature difference (K). If we have a pure countercurrent heat exchanger, where the hot stream enters at T_{h1} and leaves at T_{h2}, and the cold stream enters at T_{c1} and exits at T_{c2}, so that T_{c1} and T_{h2} are at the "cold end" C and T_{h1} and T_{c2} are at the "hot end" H of the exchanger, then ΔT_{LM} is given by:

$$\Delta T_{LM} = \frac{\Delta T_H - \Delta T_C}{\ln \dfrac{\Delta T_H}{\Delta T_C}} = \frac{T_{h1} - T_{c2} - T_{h2} + T_{c1}}{\ln \left(\dfrac{T_{h1} - T_{c2}}{T_{h2} - T_{c1}} \right)} \tag{2.5}$$

If $\Delta T_H = \Delta T_C$, ΔT_{LM} is undefined and ΔT is used in Equation (2.4). In essence, the heat exchanger area is roughly inversely proportional to the temperature difference. Hence, low values of ΔT_{min} can lead to very large and costly exchangers, as capital cost is closely related to area. Even if one end of an exchanger has a high

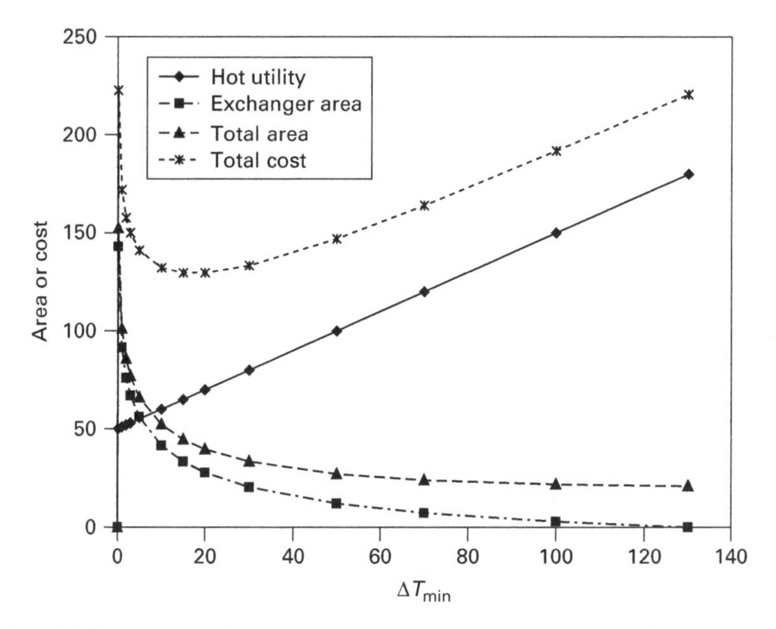

Figure 2.19 Utility use, exchanger area and cost variation with ΔT_{min}

temperature difference, the form of the expression for ΔT_{LM} means that it is dominated by the smaller temperature approach. Obviously a low ΔT_{min} value gives a low ΔT_{LM}; the precise relationship between them is discussed by Heggs (1989).

For our two-stream example, Figure 2.19 plots the utility use and heat exchanger surface area (assuming a value of $0.1\,kW/m^2K$ for the heat transfer coefficient), and also shows the effect of including heaters and coolers (assuming sensible temperature levels). Here, utility use rises linearly with ΔT_{min}, whereas exchanger area rises very sharply (asymptotically) for low ΔT_{min} values.

Now, if we assume energy cost is proportional to energy usage, and that heat exchanger cost (as a first approximation) is proportional to surface area, we can sum the operating and capital cost. Since energy cost is per hour and capital cost is a one-off expenditure, we either need to calculate the energy cost over a period (say 1–2 years) or annualise the capital cost over a similar period. The chosen timescale is known as the **payback** time. This then gives us the combined total cost graph at the top of Figure 2.19, including utility, exchanger, heater and cooler cost. We can see that there is an optimum for ΔT_{min} – in this case, about 15–20°C.

Clearly, it will be important to choose the right value of ΔT_{min} for our targeting and network design. This can be done by area and cost targeting, or **supertargeting**, based on the concept above and described in full in Section 3.7. Supertargeting is much less exact than energy targeting, because there are many uncertainties – heat transfer coefficients, total area of a exchanger network and costs are all subject to variation. However, we note that the total cost curve has a relatively flat optimum, so there is a fair amount of leeway. As long as the chosen ΔT_{min} is not excessively small or large, a reasonable design should be obtained by using a sensible "experience value" for ΔT_{min}, at least in the initial stages. Often a value of 10°C or 20°C is best, but in some industries, a very much lower or higher value is appropriate.

Summarising, the key points of area and cost targeting are:

- There is usually an optimal value for ΔT_{min} part way through the feasible range.
- The optimum is not exact and a significant error can usually be tolerated initially.
- The optimum ΔT_{min} is reduced by higher energy costs, lower capital costs or a longer payback period.

2.5 Methodology of pinch analysis

2.5.1 The range of pinch analysis techniques

This section has given us an insight into the key concepts which are the building blocks of pinch analysis. However, much more detailed analysis is needed for application in real industrial practice, and this will be covered in the following chapters. As well as refinements of the methods already described, further major techniques which will be covered are:

- Data extraction – how to take a process flowsheet, form a consistent heat and mass balance and extract the stream data needed for a pinch analysis (Sections 3.1 and 3.2).
- The Appropriate Placement principle – how hot and cold utilities, separation systems and other process items should relate to the pinch and the GCC (Section 3.3.2).
- Multiple hot and cold utility levels – integrating the heating and cooling systems optimally with the process (Section 3.4).
- Network relaxation and optimisation – modifying a network to eliminate small exchangers which are not cost-effective, or other undesirable features (Section 4.4).
- Retrofit of existing plants – adapting the techniques to deal with existing exchangers and plant-layout (Section 4.7).
- Heat and power systems, heat pumps and refrigeration systems (Sections 5.2 and 5.3).
- Process change – altering operating conditions of unit operations and other streams to maximise heat integration (Section 6.2).
- Handling batch processes (Sections 7.1–7.7) and other time-dependent situations such as startup and shutdown (Section 7.9).

2.5.2 How to do a pinch study

To apply the techniques listed above in practice, a systematic study method is required. The stages in a process integration (pinch) analysis of a real process plant or site are as follows:

1. Obtain, or produce, a copy of the plant flowsheet including temperature, flow and heat capacity data, and produce a consistent heat and mass balance (Sections 3.1 and 8.2).

2. Extract the stream data from the heat and mass balance (Sections 3.1 and 8.3).
3. Select ΔT_{min}, calculate energy targets and the pinch temperature (Chapter 3 for continuous processes, Chapter 7 for batch processes).
4. Examine opportunities for process change, modify the stream data accordingly and recalculate the targets (Chapter 6).
5. Consider possibilities for integrating with other plants on site, or restricting heat exchange to a subset of the streams; compare new targets with original one (Section 3.5).
6. Analyse the site power needs and identify opportunities for combined heat and power (CHP) or heat pumping (Chapter 5).
7. Having decided whether to implement process changes and what utility levels will be used, design a heat exchanger network to recover heat within the process (Chapters 4 and 8).
8. Design the utility systems to supply the remaining heating and cooling requirements, modifying the heat exchanger network as necessary (Chapters 3, 4 and 5).

Two particular points should be noted. Firstly, the order of operations is not the same as the order of the chapters in this book! In particular, options for changing the process configuration and conditions need to be considered in the initial targeting stage and before network design, as will be seen in Section 6.6 for example. Secondly, a considerable amount of preparatory work is required before doing the targeting analysis to form a heat and mass balance, identify our streams and derive the stream data. Extracting stream data from a process flowsheet is immensely important, yet full of pitfalls; it is one of the most difficult and time-consuming parts of a practical study, but has been largely neglected in published literature.

Exercise

Calculate the hot and cold composite curves, Problem Table, GCC, hot and cold utility targets and pinch temperature for the following problem at a global ΔT_{min} of 10°C, using a calculator, your own spreadsheet or the spreadsheet supplied with this book (Table 2.4).

Table 2.4 Stream data for five-stream problem

Stream ID	*Stream type*	*Supply temperature (°C)*	*Target temperature (°C)*	*Heat capacity flowrate (kW/K)*	*Heat load (kW)*
1	Hot	200	50	3	450
2	Hot	240	100	1.5	210
3	Hot	120	119	300	300
4	Cold	30	200	4	−680
5	Cold	50	250	2	−400

References

Heggs, P. J. (1989). Minimum temperature difference approach concept in heat exchanger networks, *J Heat Recov Syst CHP*, 9(4): 367–375.

Hohmann, E. C. (1971). *Optimum networks for heat exchangers*, PhD. Thesis, University of S. California, USA.

Linnhoff, B. and Flower, J. R. (1978). Synthesis of heat exchanger networks: Part I: Systematic generation of energy optimal networks, *AIChE J*, 24(4): 633–642. Part II: Evolutionary generation of networks with various criteria of optimality, *AIChE J*, 24(4): 642–654.

Linnhoff, B. and Hindmarsh, E. (1983). The pinch design method of heat exchanger networks, *Chem Eng Sci*, 38(5): 745–763.

Linnhoff, B., Mason, D. R. and Wardle, I. (1979), Understanding heat exchanger networks, *Comp Chem Eng*, 3: 295–302.

3 Data extraction and energy targeting

This chapter covers the practicalities of the targeting process in considerably more depth than the initial summary in Section 3.2, with the following subsections:

Forming a heat and mass balance and extracting the stream data (Section 3.1).
Individual stream contributions and threshold problems (Section 3.3).
Multiple utilities, balanced composite and grand composite curves (Section 3.4).
Zonal targeting and pressure drop targeting (Section 3.5).
Targeting for number of equipment units (exchangers, heaters and coolers), heat exchanger area and shells, and topology traps (Section 3.6).
Supertargeting for variation of capital, operating and total cost with ΔT_{min} (Section 3.7).

We also introduce a major new case study for an organics distillation unit; stream data extraction is covered in Section 3.2 and targeting in Section 3.8.

3.1 Data extraction

Extracting the stream data from the process flowsheet is an unspectacular, unexciting but absolutely crucial part of pinch analysis. It is not always clear-cut and sometimes alternative approaches present themselves. If the wrong method is chosen, one may either end up with impossible targets which cannot be achieved by realistic equipment, or conversely with a system which is so constrained that the only possible network is the one you already have!

At this stage we will simply cover the key aspects which will allow you to perform basic stream data extraction. Chapter 8 goes into more detail on the many practical problems which can be encountered in more complex situations. Details may also be found in ESDU (1987).

3.1.1 Heat and mass balance

The first step necessary is to produce a heat and mass balance for the plant. This is usually a significant challenge; on real process plant, the problem with the heat and mass balance is that it almost always doesn't! Significant data reconciliation must then take place. The most important requirement is for consistency, rather than precise accuracy. Using a "balance" with serious inconsistencies will only lead to trouble when we attempt to translate the targets into practical design projects.

For a new build, a heat and mass balance can be formed from design data on flowrates and compositions and literature data on specific heat capacity, etc. For an existing plant, even if the design data is available, it will often be significantly different to actual performance. Process design is an inexact science; settings and flows will usually be altered during commissioning to obtain stable behaviour and the desired output. Moreover, raw material composition may have altered since initial start-up, or may vary with time, while heat exchangers become fouled and their performance drops. Hence, a new heat and mass balance should be formed, reflecting current performance, possibly with a range of scenarios for different feedstocks or before/after cleaning (multiple base cases, Section 4.8).

The mass balance needs to be based on mass flowrates. Few plants have flowmeters on all streams and they can show significant errors; metres for steam (or other condensing fluids) are particularly prone to error, figures of 30% or more being not uncommon. It is always worthwhile to do a crosscheck against annual production of the various components (here, the different oil fractions and the amount of crude feed supplied from storage), bearing in mind the variation between multiple feedstocks. The mass balance can then be reconciled, though often with some difficulty; it may be necessary to modify supposed flowrates which have been taken as "gospel" for years.

The heat balance is more complicated. The raw data are temperatures, heat loads and the flowrates from the mass balance. Heat losses will have to be allowed for, in contrast to mass balances where losses from leaks are usually minuscule (except if the process includes large air or gas flows). Temperature is usually the most accurately measured parameter on a plant, often to within 1°C, although effects of fouling or location in "dead spots" must be checked for. Often, where there is no permanent instrument in a location, there is a sampling point where a thermocouple can be introduced, or the external pipe temperature may be measured (obviously less accurate). Heat loads are more difficult. Specific heat capacities and latent heats can be obtained from literature, manufacturers' data or (if necessary) measurement. Cooling loads may be found from the flowrate and temperature drop of the cooling water, and heater loads from the steam flow (often inaccurate; condensate flow measurement may be preferable). Heat transfer from furnaces is particularly difficult to measure precisely, because of the heat losses via the stack. Closing the heat balance is usually a challenge, and in some cases it may be necessary to refine the mass balance to get realistic figures.

The key point which must be remembered throughout is that our aim is to produce a reliable balance from which we can extract temperature and heat load data for streams.

3.1.2 Stream data extraction

Having obtained a reliable heat and mass balance, the next stage is to extract the hot and cold streams in the form required for pinch analysis.

A sensible criterion for a stream is that it should **change in heat load but not in composition**. Hence a flow of liquid through a heat exchanger, or a single-component liquid being evaporated, or a mixture which is being cooled without any separation of the components, can all be represented as streams. Conversely, a flow of liquid through an absorption column or a scrubber, or a mixture which is reacting, or a flow through a distillation column which emerges with a volatile component removed, should not be treated as a single stream; better methods will be found in Chapter 6.

The stream data required will be the temperature range ($T_1 \rightarrow T_2$), stream type (hot or cold) and either the heat capacity flowrate CP (kW/K) or the stream heat load ΔH (kW). These last can be obtained in several ways:

- from stream mass flows and published (or measured) specific heat capacities,
- from mass flows and specific enthalpy data,
- from heat loads measured on heat exchangers,
- by back-calculation from other stream heat loads in the heat balance.

The various quantities are simply linked by the equation:

$$\Delta H = CP \cdot (T_2 - T_1) = C_P \dot{m}(T_2 - T_1) = \dot{m}(h_2 - h_1)$$

where
$\dot{m}$ is the mass flowrate (kg/s);
C_P the specific heat capacity (kJ/kgK);
h the specific enthalpy (kJ/kg).

Other sets of units can be adopted, but either they must be mutually consistent or the correct conversion factors must be applied (see Section 8.4.5).

3.1.3 Calculating heat loads and heat capacities

The algorithms for pinch analysis were originally stated in terms of heat capacity flowrate CP. However, in most practical cases it is more convenient to work from heat loads in kW rather than heat capacity flowrates in kW/K. (For alternative sets of units, see Section 8.4.5). For a liquid mixture stream on an existing plant, the specific heat capacity may not be known, whereas the heat load can be easily found using the known loads on existing heat exchangers at known temperatures. It is then easy to back-calculate CP from this data. However there are two possible pitfalls. Firstly the quoted equipment loads may be design rather than operating loads. Secondly, the CP will tend to vary with temperature, and if only a few data points over a wide temperature range are taken, this dependence may be miscalculated. Most seriously, internal latent heat changes (e.g. from partial vaporisation or condensation), which give a very large local change in CP, may be disguised. This would often lead to serious targeting errors. When latent heat changes occur, the dew and bubble points should be fixed as linearisation points from the outset. This is because the location of the pinch is often defined by such points.

When entering data, latent heat loads again need some care. Theoretically, they can be considered to be streams with a finite heat load at a fixed temperature, so that the heat capacity flowrate *CP* is infinite. In practice, some targeting software cannot handle this and the target temperature has to be set slightly different to the supply temperature – say 0.1°C. In such cases, keep the supply temperature at its original value and alter the target temperature – upwards for a cold (vaporising) stream, downwards for a hot (condensing) stream. The reason is that the pinch is always caused by a stream beginning, so this will ensure that the pinch temperature is still exactly correct.

In the basic Problem Table method, all streams were assumed to have *CP*s independent of temperature. In real problems heat capacities are always dependent to some extent on temperature, and not all targeting software can handle this. So it is important to know when the linear approximation is valid, and when it is not. Remembering that the targets depend above all on how closely the hot and cold composite curves approach at the pinch, and their shapes in this region, it is clear that *data errors are most significant at the pinch*. It is therefore in the neighbourhood of the pinch that we must be most careful about the approximation of *CP*.

If it is not acceptable to assume that *CP* is constant, streams should be linearised in sections. This operation maintains the validity of the Problem Table algorithm (linear sections being handled in just the same way as whole linear streams), whilst improving accuracy where necessary. For variable *CP*, "safe-side" linearisation should be practised if possible, that is the hot stream linearisation should always be on the "cold side", at slightly lower temperatures than the actual data, and *vice versa* for the cold stream, as in Figure 3.1. This ensures that predicted energy targets can always be met in practice.

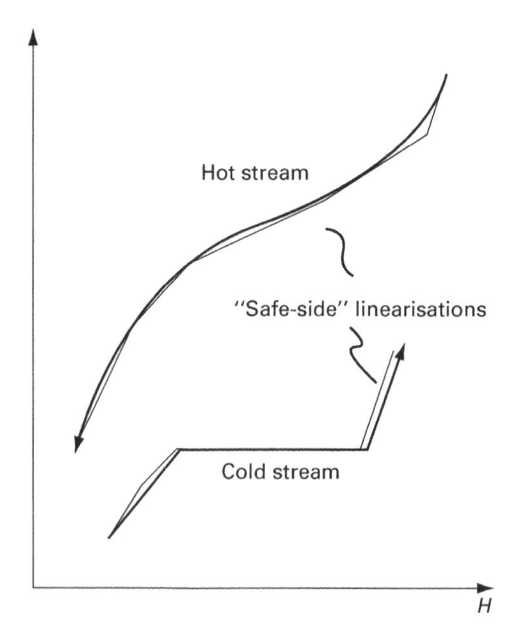

Figure 3.1 "Safe-side" linearisation

A typical situation is where one or more streams' heat capacity is given as a polynomial in temperature T, so that:

Heat capacity flowrate $CP = c_0 + c_1T + c_2T^2 + c_3T^3 \ldots$
Heat load $H = c_0T + (c_1T^2/2) + (c_2T^3/3) + (c_3T^4/4) \ldots$

Streams may be split into two or more segments, or simply left as a single stream if the variations with temperature are not great. When the initial targeting has been performed, the relevant streams should be split into two at the pinch, and the targets are then re-evaluated with the improved data. This will usually alter the energy targets but almost never the pinch temperature (since the pinch is always caused by a stream beginning or experiencing a major increase in CP). Hence, no further iteration is needed. This method can be seen in action on the case study in Section 3.8.1.

If there is a gross change in CP, for example where a stream begins to vaporise or condense and a large latent heat load is added in, it is safest to split the streams into segments or, if the targeting software does not allow this, to treat the sections as separate streams.

Summarising, the strategy for use of data should be:

- Use rough data first.
- Locate the pinch region by Problem Table targeting.
- Use better data in the neighbourhood of the pinch.

3.1.4 Choosing streams

How much should we subdivide streams when they pass through intermediate process vessels such as storage tanks and pumps?

Consider a stream which is currently heated from 10°C to 30°C, passed through a storage tank, heated to 80°C in a heat exchanger and then to 120°C in a utility heater (Figure 3.2). It could be represented in the stream data as three separate streams. This would give the correct targets, but in the network design phase we have constrained the temperatures of the stream sections so tightly that we would find each section "perfectly" matched to its original partner, that is we would be very likely to generate the original flowsheet! Suppose instead that two streams were defined, one from 10°C to the storage temperature of 30°C, and one from 30°C to the final temperature of 120°C. Now we stand a better chance of finding different matches and improving the design. However, the storage temperature of 30°C is probably not critical. If the feed is represented by one stream running right through from 10°C to 120°C, the chances of finding an improved design are greater still. The storage temperature can be fixed at the "natural" break point between two matches.

In other words, if you break up process flows into too many separate streams, you increase the apparent complexity of the network (more streams), add unnecessary constraints and are likely to conceal heat recovery opportunities (because a new exchanger may seem as if it needs to be two or three separate matches on separate streams, instead of a single match).

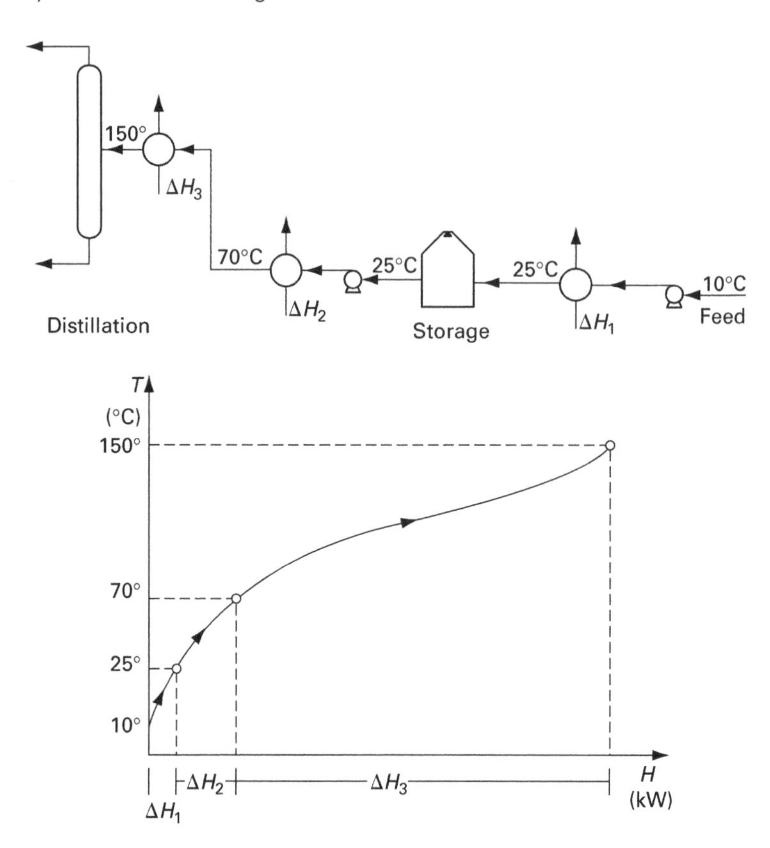

Figure 3.2 Data extraction from an existing process flow

In general, the designer should in the first instance decide which supply and target temperatures he is going to define as "hard" (invariant), and which as "soft" (with some leeway). He will then produce a base case heat exchanger network design, after which it may become apparent that by making changes to temperatures originally classified as hard, further improvements could be made. These decisions are clearly entirely dependent on the process technology and to some extent on the designer's experience. However, the designer should always be on the lookout for opportunities of improving his networks by modifying the base data. He therefore needs to know which constraints on his plant are real, and which areas are flexible; and if the latter, by how much.

For example, in crude oil distillation trains such as that described in the case study in Section 9.2, there is usually a desalter vessel on the crude oil feed stream. The mass of salt removed has a negligible effect on the heat balance and it seems attractive to treat the feed before and after the desalter as a single stream. However, there is generally a tight temperature constraint on the desalter – it will only operate effectively at a temperature between, say, 125°C and 130°C. If the crude oil feed is treated as a single stream, a match may not finish at the correct temperature to allow effective desalter operation. So it is better to have two separate streams in this instance, but noting that there is a little flexibility on the break-point temperature if necessary.

Again, we see that pinch analysis is not a substitute or replacement for the skill and experience of the designer, but a supplement, enabling him to exploit his knowledge in the most effective way.

One can also have "optional" streams. Typically, these are hot products or waste streams which could be allowed to cool down naturally (or discharged to atmosphere when warm – a so-called "national cooler"), or passed through a heat exchanger so that heat can be recovered. Different targets will be obtained with and without these streams. It is up to the designer whether or not to include them in the main "base case"; the other option can then be treated as an alternative case. Process change analysis (Chapter 6) is useful, particularly the plus–minus principle.

It is also vital that for a vessel in which a significant change in composition takes place – for example a distillation column, evaporator or reactor – the input and output streams are kept separate, not combined. Again, the temperature of the unit operation can be optimised by the methods given in Chapter 6, but it cannot be ignored – it is a vital processing step.

3.1.5 Mixing

Mixing and splitting junctions can also cause problems in stream data extraction. Consider two process flows of the same composition leaving separate units at different temperatures, mixing and then requiring heating to a common final temperature. This could be considered as one stream, and the heating duty could be performed by a single heat exchanger. However, mixing degrades temperature. Consider what may happen if the system is regarded as only one stream for energy targeting. If the mixing temperature lies below pinch temperature, then the "cooling ability" of the cold stream below the pinch is degraded. More heat must therefore be put to utility cooling, and by enthalpy balance, heat must be transferred across the pinch increasing hot utility usage. To ensure the best energy performance at the targeting stage, the mixing should be assumed isothermal. Hence, heat each stream separately to its final temperature, or heat/cool one stream to the temperature of the other, then mix, and then heat/cool the resulting mixture to its final temperature.

Figure 3.3 shows a numerical example, this time for two hot streams. The correct method is to assume that the streams mix isothermally, here at the target temperature of 30°C. If the original layout were retained and the pinch corresponded to a hot stream temperature between 70°C and 100°C, energy would be wasted. However, whatever the pinch temperature is, mixing the streams will degrade temperatures and reduce the driving forces in heat exchangers, giving increased capital cost. This is illustrated by the hot composite curves in Figure 3.4; the solid line is for mixing and the broken line for the situation where the streams are kept separate and run down to the final temperature.

3.1.6 Heat losses

Heat losses, like taxes, are an annoying and unavoidable fact of life. In pinch analysis, they make stream data extraction more complicated by causing a mismatch

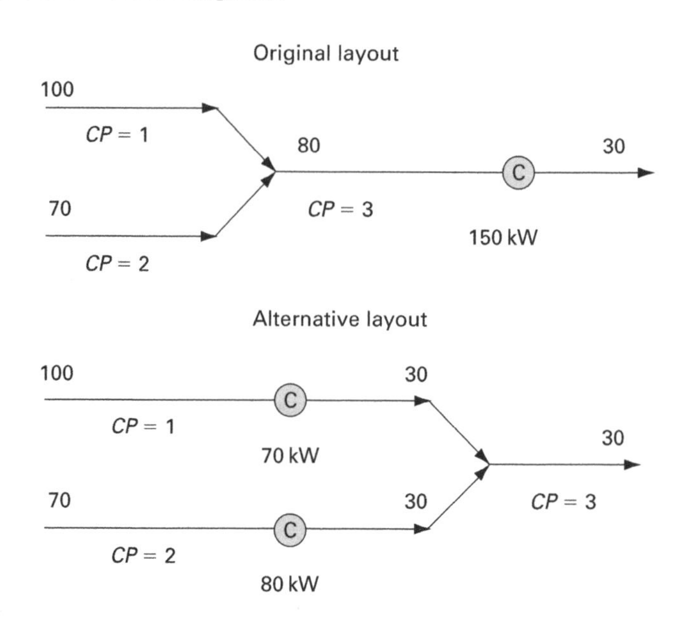

Figure 3.3 Data extraction for mixing streams

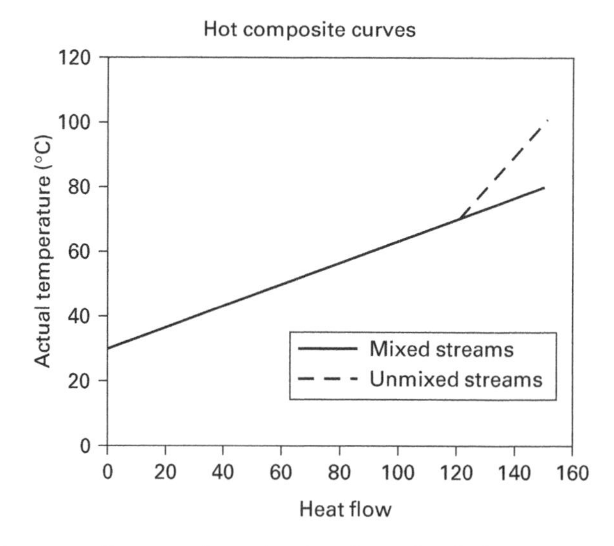

Figure 3.4 Hot composite curve for mixed and unmixed streams

between streams' inherent heat capacities and the actual amount of heat which must be supplied or can be extracted.

One obvious example is a heat exchanger which exchanges 100 kW of heat between hot and cold streams but also experiences 10 kW heat losses from the outer surface of the shell. If the hot stream is on the shell side, its net enthalpy will fall by 110 kW instead of 100, and the temperature drop will be greater than expected. A stream heat load calculated from specific heat capacity data, mass flowrate and overall temperature drop will overestimate the heat recoverable in practice from the stream. Another obvious source of heat losses is from long pipe runs.

The case study in Section 3.2 illustrates heat losses being allowed for in practice.

3.1.7 Summary guidelines

Some useful overall guidelines to bear in mind when extracting data are:

1. keep your hot streams hot and your cold streams cold,
2. avoid over-specifying the problem; don't break up streams unnecessarily,
3. avoid non-isothermal mixing at the energy targeting phase,
4. check and refine data in the neighbourhood of the pinch (or other areas of low net heat flow),
5. identify any possible process constraints and find targets with and without them, to see what energy penalty they impose (the same technique as is used for targeting two plants separately or thermally linked together).

3.2 Case study: organics distillation plant

3.2.1 Process description

The two- and four-stream examples used so far have been relatively simple situations. We will now introduce a third example which is based closely on a real process – a small-scale fractional distillation of an organic mixture – and is an existing plant, rather than a new design. In some respects this is similar to a smaller-scale version of the crude distillation unit described in Section 9.2. This will allow us to see stream data extraction at work in a practical situation.

The crude feed is supplied at ambient temperature and fed to a distillation column at atmospheric pressure, where it is split into three fractions: light oil, middle oil and residue. Naturally, the feed has to be heated up to the operating temperature of the column and in this case it must also be partially vaporised, as there is no separate reboiler; some flashing also occurs as the hot liquid enters the column. The light oil comes off the top as vapour (overheads), is condensed and the majority is recycled to provide the top reflux; the remainder is cooled and a small amount of water is removed in a gravity separator. The various products are cooled to different levels, depending on their viscosity. The crude feed passes through two heat exchangers, and is heated by the overheads and middle oil, before entering a furnace which brings it up to its final feed temperature. All other heating and cooling duties are performed by utilities. The overall flowsheet is shown in Figure 3.5.

3.2.2 Heat and mass balance

Information is available on the production rates of the various fractions and their specific heat capacities, and several of the temperatures are recorded by thermocouple,

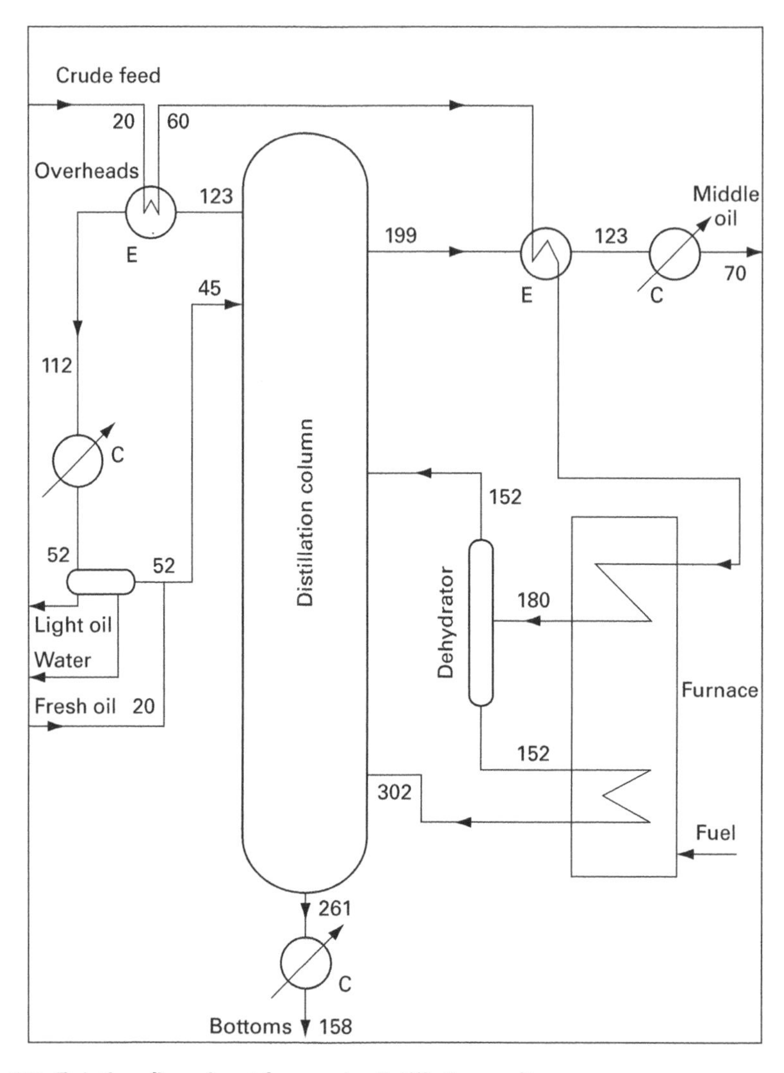

Figure 3.5 Existing flowsheet for crude distillation unit

as shown on the process flowsheet. Hence it is possible to form a preliminary heat and mass balance, as shown in Table 3.1. However, some information is unknown, including a number of temperatures and the exact reflux ratio (it is thought to be approximately 5:1). Some units conversion has been necessary to give a consistent set. The specific heat capacity for the crude feed is known to be significantly temperature-dependent, at $(2 + 0.005T)$ kJ/kg K; since the mass flow is 10 kg/s, $CP = (20 + 0.05T)$ kJ/K and, integrating, enthalpy (relative to a 0°C datum) is $(20T + 0.025T^2)$ kJ/s (kW).

No data are available on heat transfer coefficients. However, the sizes of the two existing heat exchangers are known and information can be deduced from them to obtain the HTC's and also provide further information on the heat balance. The heat

Table 3.1 Heat and mass balance for organics process

Flow	*Production rate (te/h)*	*Mass flow (kg/s)*	*Specific heat (kJ/kg K)*	*CP (kW/K)*	*Initial temperature (°C)*	*Final temperature (°C)*	*Heat flow rate (kW)*
Crude feed	36	10	2 + 0.005T	25 (mean)	20	180	4,000
Dehydrate	?	(9.67)	3.1	30	152	302	4,500
Bottoms	14.4	4	2.5	10	261	158	1,030
Middle oil	18	5	2	10	199	70	1,290
Light oil	9.6	2.67	2	5.33	52	52	0
Overheads	?	?	?	?	112	45	?
Fresh oil	7.2	2	2	4	20	45	100
Water	1.2	0.33	4.19	1.4	52	52	0

Note: Question marks denote unknowns, bracketed values were found by back-calculation.

Table 3.2 Data for existing heat exchangers

Streams	*Exchanger area (m^2)*	*Hot stream temperatures (°C)*	*Cold stream temperatures (°C)*	*Log mean temperature difference (°C)*	*Calculated heat load (kW)*	*Calculated overall HTC (kW/m^2K)*
CF/Ohds	57.5	123–112	20–60	77	880	0.20
CF/MO	73.2	199–123	60–(92)	83	760	0.125

load on the crude feed–overheads exchanger can be calculated from the known temperatures and heat capacity flowrate of the crude feed; likewise, that for the middle oil–crude feed exchanger can be calculated from the details of the middle oil, and the exit temperature of the crude feed can then be back-calculated as 92–C. The overall heat transfer coefficients U for the exchangers are then calculated to be 0.20 and 0.125 kW/m^{2}K, respectively. We remember that the basic relationship of U to the film heat transfer coefficients h (ignoring fouling and wall resistances) is:

$$\frac{1}{U} = \frac{1}{h_1} + \frac{1}{h_2}$$

Assuming that all organic liquids have the same film HTC, setting $h_1 = h_2$ for the crude feed/middle oil exchanger gives h as 0.25 kW/m^{2}K. Back-calculation for the crude feed/overheads exchanger gives a film HTC for the overheads of 1.0 kW/m^{2}K, which is reasonable for a condensing stream (Table 3.2).

The *CP* for the condensing overheads over this range can be calculated as 80 kW/K. This will not, however, apply to the rest of the stream. Measurements of cooling water flow and temperature drop indicate that the heat load on the following overheads cooler is 1.8 MW, and since the temperature drop is 60°C this gives a *CP* of 30 kW/K.

The current level of heat recovery in these two exchangers is 1,640 kW.

Table 3.3 Process stream data

Stream name	*Stream type*	*Initial temperature (C)*	*Target temperature (C)*	*Film HTC (kW/m^2K)*	*Heat capacity flow rate (kW/K)*	*Heat flow rate (kW)*
Bottoms	Hot 1	261	158	0.25	10	1,030
Middle oil	Hot 2	199	70	0.25	10	1,290
Overheads	Hot 3A	123	112	1	80	880
	Hot 3B	112	52	1	30	1,800
Crude feed	Cold 1	20	180	0.25	20 + 0.05T	−4,000
Dehydrate	Cold 2	152	302	0.25	30	−4,500

3.2.3 Stream data extraction

With a consistent heat and mass balance available, the stream data can be deduced. Only flows which require heating or release heat need to be considered, so the light oil and water from the separator can be ignored. The overheads emerging from the separator are mixed directly with the fresh oil and direct heat exchange of 100 kW takes place. Theoretically, these two streams should be added to the analysis; in practice, the amount of heat involved is so small and the temperatures so low that they can be safely ignored. This leaves five actual streams, whose characteristics are listed in Table 3.3.

Note that the total heat load for the cold streams is 8,500 kW and for the hot streams is 5,000 kW. Since the current level of heat recovery is 1,640 kW, this implies that the current hot utility and cold utility demands are 6,860 kW and 3,360 kW, respectively.

3.2.4 Cost data

The final information to be collected is the cost of heating and cooling and the capital cost of new heat exchangers. The hours worked per year are also needed; here the figure is 5,000.

Heating is provided in the coal-fired furnace whose mean temperature is approximately 400°C; the fuel costs £72/ton, has a gross calorific value of 28.8 GJ/ton and the furnace has a gross efficiency of 75%. This equates to a cost of £3.33/GJ of useful heat delivered, or £12/MWh.

Cooling is much cheaper in this case; cooling water is recirculated to a cooling tower which works in the range 25–35°C. Charges for additives, maintenance, make-up water and treatment of a small amount of effluent average out at £0.5/MWh.

Exchanger costs are notoriously difficult to estimate accurately, but in this case we can assume the cost in pounds sterling to be £$(10{,}000 + 300A^{0.95})$. The only exception is that coolers can be cheaper to construct and the first constant can be reduced to 5,000. All of these are in any case only "ballpark" figures and, if any exchangers are proposed for installation, a proper budget price must be obtained from a manufacturer.

Targeting can now be performed and is described in Section 3.8.

3.3 Energy targeting

Basic energy targeting was described in Chapter 2. The Problem Table calculation finds the hot and cold utility requirements, pinch temperature and relationship between net heat flow and temperature for a chosen ΔT_{min} value. We will now take this further to look at more detail and a variety of special cases.

Step-by-step algorithms for calculating the Problem Table and composite curves are given in the Appendix to this chapter, Section 3.11.

3.3.1 ΔT_{min} contributions for individual streams

We can see from the simple equation for countercurrent heat exchange ($Q = UA\,\Delta T_{LM}$) that the heat exchanger area A required is inversely proportional to both the overall heat transfer coefficient U and the temperature difference ΔT on a match. So far, we have assumed that any streams with given temperatures can be matched, irrespective of their characteristics. However, some streams may be gases or viscous liquids with poor heat transfer coefficients, or may be prone to fouling heat exchanger surfaces. Heavy oils and waxes are particular culprits on both grounds. The U-value on matches involving these streams will be low, and hence the corresponding area of a match involving these streams will be alarmingly high.

Later, when we consider area targeting (in Section 3.6.3), we will see how this can be allowed for explicitly, but can we take it into account when calculating basic energy targets?

There is a simple method, which involves allocating an individual **ΔT_{min} contribution** to each stream. Normally this is $\Delta T_{min}/2$, as with other hot and cold streams. However, for an "awkward" stream, we can allocate a higher value, entitled ΔT_{cont}. The shifted temperature for that stream will then be given by $S = T \pm \Delta T_{cont}$. The Problem Table calculation using shifted temperatures can now proceed exactly as before, but the ΔT on any match involving this stream can now be higher. This compensates for the low U-value, so that the area of the new match will now be acceptable. As an example, suppose that liquid streams are assigned a contribution of 5°C and gas streams 10°C. Then a liquid/liquid match has a ΔT_{min} of 5 + 5 = 10°C, a liquid/gas match has a ΔT_{min} of 5 + 10 = 15°C and a gas/gas match has a ΔT_{min} of 10 + 10 = 20°C.

The calculation is an easy one and all commercial targeting software allows for the possibility of ΔT_{cont} on individual streams.

In the same way, a ΔT_{cont} can be assigned which is lower than $\Delta T_{min}/2$, in situations such as:

a boiling or condensing stream with high heat transfer coefficients,
a below-ambient cryogenic stream where economics favour maximum heat recovery to reduce expensive refrigeration costs,
a stream which is likely to undergo direct contact heat exchange (for which ΔT_{cont} may be set to zero, or even negative if the other stream involved has a positive ΔT_{cont}).

3.3.2 Threshold problems

In many processes, both hot and cold utilities are always required. For instance, in our two-stream example, even if ΔT_{min} were reduced to zero, and hence capital cost increased to infinity, the need for both hot and cold utilities remains. However, this is not true of all problems. In the four-stream example, Figure 3.6 shows that as ΔT_{min} is reduced, a point is reached (at 5.55°C) where no hot utility is required; at all lower values of ΔT_{min}, the only utility needed is 40 kW cold utility. The value of ΔT_{min} at which one utility target falls to zero is termed "$\Delta T_{threshold}$", and a situation where only one utility is required is called a *threshold problem*. If the composite curves are shifted further together, reducing ΔT_{min} further, this does not cause a further change in utilities requirements, although it does mean that, if desired, part of the hot utility could be supplied at the low temperature end of the problem, or any intermediate value.

Threshold problems fall into two broad categories, which can easily be distinguished by looking at the composite and grand composite curve (GCC). In one type, the closest temperature approach between the hot and cold composites is at the "non-utility" end and the curves diverge away from this point, as shown in

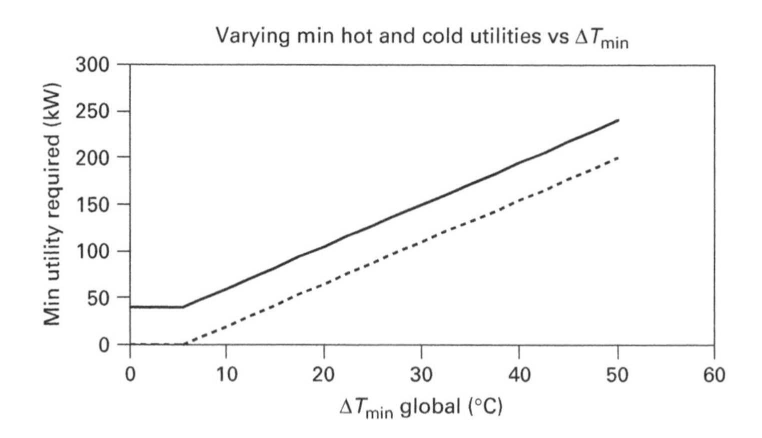

Figure 3.6 Energy–ΔT_{min} plot for four-stream example

Figure 3.7(a). In this case, design can be started from the non-utility end, using the pinch design rules; the GCC in Figure 3.8(a) shows the similarity to a true pinch, with zero heat flow. In the other type, there is an intermediate near-pinch, which can be identified from the composite curves as a region of close temperature approach and from the grand composite as a region of low net heat flow. This is the case for our four-stream example with $\Delta T_{min} = 5°C$, as shown by Figure 3.7(b); there is a near-pinch at 85°C for hot streams, 80°C for cold streams, and the GCC (Figure 3.8(b)) shows the net heat flow is only 2.5 kW at this point. Here it is often advisable to treat the problem like a "double pinch" and design away from both the near-pinch and the non-utility end. Network design for threshold problems is given in more detail in Section 4.5.1.

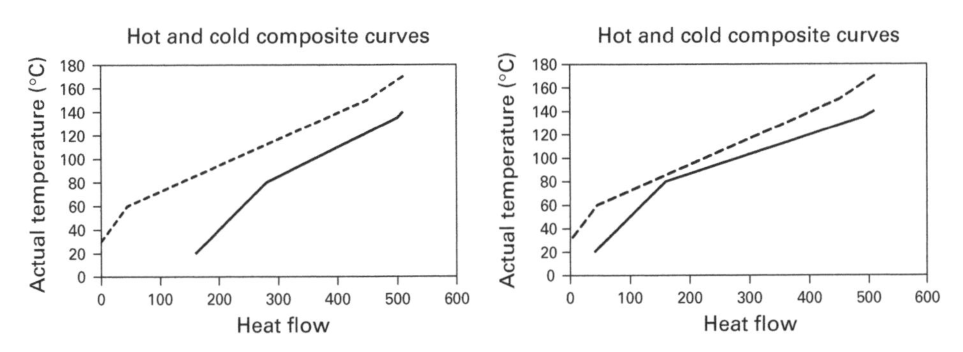

Figure 3.7(a/b) Composite curves for different types of threshold problem

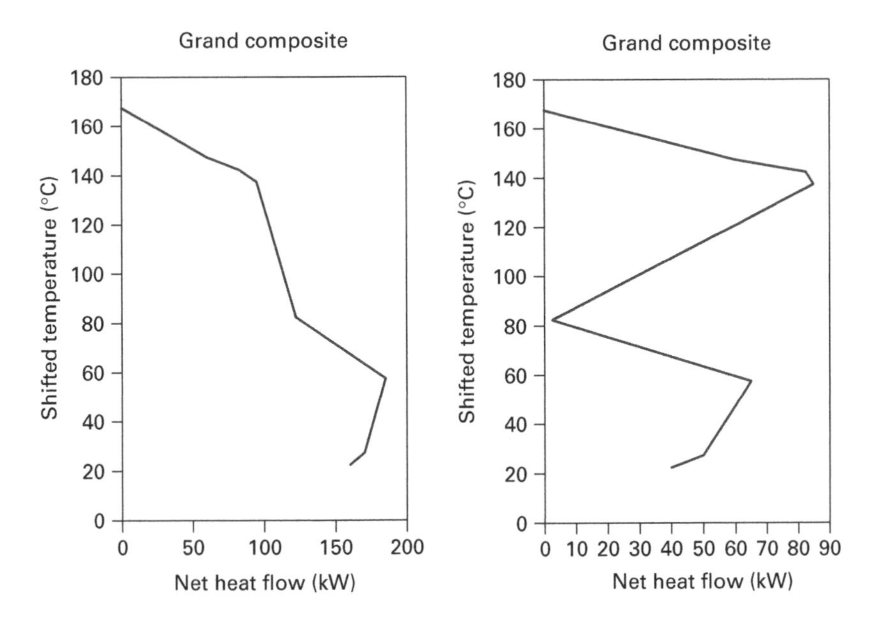

Figure 3.8(a/b) GCC for threshold problems

3.4 Multiple utilities

So far, we have just assumed that the external heating and cooling requirements are supplied as a single hot utility and single cold utility, at an unspecified temperature sufficient to fulfil the duty. In practice, more than one utility may be available, and there are often price differentials between them. Conversely, the needs of the process may help us to choose the utility levels. The GCC is our key weapon in this area.

3.4.1 Types of utility

Hot utilities, supplying heat to a process, may include:

1. furnaces,
2. steam heaters,
3. flue gas,
4. heat rejected from heat engines,
5. thermal fluid or hot oil systems,
6. exhaust heat from refrigeration systems and heat pump condensers,
7. electrical heating.

Likewise cold utility systems remove heat from the process. They may include:

1. cooling water systems,
2. air coolers,
3. steam raising and boiler feedwater heating,
4. chilled water systems,
5. refrigeration systems and heat pump evaporators,
6. heat engines below the pinch.

It is also useful to distinguish between constant- and variable-temperature utilities. For example, condensing steam (providing latent heat at a single temperature) is a **constant-temperature utility**, while hot flue gas (giving up sensible heat over a temperature range) is a **variable-temperature utility**. Some utilities are a mix of both types; for example, a furnace chamber gives out radiant heat at effectively a constant high temperature, whereas the exhaust gases can release further heat as a variable-temperature utility. The hot and cold utility needs may also be integrated with the power requirements for the site, as described in Chapter 5.

Often a wide range of hot and cold utilities can be used, and some will be more convenient than others. In particular, low-temperature heating may be cheaper than high-temperature heating; for example, low-pressure (LP) steam can be cheaper than high-pressure (HP) steam, especially in a combined heat and power system. Heat loads at high temperature may need to be supplied by a dedicated furnace. Correspondingly, some of the heat released below the pinch could be used at relatively high temperature to raise steam or preheat boiler feedwater, rather than being thrown away in cooling water. The biggest effect of temperature on unit cost of utilities occurs in *refrigeration systems*; below-ambient cooling needs heat pumping to ambient temperature, and the work requirement and cost increase steeply as the required temperature falls (see Section 5.3.2, Table 5.4).

3.4.2 The Appropriate Placement principle

Let us imagine that we have a heat source which can provide a fixed quantity of heat in kW but at any temperature we choose. Where should we place it to provide heat to the process? Clearly it should not go below the pinch, because it would then break one of our golden rules – don't heat below the pinch. Therefore it should be placed above the pinch. But furthermore, the process must be able to absorb the heat provided by the source at that temperature – otherwise some of it will again be wasted. The net heat flow at any shifted temperature is given by the Problem Table or GCC, and any amount of external heat up to this value can be provided at that temperature. Therefore the source should be placed not only **above the pinch**, but also **above the GCC**.

In the same way, a heat sink which removes heat from the process should do so **below the pinch** and **below the GCC**. If this condition is not met, not all the heat absorbed by the sink can come from the process, and additional hot utility must be supplied, which is clearly wasteful.

This, in essence, is the Appropriate Placement principle, originally introduced by Townsend and Linnhoff (1983). It is obvious that it provides the key to choosing the correct levels and loads for the various hot and cold utilities. Less obviously, it also applies to other process items such as reactors and separation systems, and this will be discussed in Chapter 6.

The penalty for violating either the Appropriate Placement principle or one of the three golden rules of network design in Chapter 2 is the same; heat is transferred across the pinch and both the hot and cold utility requirements go up. This has been stated before, but the GCC provides an elegant representation.

Figure 3.9 shows the GCC of the organics distillation unit. Suppose we have a heat source available at a shifted temperature of 200°C (e.g. medium pressure steam). Reading from the curve, we see that just under 2 MW can be provided to the process at this level. The remaining 2.8 MW can still be provided at or above the maximum process temperature of 330°C (e.g. using a furnace).

Now suppose we tried to supply more heat at 200°C – say 4 MW. 2 MW can be absorbed by the process successfully. However, the remainder cannot be used to heat cold streams above 200°C. It can heat cold streams below 200°C, but it will be substituting for hot streams which could also have done this. So the net effect is that the additional 2 MW cannot be used effectively. This heat will be transferred across the pinch and end up as additional cold utility. In effect, the grand composite is pushed to the right. It is now obvious that heat transfer across the pinch and inappropriate placement of utilities (or of any other heat source or sink) have the same effect; the hot and cold utilities are both increased by the amount of the violation, and the process no longer achieves its energy targets.

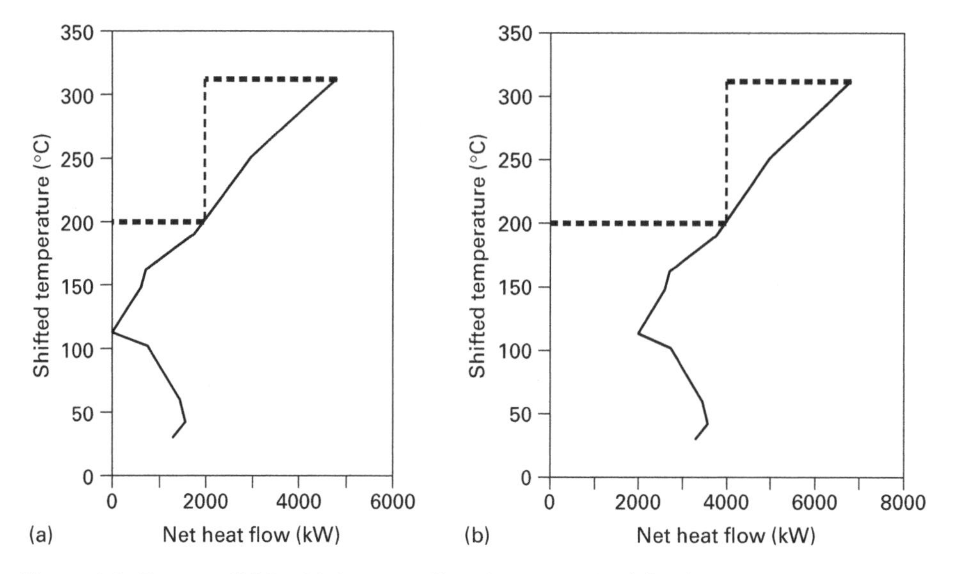

Figure 3.9 Process GCC with intermediate heat supply: (a) achieving target, (b) off target

3.4.3 Constant-temperature utilities

If we want to supply some of our utility requirements at a lower temperature, we can find the potential easily from the GCC or the Problem Table. Thus, for the organics distillation unit, we can see from Figure 3.9 that if we have a utility available at 200°C, we can use it to supply just under 2,000 kW without incurring an energy penalty (we can obtain a more accurate figure by interpolating from the Problem Table). Note that 200°C is a *shifted* temperature – as the ΔT_{min} for this problem is 20°C, the actual utility temperature is 210°C (or 200 + ΔT_{cont} if the utility is given a different ΔT_{min} contribution to the process streams).

Often, it is not even necessary to supply utility at the highest process stream temperature. Consider the four-stream problem from Chapter 2, whose GCC is shown in Figure 3.10. The highest shifted temperature reached by a process stream is 165°C and all the heat requirements could, if desired, be supplied at this temperature. However, this stream is a hot stream 2 and the upper temperature intervals have a heat surplus rather than a deficit, and this heat can be used to heat process streams at lower temperatures. Although cold stream 3 requires heating to $T = 140$°C (shifted temperature $S = 145$°C), we do not even have to supply utility heating at this level. Not until just below 100°C do we actually encounter a **net** deficit, and the 20 kW of hot utility could all be supplied at this temperature. Likewise, below the pinch, the 60 kW of cooling can be removed at 61°C (corresponding to an actual utility temperature of 56°C) instead of the lowest shifted temperature of 25°C. If any difficulty is experienced in reading the relevant temperatures from the graph, they are easy to find by interpolation from the Problem Table.

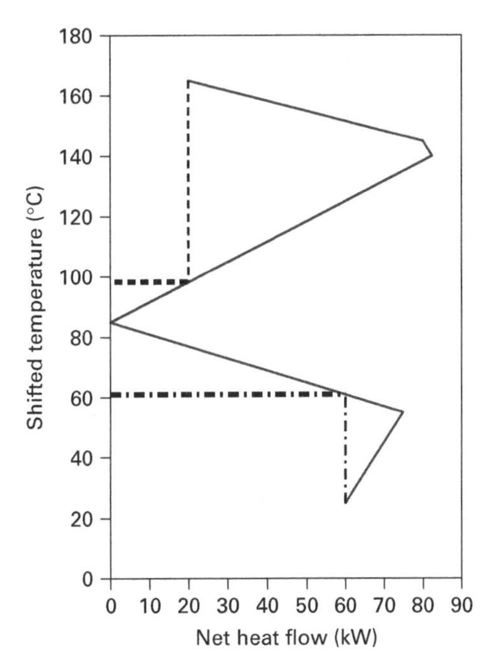

Figure 3.10 Fitting constant-temperature utilities to the GCC for the four-stream example

The regions where the GCC bends back on itself and net heat can be exchanged between different temperature intervals are known as **pockets** or re-entrants. Note that they do **not** represent all the heat exchange taking place between hot and cold streams, which is only revealed by the composite curves. For the four-stream example, the total heat exchange in the process is 450 kW but only 77.5 kW occurs in the pockets (62.5 above and 15 below the pinch). It is relatively unusual to find such large pockets, especially at the upper end of the temperature range. Many plants are like the organics distillation unit in Figure 3.9, where above the pinch the net heat flow increases monotonically with temperature, although there is a small pocket below the pinch. However, there are good opportunities to use multiple utility levels even without pockets.

3.4.4 Utility pinches

If it is decided to use a utility at a specified temperature, this introduces an additional constraint into the design problem. In fact, whenever a utility profile touches the GCC, a new pinch is created. We can designate these as **utility pinches** to differentiate them from the original **process pinch**. However, transferring heat across a utility pinch does not automatically lead to an energy penalty; it simply substitutes high-temperature utilities for low-temperature ones. Whether this is a serious problem depends on the relative cost of the utilities. For example, if both HP and LP steam are being generated in package boilers, there is little difference in fuel requirement or in cost to produce 1 ton of either. This is also the case if all the steam is generated in a HP boiler and LP steam is obtained by simple letdown. However, if a combined heat and power system is being used and LP steam is obtained by passing HP steam though a turbine and generating useful power, there is great benefit from substituting LP for HP steam. Indeed in some cases the apparent cost of LP steam can be virtually zero (if the fuel cost is comparable to the value of the power generated). This can lead to the paradox that it is hardly worthwhile to reduce heat transfer across the actual process pinch, but very important to avoid any violation of the utility pinch. However, these situations are rare.

As described in Section 3.6.1, the presence of a pinch increases the number of units required to solve the network design problem above u_{min}. The addition of utility pinches over and above the process pinch can therefore cause a marked increase in the number of units required (as will be seen in Section 4.6.2). It is often better to "relax" either load or level of the intermediate utility in order to simplify the network, bearing in mind that there is no immediate energy penalty for transferring heat across a utility pinch and that in most situations the cost penalty for transferring heat across the utility pinch is much less than for violations of the process pinch.

3.4.5 Variable-temperature utilities

Variable-temperature utilities gain or lose sensible heat and therefore change in temperature as they do so. Examples are hot oil circuits, flue gases from boilers, exhaust gases from CHP systems, boiler feedwater being preheated and cooling water. The Appropriate Placement principle still applies – hot utilities should lie above the pinch and the GCC, cold utilities below – but the utility itself plots as a sloping line instead of a horizontal one (although, in the case of cooling water, the temperature range is usually so small and the slope so flat that it can often be treated as a constant-temperature utility).

Again, there may be a choice of temperatures, and there are other trade-offs to be explored. One important question is whether the utility stream is used once only (e.g. flue gas which is then rejected to atmosphere) or recirculated (e.g. hot oil circuits where the oil is returned to a furnace, cooling water systems where the heat is rejected in a cooling tower). Also, as many variable-temperature utilities are hot gas streams, it is common for the ΔT_{min} on these matches to be given a higher value – typically 50°C.

3.4.5.1 *Once-through streams*

Returning to the organics distillation unit, consider a hot exhaust or flue gas stream which is providing sensible heat above the pinch as shown in Figure 3.11. The stream has its own mass flowrate, *CP*, supply and target temperatures. We will use a ΔT_{min} of 50°C on these matches; as the global ΔT_{min} is 20°C and process streams have a ΔT_{cont} of 10°C, the ΔT_{cont} for the utility must be 40°C. If we choose a supply temperature of 400°C ($S = 360$°C) and a *CP* of 24 kW/K, we get a stream which is easily

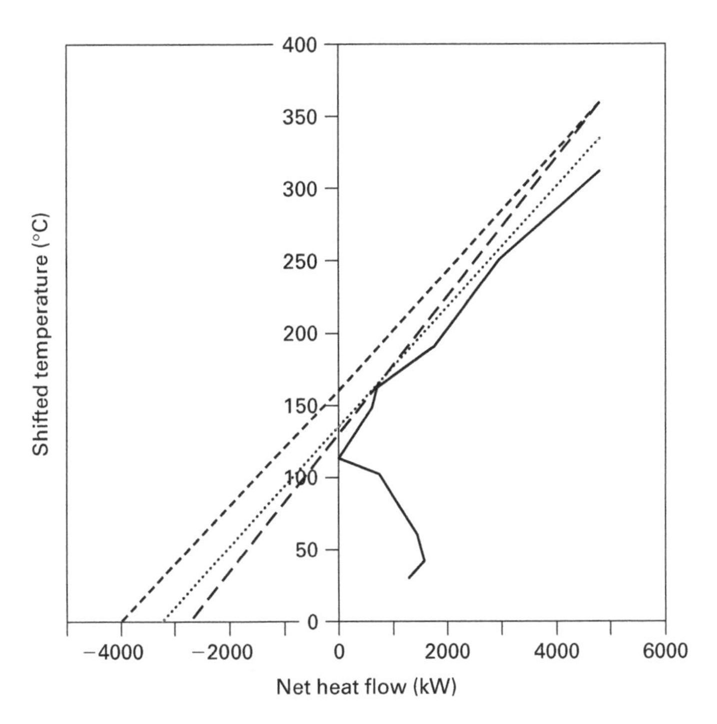

Figure 3.11 Fitting variable-temperature once-through utilities to the GCC

capable of supplying the entire 4,795 kW hot utility requirement, as it releases 4,800 kW over the temperature range down to 200°C (S = 160°C). However, the stream contains a further 4,800 kW above the datum level of 0°C, and this heat is being thrown away. Can we reduce this? There are two possibilities: (a) lower the supply temperature or (b) lower the *CP* (by reducing the mass flowrate of the stream). As we do so, the locus of the utility stream moves down, and the limit comes when it touches the GCC. From Figure 3.11, we see that this happens if the supply temperature is reduced by about 25°C in (a), which will save about 600 kW, and the new utility pinch occurs at $S = 162$°C. For (b) the utility pinch is in the same place, and from the Problem Table we can read off the net heat flow at this temperature as 697 kW. We can thus calculate the new *CP* as (4,795–697)/(360–162), or 20.7 kW/K, and the heat content at 0°C is 3,485 kW, or 1,310 kW less than before.

In this case, not only do we have a new utility pinch, but the process pinch has disappeared completely, as will be seen from the balanced grand composite curve (Section 3.4.6, Figure 3.15). However, the criterion still applies that the pinch must occur where a stream (either process or utility) *begins* or *increases CP*. For the constant-temperature utility, the utility stream itself caused the pinch. Here, however, the pinch is caused by the dehydrate (cold process stream) beginning at 152°C. We can also see a further near-pinch at $S = 251$°C, caused by the bottoms stream starting at 261°C. This will give a completely different network design, as will be shown in Section 4.9.4.

The actual fuel consumption Q_{fuel} of the furnace will depend on the supply temperature of the air input stream (typically ambient) and the heat losses:

$$Q_{fuel} = m_{fuel}\Delta H_{com} = \frac{C_{PG} m_{air} (T_{final} - T_{ambient})}{\eta_{furnace}} \quad (3.1)$$

Here $n_{furnace}$ is the furnace efficiency (fractional), including heat losses.

The furnace requirement is significantly greater than the heat released from the hot flue gas to the process, which is $C_{PG}\, m_{air}(T_{final} - T_{outlet})$. The air had to be heated up from ambient temperature to the furnace exhaust temperature; it releases heat to the process above the pinch but any remaining heat in the stream is wasted if the flue gas is then discharged to atmosphere. Hence, if the flue gas final temperature is well above the pinch and large air flowrates are used, energy is wasted and the heat load on the furnace is increased. Because of this additional penalty, it is usually desirable to minimise the exhaust gas flowrate from a dedicated furnace or direct-fired burner and to make the profile touch the pinch. However, if the exhaust gas is simply a waste by-product of a furnace used for some other purpose (e.g. a steam-raising boiler or a gas turbine cogenerating power) this constraint does not always apply.

The energy penalty may also be reduced if the flue gas below the pinch is used to preheat the incoming air to the furnace. This is best investigated by including the utility streams (air and flue gas) in the pinch analysis, as described in Section 3.4.6.

3.4.5.2 *Recirculating systems*

Now consider a hot oil circuit which is providing heat above the pinch as shown in Figure 3.12. The cool oil is returned to a furnace and reheated; Figure 3.13 shows the basic system layout. In this case, no heat is thrown away from exhaust gas below the pinch; the heat duty on the furnace is the same as the heat released to the process, and is independent of the oil circulation rate. The hot oil simply plots as a straight line on the GCC between its supply and target temperatures.

Capital costs of pumps and pipework in the hot oil circuit (excluding the process duty exchangers) are minimised by minimising oil flowrate. This means maximising the oil supply temperature and minimising the oil return temperature. Costs in the process-utility exchangers are minimised by maximising driving force (i.e. by maximising both the oil supply and return temperatures). For the optimum total system we therefore maximise the oil supply temperature (usually dictated by oil stability) and optimise the oil return temperature. The minimum flow for given supply temperature is predicted when the oil cooling profile just touches the process sink profile at some point, as shown in Figure 3.12. The trade-off between circuit cost and exchanger cost can then be explored by increasing this flowrate.

3.4.6 Balanced composite and grand composite curves

In order to get a clearer overall picture, it is often useful to include the utility-related streams, such as the cold air being heated in the furnace, in the stream data

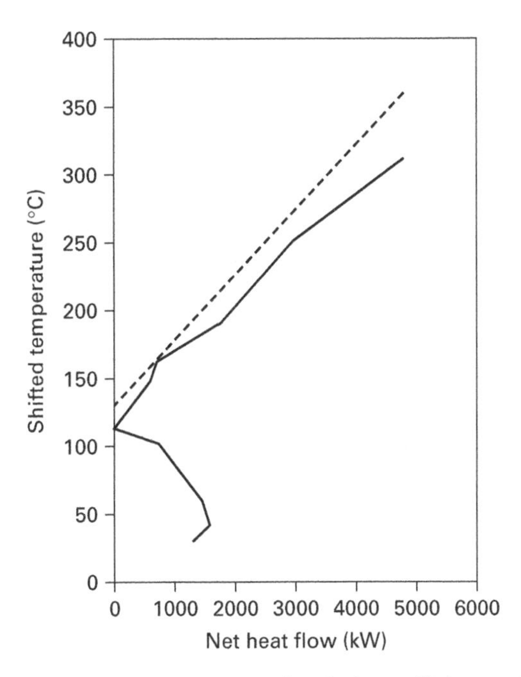

Figure 3.12 Fitting variable-temperature recirculating utilities to the GCC

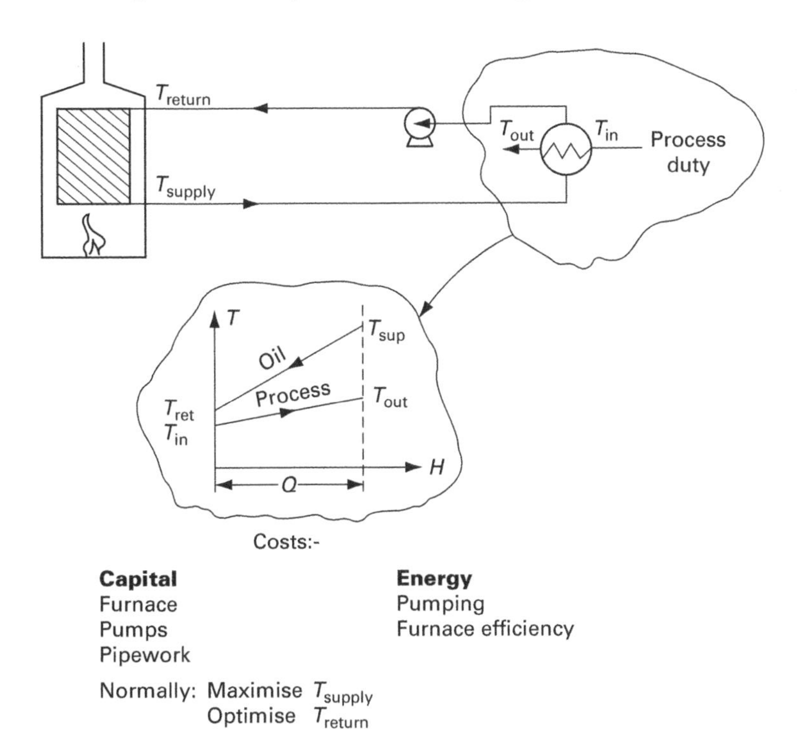

Figure 3.13 Schematic diagram of hot oil heating system

set. The utility system and the process can then be studied as a whole, as well as separately. It may then become easier to evaluate possibilities for preheating air or feedwater to the boilers with waste heat from the process below its pinch.

The composite curves and GCC can be re-plotted including the utility streams (at their target heat loads). The resulting curves should have no unbalanced "overshoot" at either end, because the utilities should precisely balance the process net heat loads. As a result they are known as **balanced composite curves (BCC)** and the **balanced grand composite curve (BGCC)**. They are particularly useful for showing the effect of multiple utilities, multiple pinches and variable-temperature utilities on temperature driving forces in the network, thus revealing constraints on network design more clearly.

The BCC and BGCC give a clear visualisation for constant temperature utilities, where the utility streams are clearly separate and identifiable. However, for variable-temperature utilities, the process streams and utility streams cannot be visually distinguished where they coexist over the same temperature range. This can be seen by comparing Figures 3.14 and 3.15. Figure 3.14 shows the BCC and BGCC for the

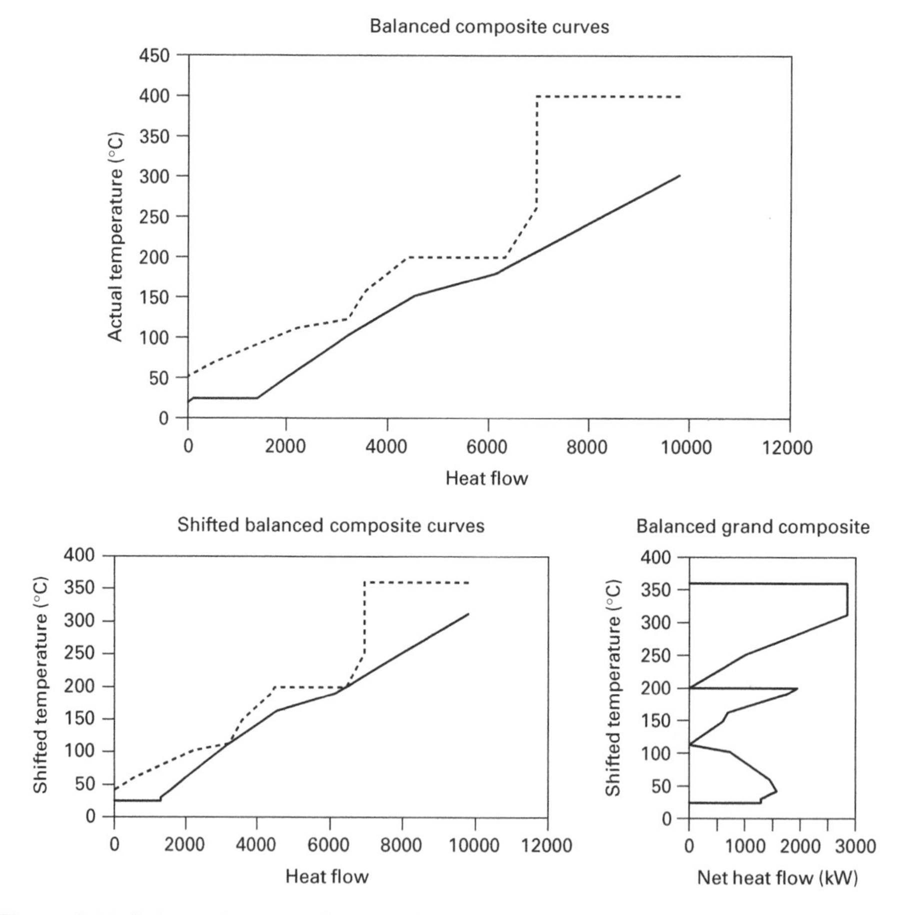

Figure 3.14 Balanced composite and grand composite curves for organics unit, with intermediate steam

organics distillation unit with heating provided by fuel (nominally at $S = 360°C$), intermediate steam at 200°C and cooling water at 25°C. The utility streams are easily identifiable and the additional utility pinch and its effect on local driving forces can be clearly seen. In contrast, Figure 3.15 shows the BCC and BGCC with heating from flue gas above the pinch (supplied at 400°C ($S = 360°C$) – option (b) in Section 3.4.5.1), and cooling water at 25–35°C. The cooling water stream stands out clearly as there is no temperature overlap, but the flue gas is combined with the utility streams; however, the major change in driving forces and the "squeezing" throughout the pinch region can be clearly seen.

Including the utility streams in the analysis may radically alter the overall assessment. For example, suppose we have a flue gas stream from an efficient modern boiler which has been brought down to 120°C by preheating the inlet air. It would be more useful for process needs if it were available at 200°C, but then the air preheating would be lost and the furnace fuel requirement would rise, so it seems that

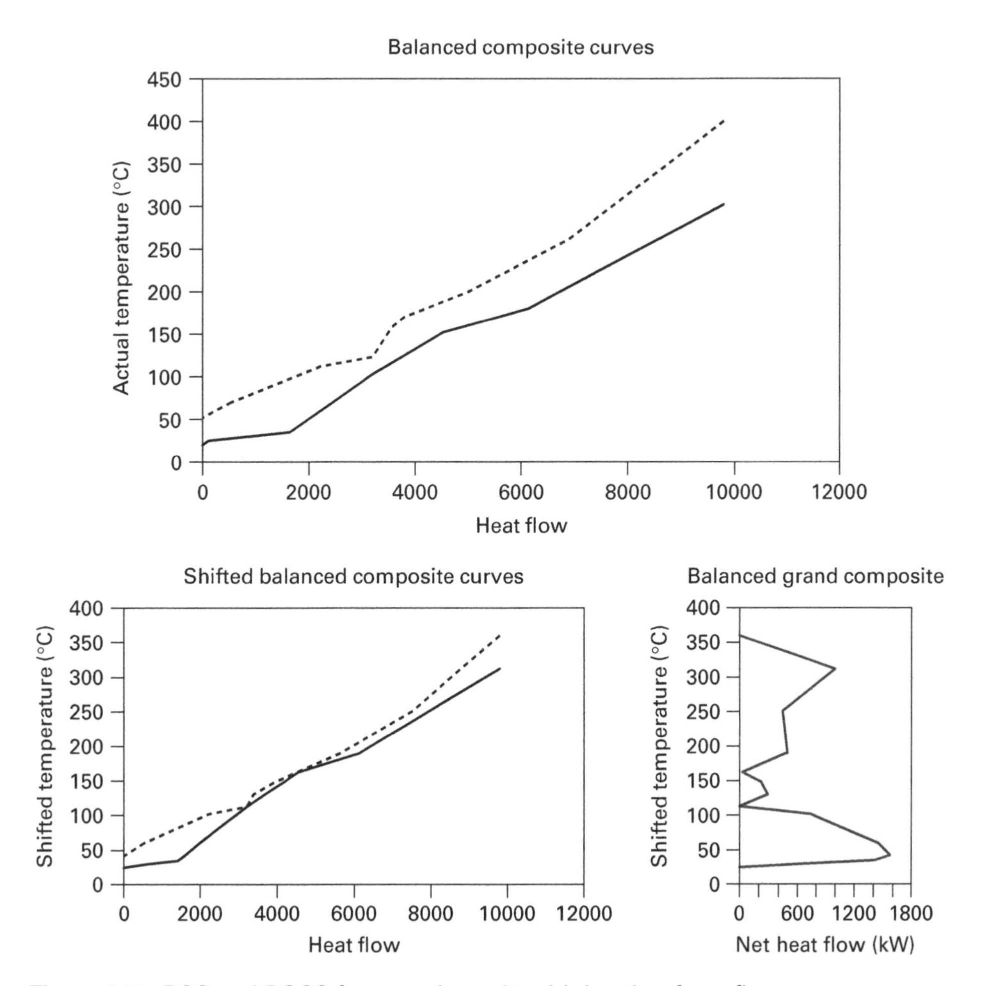

Figure 3.15 BCC and BGCC for organics unit, with heating from flue gas

you lose on the swings what you gain on the roundabouts. However, if the preheat air were included in the stream data, it might well be found that it is possible to heat it with "free" below-pinch waste heat from the process instead! Indeed, the organics distillation unit case study shows precisely this happening (in Section 3.8.5).

Linnhoff (1993) commented that the most common mistake made even by otherwise experienced users of pinch analysis is the failure to look at the BCC and BGCC, often leading to consideration of utility choices only after preliminary network design. Design should be carried out based on balanced composites and using a balanced network grid (Section 4.6). This ensures it is carried out after proper targeting of utilities, consideration of process changes and optimisation. Otherwise, an unnecessary iteration may be needed in the design process, when it turns out that the original network designed in isolation is non-optimal when utilities are included.

The BCC and BGCC also explicitly reveal the location of utility pinches. Consider our four-stream example. We know where the process pinch is, but if we supply steam at some intermediate level, inside the pocket, it is not intuitively obvious where the utility pinch is. For a steam level at $S = 120°C$, the BGCC in Figure 3.16(a) shows that the utility pinch is in fact at the highest process stream temperature, not the utility temperature (though the latter still causes a region of low net heat flow). However, if the steam is supplied at the minimum practicable level, where the utility level just touches the pinch, we see from Figure 3.16(b) that we now have *two* utility pinches, one at the top and one at the utility temperature. Since we have 20 kW of utility and we know from the Problem Table that there is a net CP of 1.5 in the region above the pinch at 85°C, we can calculate that the minimum allowable shifted temperature for the steam is 98.3°C.

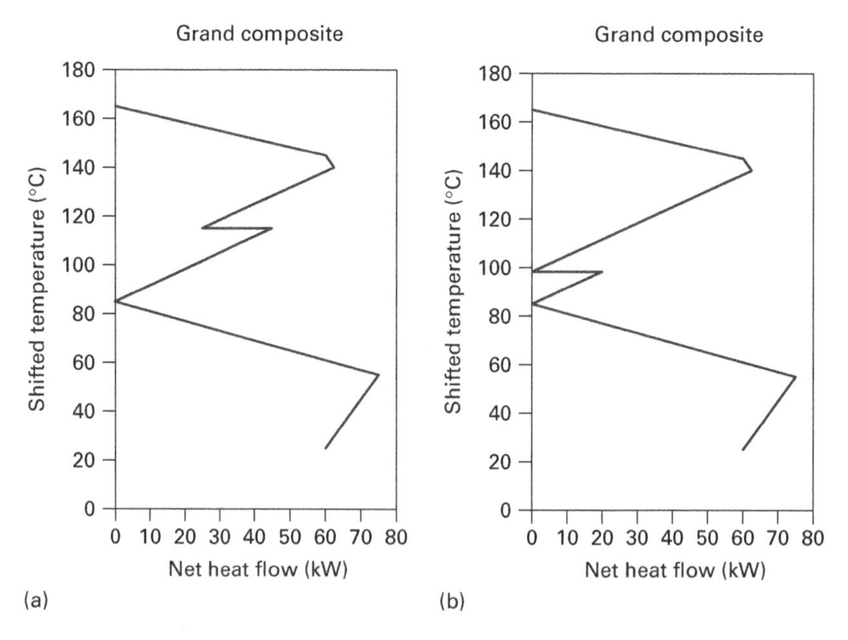

Figure 3.16 BGCC for four-stream example with different steam supply levels

A final important point is that the heat loads of utility streams are linked to the energy consumption. Thus, for example, if steam use is reduced, so is the mass flow of steam needing to be raised, and therefore the boiler feedwater heating duty also falls.

3.4.7 Choice of multiple utility levels

Normally, if there are cost differentials between utilities, we naturally want to maximise the use of the least expensive ones. This usually means that we want to maximise the use of the coldest hot utility and the hottest cold utility. The shape of the grand composite often dictates the most appropriate choice of levels and loads, but in many processes there is no "natural" choice of utility levels and loads, as in the above-pinch region of the organics distillation unit (Figure 3.29). Several choices are possible; the designer could use a single utility to supply the 4,795 kW at or above 311°C shifted temperature, an intermediate level supplying 2,000 kW at just over 200°C (as in Section 3.4.2), or two additional levels at, say, 250°C and 200°C, or any higher number. The trade-offs from using multiple utilities are as follows:

1. steam can often be raised more efficiently at a lower temperature level, as more heat can be recovered from boiler flue gases;
2. power can be generated if the steam is let down from a higher pressure through a passout turbine (see Section 5.2.1);
3. lower-temperature steam is at lower pressure and hence the capital cost of the system is less;
4. the driving forces between utilities and process are reduced, so that the surface area and capital cost of heaters/coolers will increase;
5. each extra level increases complexity of design and incurs additional capital cost for boilers and pipework.

The designer needs to balance the reduction in running cost (if any) against the increased capital cost brought about by increasing the number of levels.

3.5 More advanced energy targeting

3.5.1 Zonal targeting

So far, we have assumed that all streams are equally free to exchange heat and that there is no preference on pairings except on temperature grounds. Often, this is not the case. Two specific situations are:

1. Where two parts of a plant are physically separated and exchanging heat between the two involves long pipe runs with high capital cost, heat losses and pressure drop.
2. Where part of the plant forms a natural "subset" which can be operated on its own, for example if the rest of the plant is shutdown for maintenance.

In such situations, we can make use of **zonal targeting**. The different streams are allocated to subsets or zones, and targets for each of these are obtained separately. The targets for various combinations of zones can then be obtained by combining the relevant sets of stream data, and the difference in the targets between the separate and combined data is found. If the difference is small, the zones can be treated as independent entities. If the difference is large, and is a high proportion of the total energy savings available, this gives a clear incentive for heat exchange between the zones, which must then be traded off against the extra cost and operating inconvenience.

This technique has been used since the early days of pinch analysis, but was developed into a formal zoning concept by Ahmad and Hui (1991) and modified by Amidpour and Polley (1997). It is closely related to total site analysis (Section 5.4), which includes evaluation of the benefits from exchanging heat between different processes on site, by various methods. Good examples of zonal targeting are shown in the organics distillation unit (Section 3.8.4) and the case study in Section 9.5.

3.5.2 Pressure drop targeting

Additional heat exchangers and pipework on a stream will tend to increase its pressure drop. At best, this will increase pump power consumption. More seriously, the higher pressure drop may be beyond the capabilities of the current pumps, so that additional pumps are needed, or the project may become completely infeasible. For example, furnaces and pumps are often limiting in the revamp or debottleneck of crude preheat trains. The challenge is to make the unit more fuel efficient and debottleneck the furnace by adding and reconfiguring surface area *while at the same time avoiding the installation of new pumps*. In other words, revamp or debottlenecking objectives need to fit in with pressure drop limits set by existing pumps. This is a complicated design problem.

Early pinch analysis methods ignored pressure drop. Heat exchangers were designed in the context of temperature and heat load considerations. Pressure drop was considered as an afterthought. It became apparent during the first practical applications of pinch analysis that pressure drop could not be treated in this fashion. An optimised network would settle heat exchangers at given sizes. Subsequently, optimisation of heat exchanger surface area against pressure drop might double certain exchangers in size, while others became much smaller. This clearly invalidated any optimisation that had taken place during the initial design. Worse, in retrofits, the initial design would often exceed the available pressure drop limits, rendering the design impractical. The conclusion was soon reached that the optimisation of heat exchanger surface area vs. thermal energy is inextricably linked with the optimisation of heat exchanger surface area vs. pressure drop. There is a three-way trade-off.

This problem led, after several years of work, to the concept of "pressure drop targeting" and of "three-dimensional supertargeting" (Polley *et al.* 1990). Three-dimensional supertargeting takes on board pressure drop related costs or pressure drop limits for hot and cold streams. Targeting is then carried out consistent with the cost of (or limits on) pressure drop, the cost of fuel and the cost of heat exchanger surface area. Prior to design, streams are set at optimised pressure drop, or at the available pressure drop limit, and utilities are set at the optimised level of thermal recovery. Heat exchangers are placed in this context. If an attempt is made to optimise a design so initialised for fuel, capital cost or pressure drop, little benefit is found. Pressure drop targeting and three-dimensional supertargeting have made an important contribution in making pinch analysis more practical, particularly in the context of oil refinery applications. They join with zonal targeting in allowing effectively for practical real-life constraints (Polley and Heggs 1999).

3.6 Targeting heat exchange units, area and shells

3.6.1 Targeting for number of units

The capital cost of chemical processes tends to be dominated by the number of items on the flowsheet. This is certainly true of heat exchanger networks and there is a strong incentive to reduce the number of matches between hot and cold streams.

Referring back to the flowsheet in Figure 1.4, three exchangers, two heaters and one cooler are used in the design, making six units in all. Is this the minimum number, or could the designer have managed with fewer units? If so, can we find a rapid way of identifying this?

As previously described, the bare flow diagram in Figure 1.5 shows that there are four separate process streams to consider. The target energy performance for this system as calculated by the Problem Table method shows that only heating is required, and no cooling. Straight away then, we know that the cooler is surplus to requirement! Figure 3.17 shows the heat loads on the one hot stream and three cold streams written within circles representing the streams. The predicted hot utility load is shown

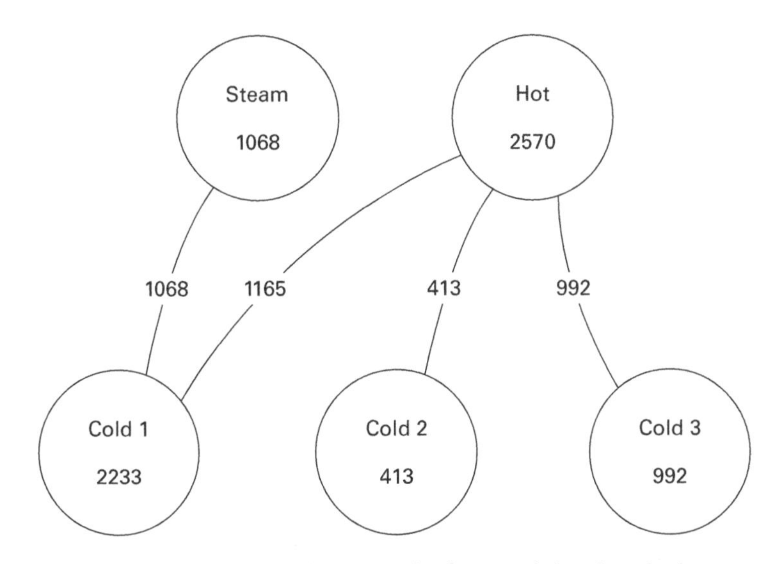

Figure 3.17 Schematic of heat exchange units for specialty chemicals process

similarly. Note that the total system is in enthalpy balance (i.e. the total hot plus utility is equal to the total cold). If we assume that temperature constraints will allow any match to be made, then we can match the whole of cold streams 2 and 3 (total 1,405 units) with hot stream 1, leaving a residual heat load of 1,165 units on Hot 1. Matching Cold 1 with Hot and maximising the load on this match so that it "ticks off" the 1,165 residual requirement on Hot 1 leaves Cold 1 needing 1,068 units, which can be exactly supplied by Steam. So following the principle of maximising loads, that is "ticking off" stream or utility loads or residuals, leads to a design with a total of four matches. This is in fact the minimum for this problem. Notice that it is one less than the total number of streams plus utilities in the problem.

Thus:

$$u_{min} = N - 1$$

where u_{min} = minimum number of units (including heaters and coolers) and N = total number of streams (including utilities) (Hohmann 1971).

It is possible to produce a design for this system with four units, as shown in Figure 1.1(b) in the Introduction. In fact, it is normally possible in heat exchanger network design to find a u_{min} solution, as will be shown.

Certain refinements to this formula are required, however. In Figure 3.18(a), a problem having two hot streams and two cold streams is shown. In this case, both

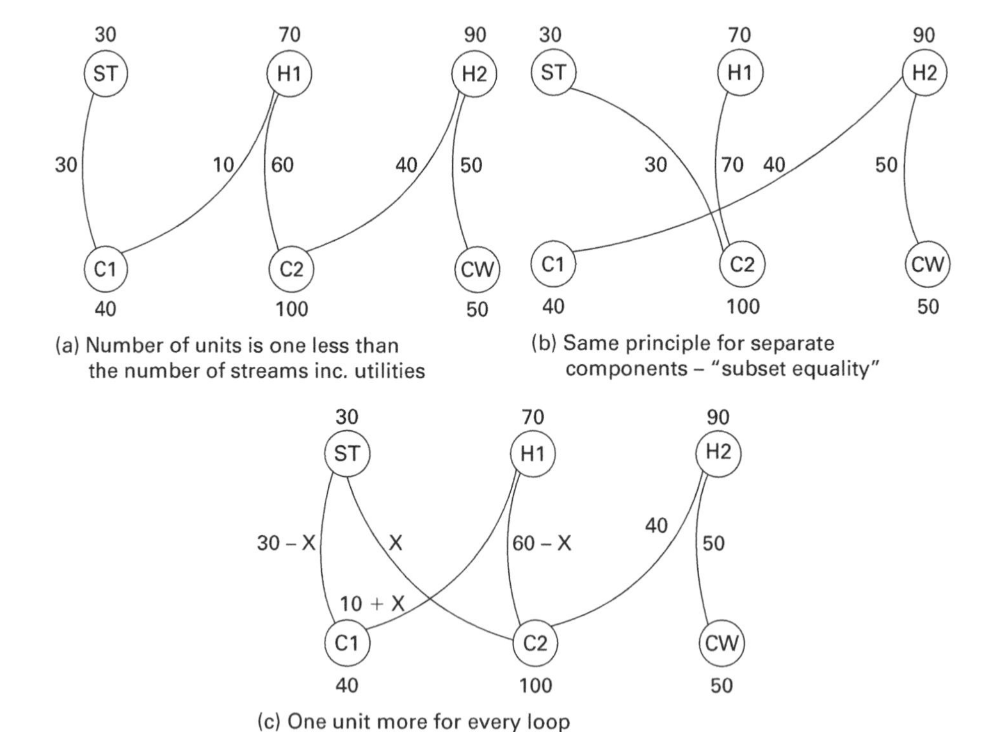

Figure 3.18 Principles of subsets and loops

utility heating and utility cooling are required. Putting in matches as before by ticking off loads or residuals leads to a design with $N-1$ units. However, in Figure 3.18(b) a design is shown having one unit less. The reason why the number appears less than minimum is not hard to see. Whilst overall the problem is in enthalpy balance, the subset of streams H2, C1 and CW is itself in enthalpy balance. Similarly ST, H1 and C2 are in enthalpy balance (which they must be if the total problem is in balance). What this means is that for the given data set we can design two completely separate networks, with the formula $u_{\min} = N-1$ applying to each individually. The total for the overall system is therefore $(3-1)+(3-1)=4$ units, or one less than in Figure 3.18(a). This situation is termed "subset equality", that is for the given data set it is possible to identify two subsets which by enthalpy balance can form separate networks. The data set is said to comprise two "components". Since the flowsheet designer is in control of the size of the heat loads in his plant, it is sometimes possible to deliberately change loads so as to force subset equality and thus save a unit.

Finally, in Figure 3.18(c) a design is shown having one unit more than the design in Figure 3.18(a), the new unit being the match between ST and C2. The extra unit introduces what is known as a "loop" into the system. That is, it is possible to trace a closed path through the network. Starting, say, at the hot utility ST, the loop can be traced through the connection to C1, from C1 to H1, from H1 to C2, and from C2 back to ST. The existence of the loop introduces an element of flexibility into the design. Suppose the new match, which is between ST and C2, is given a load of X units. Then by enthalpy balance, the load on the match between ST and C1 has to be $30-X$, between C1 and H1 $10+X$, and between H1 and C2 $60-X$. Clearly X can be anything up to a value of 30, when the match between ST and C1 disappears. The flexibility in design introduced by loops is sometimes useful, particularly in "revamp" studies.

The features discussed in Figure 3.18 are described by a theorem from graph theory in mathematics, known as Euler's General Network Theorem. This theorem, when applied to heat exchanger networks (Linnhoff *et al.* 1979), states that:

$$u = N + L - s$$

where u = number of units (including heaters and coolers);
N = number of streams (including utilities);
L = number of loops and
s = number of separate components.

Normally we want to avoid extra units, and so design for $L=0$. Also, unless we are lucky, there will be no subset equality in the data set and hence $s=1$. This then leads to the targeting equation:

$$u_{\min} = N - 1$$

introduced previously.

3.6.2 Targeting for the minimum number of units

Figure 3.19(a) shows how the targeting equation is applied to a "maximum energy recovery" (MER) design. The pinch divides the problem into two thermodynamically independent regions. Since the regions are independent, the targeting formula must be applied to each separately as shown (Linnhoff and Hindmarsh 1983).

The total for the whole problem, "$u_{\text{min, MER}}$", is then the sum of the u_{mins} for each region. Suppose, however, that α units of heat are transferred across the pinch as shown in Figure 3.19(b), thus increasing the hot and cold utilities by α. Now, the regions are no longer thermodynamically independent, and we have a single problem. Re-applying the targeting formula, that is ignoring the pinch, leads to the conclusion that:

$$u_{\text{min}} \leqslant u_{\text{min MER}}$$

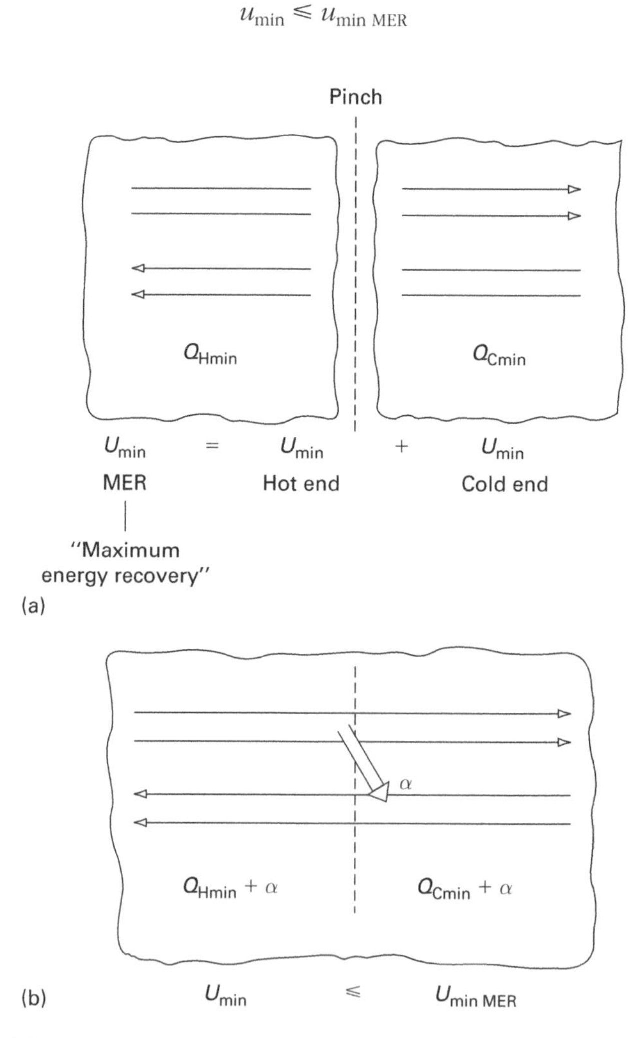

Figure 3.19 Units targeting with and without subdivision at the pinch

This is because in targeting for the MER design, streams that cross the pinch are counted twice. The conclusion is that there is a trade-off between energy recovery and number of units employed.

Referring back to the four-stream example problem shown in Figure 2.13 and applying the targeting formula to the hot and cold ends, with three process streams and hot utility above the pinch and four process streams and cold utility below the pinch, we obtain:

$$u_{\text{min MER}} = (4 - 1) + (5 - 1) = 7$$

However, the final design shown in Figure 2.18 has only six units. The reason is the coincidence of data mentioned in the description of the hot-end design. Above the pinch as shown in Figure 2.20(c), streams 2 and 3 form a subset, allowing the hot end to be designed with three units rather than four ($s = 2$ in Euler's equation). Conversely, applying the targeting formula to the whole problem ignoring the pinch gives:

$$u_{\text{min}} = (6 - 1) = 5$$
$$\text{(4 streams + 2 utilities)}$$

Hence by transferring energy across the pinch, the scope for reducing the number of units is 1. This can be seen in the "commonsense" network design in Figure 2.14, which uses 5 units – two exchangers, two heaters and one cooler.

Similar techniques may be used to allow a small energy penalty at various points in the network to reduce the number of heat exchangers, thus trading off energy against capital cost. This is known as **network relaxation** and is described in detail in Section 4.4.

3.6.3 Area targeting

The area of a single countercurrent heat exchanger is defined by Equation (2.4):

$$A = \frac{Q}{A\Delta T_{\text{LM}}}$$

The log mean temperature difference is defined by Equation (2.5). The product (UA) is also often useful.

For a multi-stream problem with several exchangers, we can estimate the total heat exchanger area in a similar way. We divide the composite curves into segments, based on heat load, calculate the area value for each segment k, and sum them together to give a total area for the heat exchangers in the network. The segments should be chosen to start and finish at heat loads corresponding to gradient changes on the hot and cold composite curves (where streams start, finish or change CP, as with the temperature intervals when calculating the Problem Table). By using the BCC, including the hot and cold utilities (Section 3.4.6) we can also include the area of heaters and coolers; the balanced curves should always be used where multiple utilities and utility pinches are involved. The total network area is given by summing over all the segments (K):

$$A_{\text{total}} = \frac{1}{U} \sum_{k=1\ldots K}^{\text{Intervals}} \frac{\Delta H_k}{\Delta T_{\text{LM}.k}} \tag{3.2}$$

This gives an area (or UA) total for "vertical" heat transfer if the energy targets are met, as shown in Figure 3.20(a). This is of general usefulness, as Hohmann (1971) observed that all networks featuring MER show similar surface area requirements, and that this area is approximately equal to the "total minimum area" calculated from the composite curves. However, two constraints should be noted. Firstly, the heat transfer coefficients U are assumed to be the same for all streams, which is often not the case. To overcome this, Equation 3.2 can be extended to allow for individual film heat transfer coefficients h on each stream (Townsend and Linnhoff 1984; Linnhoff and Ahmad 1990):

$$A_{\text{total}} = \sum_{k=1\ldots K}^{\text{Intervals}} \frac{1}{\Delta T_{\text{LM}.k}} \left(\sum_{i=1\ldots I}^{\text{Hotstreams}} \frac{q_{i,k}}{h_{i,k}} + \sum_{j=1\ldots J}^{\text{Coldstreams}} \frac{q_{j,k}}{h_{j,k}} \right) \tag{3.3}$$

Here $q_{i,k}$ and $q_{j,k}$ are the individual heat loads on hot stream i or cold stream j in segment k; likewise $h_{i,k}$ and $h_{j,k}$ are the individual film heat transfer coefficients.

Secondly, to achieve fully "vertical" matching, the CPs of the hot and cold streams must be matched in exact proportion throughout the segment. This means in effect that a different pattern of matches is required for each segment, and these will usually entail substantial stream splitting, so that an overall network achieving this is usually impracticably complex. The usual practical situation is that the ratio of CP_{H} to CP_{C} will vary for different matches in a segment, so the temperature changes of the different streams do not exactly correspond and there is some "criss-crossing"

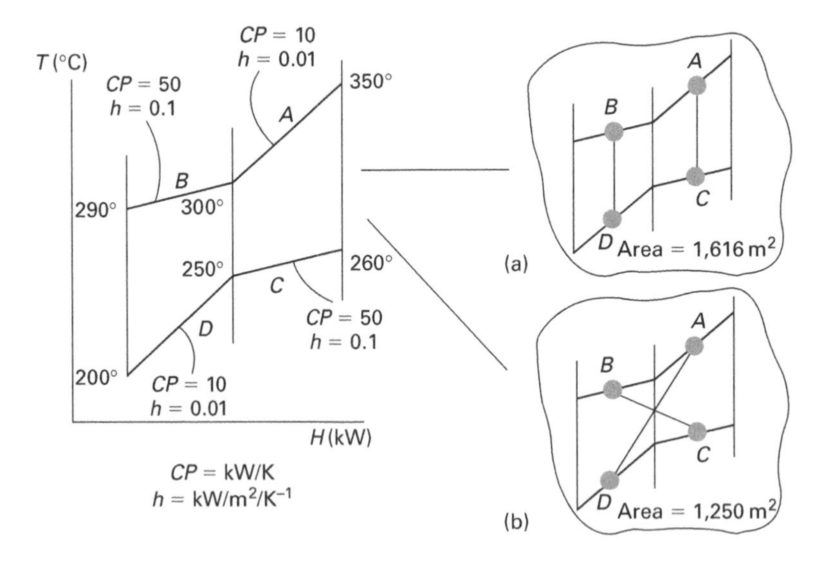

Figure 3.20 Vertical matching and "criss-crossing"

of the heat exchange, as shown in Figure 3.20(b). If all the heat transfer coefficients are the same, this will give an area penalty (usually small), because the extra A due to reduced ΔT on some matches outweighs the reduction in A on the matches with higher ΔT. However, this does not necessarily apply if the heat transfer coefficients are different. In fact, if higher ΔTs are used on matches with low U and vice versa, the area target from "criss-crossing" can actually be lower than that from "vertical" matching! A linear programming algorithm could be used to calculate a true minimum area allowing for this (Saboo *et al.* 1986; Ahmad *et al.* 1990). However, in practice, Equation (3.3) gives an area accurate to within 10%, unless film heat transfer coefficients differ by more than an order of magnitude.

All these methods require a knowledge of film heat transfer coefficients. These are very rarely available in practice. They may be back-calculated where heat exchangers already exist, or estimated from rough sizing calculations, or deduced from stream pressure drops as described by Polley *et al.* (1990).

It is clear that area targeting is less precise and more complex than energy targeting. Hence, although it is conceptually important, it is less useful in practice. Actual heat exchanger areas can be obtained during the network design phase. If the network is relaxed significantly, the area targeting algorithm will no longer be exact. The basic area or *UA* target is helpful in giving an indication of how many exchangers will be needed, or whether a lot of area needs to be added to an existing network, and hence an indication of likely capital cost. However, the cost optimum is generally flat and, where there is a sharp discontinuity due to a topology change, it can be identified by other means.

The area target can also give a good visualisation of the energy saving strategy on a plant, especially if the energy-area relationship can be plotted (this is usually similar to the area–ΔT_{min} graph). An existing plant usually lies in the non-optimal region. An improvement strategy will seek to move closer to the target line; energy will be reduced, but normally additional area will need to be added. Typically, options identified by a retrofit strategy will therefore follow an oblique line or curve, as in Figure 3.21.

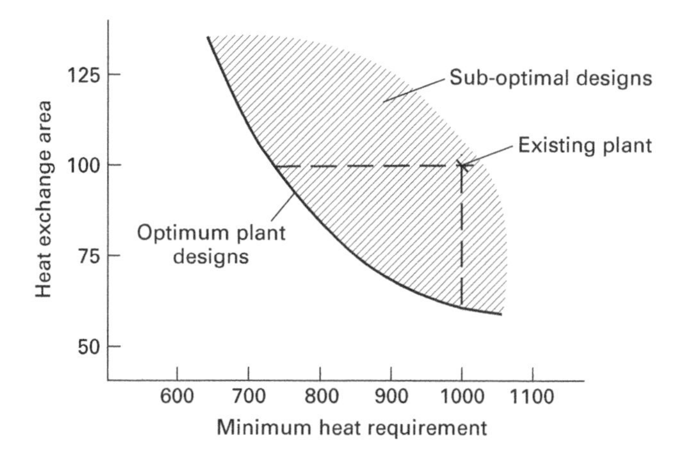

Figure 3.21 Retrofit strategy plotted on the energy–area graph

3.6.4 Deviations from pure countercurrent flow

The energy and area targeting algorithms presented so far have assumed that heat exchange matches are in pure countercurrent flow. However, as will be seen in Section 4.1, practical heat exchangers may be in crossflow, cocurrent flow (rare) or partially mixed flow. Hence, the temperature driving forces along the match may be less than the predicted levels.

For many streams this is unimportant, for example where one stream is at almost constant temperature due to latent heat loads, or where the average ΔT across the match is high. However, for a "long" duty, where the temperature range of both hot and cold streams through the exchanger is large compared to the temperature driving force, a real exchanger (e.g. U-tube type) will often have squeezed internal temperature driving forces, or even a temperature cross.

Considerable research has been done to refine area targeting algorithms to allow for this situation. However, the practical benefit of all this is limited, as the main benefit of area targeting is to get a rough idea of how much extra area will need to be installed, and a first estimate of capital cost for optimisation. The capital cost optimum is normally very flat. Where it is not, this is generally due to sharp discontinuities or "topology traps", which can be identified from the stream temperatures as described in Section 3.6.7. It is more important to get the basic configuration of the network and then to perform detailed design of the individual exchangers to get exact sizes and costs.

3.6.5 Number of shells targeting

For shell-and-tube exchangers, the number of shells required may be greater than the minimum number of units, for two reasons:

1. The area required for a single heat exchanger may be inconveniently large.
2. There may be a temperature cross in a single shell, especially if it is of crossflow or "1–2" type.

Both these are most likely to occur in "long" duties, with low temperature driving forces and large temperature differences on the streams, which are most likely to occur near the pinch.

Each additional shell incurs extra capital cost, so targets for the minimum number of shells are useful; these can be calculated as described by Ahmad and Smith (1989) and Smith (2005).

The number of shells plot is not a smooth curve like the previous ones, as it can only proceed in integer steps. An example can be seen in Figure 3.32.

3.6.6 Performance of existing systems

An existing network can be evaluated for how efficiently its heat exchange area is currently distributed by plotting the energy targets against the area calculated to be required for these targets (at a given ΔT_{min}). The area target can then be compared with the actual area used in the current network.

Few software packages generate an energy–area plot automatically, so a good alternative is to plot the energy–ΔT_{min} and area–ΔT_{min} graphs separately. The ΔT_{min} corresponding to the current energy consumption (if energy targets were met) can be found, and the area target for this ΔT_{min} is compared with the current area. Alternatively, we can read off the ΔT_{min} corresponding to the current installed area, and see what the energy target should be.

In most cases we will find that we have to install a significant amount of additional area to reach the energy targets at our preferred ΔT_{min}.

3.6.7 Topology traps

So far, the variations of energy and area with ΔT_{min} have been smooth curves, and a modest error will not greatly affect the results. However, it is possible for a sharp discontinuity to exist, where the energy targets, area requirement and optimal network change very sharply at a critical ΔT_{min}. These "topology traps" were first identified by Tjoe and Linnhoff (1986). They will be revealed by graphs of area or cost against ΔT_{min}.

Kemp (1991) suggested a simpler way of identifying topology traps. The discontinuity usually occurs where there are significant latent heat loads, giving a major change in net heat flow over a narrow temperature range. The following dataset illustrates a topology trap.

Stream ID	*Supply temperature (°C)*	*Target temperature (°C)*	*Heat capacity flowrate* CP *(MJ/K)*	*Heat flow (MW)*
h1	70	30	0.1	4
h2	70	60.1	0.101010101	1
h3	60.1	60	30	3
c4	20	60	0.1	−4
c5	40	40.1	60	−6

The discontinuity is clear from the graphs showing the variation with ΔT_{min} of total cost (Figure 3.22). However, to obtain this, the exchanger heat transfer coefficients and cost factors had to be known. What would happen if we did not have this data, or did not have an area targeting program available?

In fact, plots of utility use and pinch temperature against ΔT_{min} show that an anomaly exists, and these can easily be generated simply by repeating the energy targeting calculation several times over a range (Figure 3.23).

In practice, if a topology trap exists, it is always found to be at a discontinuity in utility use or pinch temperature. The reverse does not apply; sharp changes in energy targets or pinch temperature often makes no real difference to energy and area targets

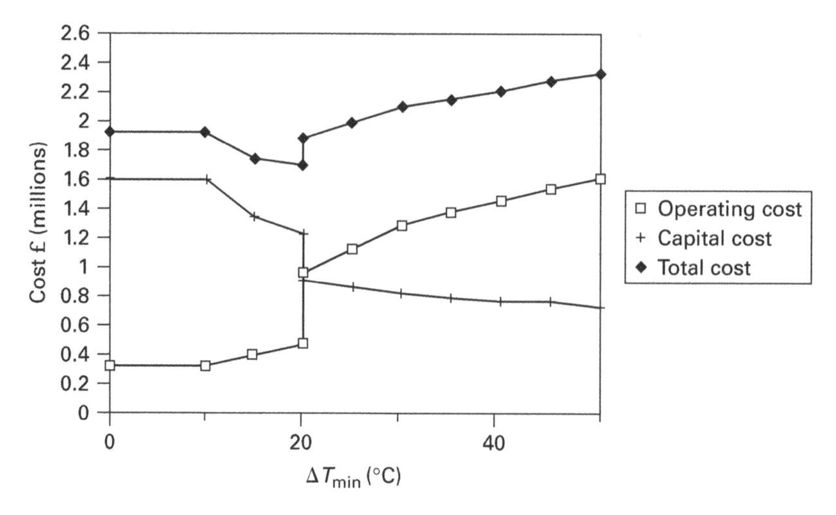

Figure 3.22 Cost–ΔT_{min} plot for topology traps example

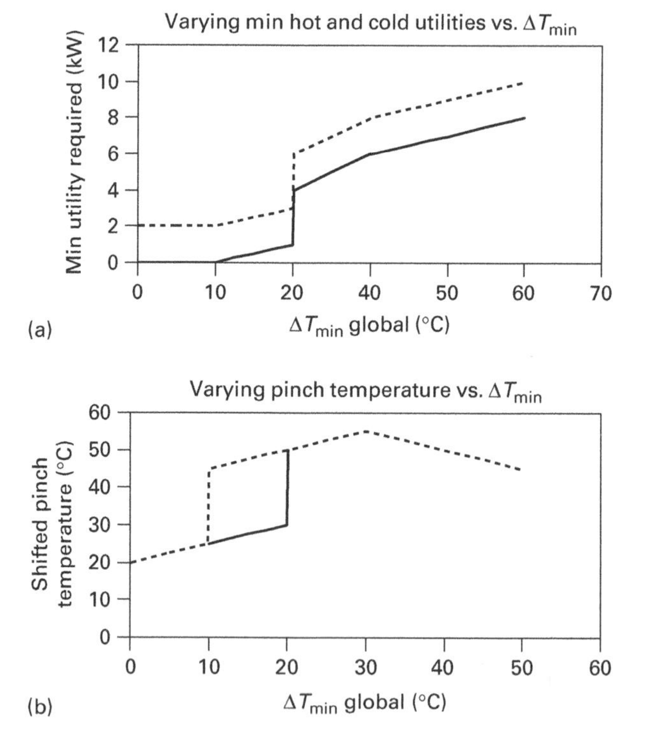

Figure 3.23 Plots of energy and pinch temperature against ΔT_{min} for topology traps example

(see e.g. our four-stream example, Figure 3.6), but they do identify where a problem will occur if there is one.

The final question is, can we home in quickly on values which are likely to give us discontinuities, rather than needing to make a very large number of retargeting calculations? We can make use again of the knowledge that the pinch is always caused by a new stream starting (or increasing sharply in *CP*). Hence, ΔT_{min} values of interest will be those between a supply temperature and the target (or supply) temperature of another stream. For our example, the key ΔT_{min} values can be read from the table as:

0, 0.1, 9.9, 10, 20, 20.1, 29.9, 30, 30.1, 40, 40.1, 50.

We could simply calculate targets for these values of ΔT_{min} and would know that intermediate variations would simply be obtained by linear interpolation. In the event, pinch discontinuities occur at only 3 of these 12 temperatures, and only 1 of these gives a significant discontinuity in the energy, area and cost curves. In several cases there is a multiple pinch or pinch region. This is a drawback to using the pinch temperature plot, as the discontinuity may only occur at one end of the pinch region (as indeed happens here) and may not be picked up by simple software which only shows one pinch. Hence, using both the energy and pinch temperature plots in tandem is the safest approach.

3.7 Supertargeting: cost targeting for optimal ΔT_{min}

3.7.1 Trade-offs in choosing ΔT_{min}

We can calculate energy targets for a given ΔT_{min}, but what if our chosen value of ΔT_{min} is wrong? For a different ΔT_{min}, the streams present at the pinch may change, and we could get different pinch matches and a completely different heat exchanger network. How can we find the optimal ΔT_{min}?

We have seen that a lower value of ΔT_{min} usually gives a reduction in energy use but needs more heat exchanger area; there is a trade-off. Since energy and area are in different units, however, we need to find a basis to compare their importance, and find an optimum ΔT_{min}. The obvious basis is cost. Energy requirements will affect the operating cost, while the size of heat exchangers, heaters and coolers will affect capital cost.

In fact, ΔT_{min} affects cost in several ways. If we try to reduce our energy costs by choosing a lower ΔT_{min}:

1. The hot and cold utility usage falls, so energy costs fall.
2. The amount of heat exchange is greater, so larger heat exchangers will be needed and their capital cost is increased.
3. Temperature driving forces in heat exchangers are lower, again requiring larger heat exchangers and higher capital cost.
4. Heaters and coolers have a lower heat load, so they can be smaller and their capital cost will fall.

In practice, the increased size and cost of the heat exchangers almost invariably outweighs any reduction in heater and cooler sizes. Hence the main trade-off as ΔT_{min} is reduced is between energy cost reduction and capital cost increase. The balance will depend on the cost per unit area of exchangers, the cost of heating and cooling, and the period over which the capital cost is to be regained (payback period).

Capital cost can be expressed in £, $, € or any other currency units and are a one-off expenditure when the plant is built or revamped. However, energy and other operating costs are cumulative over a time period, and are therefore expressed as £/h, $/yr, etc. To get an overall cost, capital and operating cost must be put on the same basis. This can be done in two ways:

1. Evaluate the operating cost over a fixed period (e.g. 1 or 2 years). Often, the desired payback time on capital investment is used as the evaluation period. The total cost graph will then be in £.
2. Annualise the capital cost, dividing it by a time period (again, often the payback time). The total cost graph can then be in £/yr, £/h or any desired unit of "cost per unit time".

The first option is more commonly used, and we will use it in our examples.

3.7.2 Illustration for two-stream example

We will use the two-stream example from Section 2.1.1 as it gives the simplest calculations to illustrate the key points.

Figure 3.24 plots the calculated surface area of (i) heat exchangers and (ii) heat exchangers plus heaters and coolers for our two-stream example, together with hot utility usage. Heat transfer coefficients have been taken as 100 W/m^{2}K (0.1 kW/m^{2}K) throughout, and utility temperatures have been taken as 250°C for steam and 20°C for cooling water. Taking the concept to its limits, at a ΔT_{min} of zero an infinitely large heat exchanger would be required which would be infinitely expensive. On the other hand, if ΔT_{min} and utility use are high, if the cost is summed over a long enough period, this will also be an extremely expensive option. Somewhere between the two should be an economic optimum.

For this example, we will use arbitrary cost units. The cost of heat exchangers is (10 + A) where A is the area in m^2; for heaters and coolers, cost is (5 + A). Likewise, the cost of 1 kW of heating and cooling over the evaluation period is taken as 1 cost unit. (The ratios are realistic; in the case study in Section 3.8, real cost data will be used and it will be seen that the trade-offs are very similar.)

Figure 3.25 shows indicative values for utility cost, exchanger cost (with and without heaters/coolers) and total cost (graph 2 with heaters and coolers included, graph 1 without). The drop in cost for a ΔT_{min} of 130°C (where no heat exchange takes place) is because the cost of the exchanger shell disappears. However, the cost is still

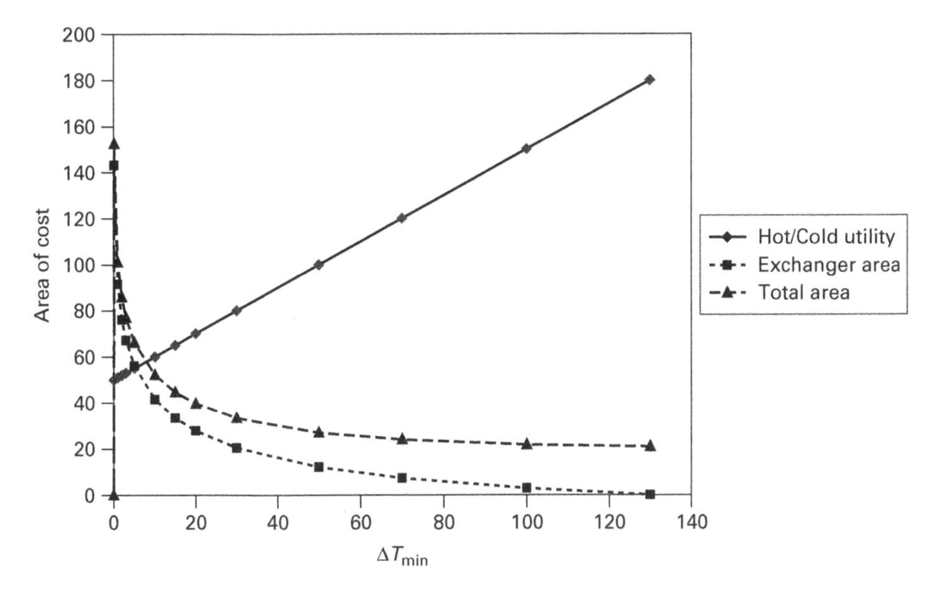

Figure 3.24 Variation of utility usage and heat exchanger area with minimum temperature difference ΔT_{min}

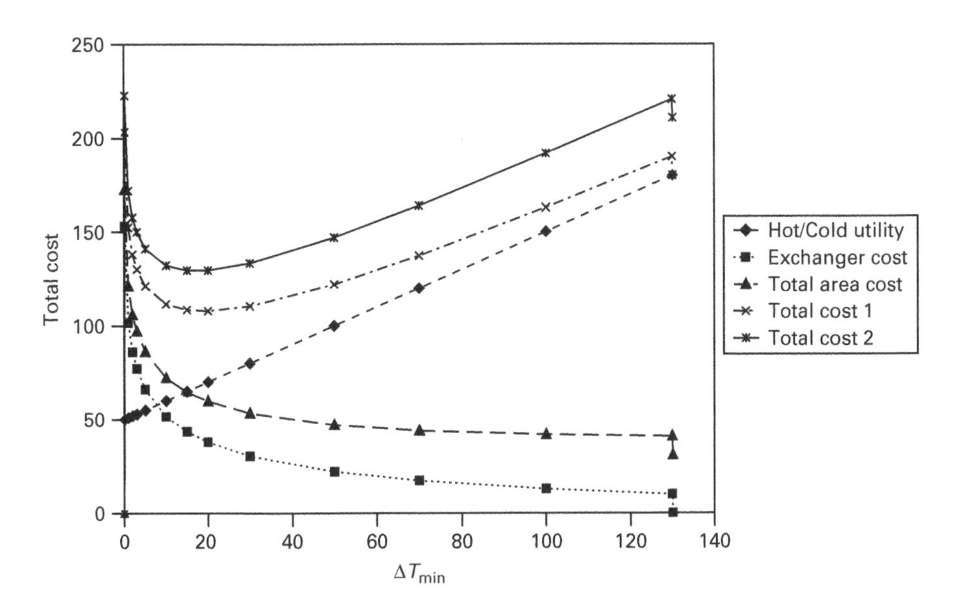

Figure 3.25 Variation of capital, energy and total cost with minimum temperature difference ΔT_{min}

higher than that with heat exchange. The optimum ΔT_{min} is approximately 15–20°C. Note, however, that the optimum is relatively flat; the total cost is within about 10% of the optimum for a range from about 5°C to 50°C.

3.7.3 Factors affecting the optimal ΔT_{min}

How sensitive is the calculation? Figure 3.26 compares the graph with one for a situation where the overall heat transfer coefficient has been halved (the same effect would be obtained by doubling utility cost, or halving hours worked per year, or doubling the cost per unit area of the heat exchanger). The optimal ΔT_{min} has changed from roughly 20°C for "$U = 1$" to 30°C for "$U = 0.5$", but the curve is still very flat in this region, and choosing one value rather than the other will only give a change of 2–3% – far less than the limits of error on the area and cost estimates. Hence the economic optimum is relatively insensitive to small variations. However, if the heat transfer coefficient is reduced by a factor of 5 (e.g. a heavily fouling or highly viscous stream) or the capital cost increases 5 times (e.g. a highly corrosive stream requiring a heat exchanger in an exotic alloy) a third curve ($U = 0.2$) results. Now the optimum ΔT_{min} has shifted substantially, from 15–20°C to 50°C.

Nevertheless, for typical bulk duties of reasonably free-flowing liquids, 20°C is a reasonable ΔT_{min}. For high throughputs or continuous three shift operation, a value of 10°C may be more appropriate. For cryogenics, as the cost of refrigeration is extremely high, ΔT_{min}s of 2–3 K are common, and plate-fin exchangers are often used. Plate heat exchangers may also allow lower ΔT_{min}.

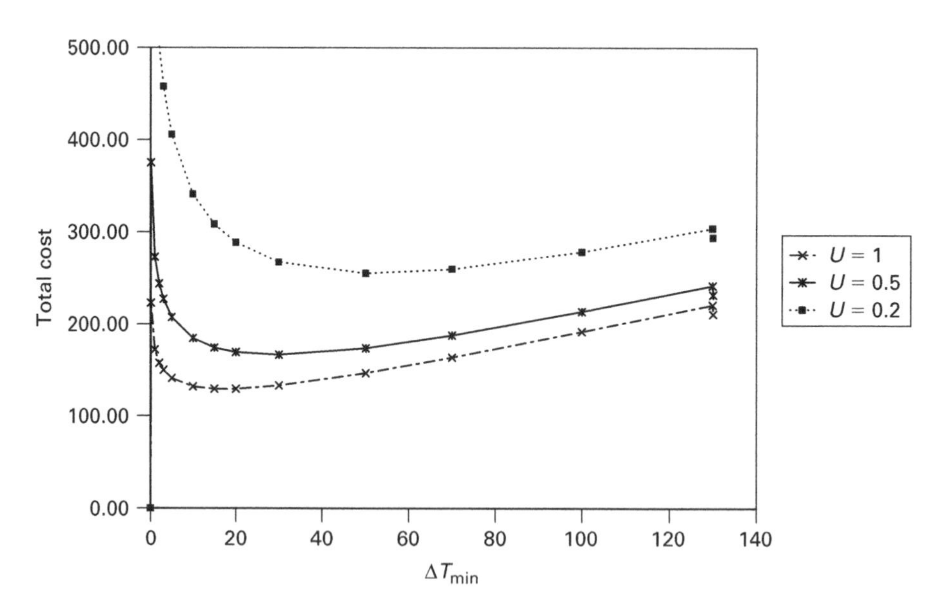

Figure 3.26 Variation of Cost–ΔT_{min} chart with different cost parameters

Although we explained this in terms of changing heat transfer coefficients, the same trends in shifting optimal ΔT_{min} will occur for anything which changes the *balance* of capital cost and energy cost. If capital becomes more expensive, or energy becomes cheaper, the best economic ΔT_{min} will increase. This can be caused by any of the following:

1. viscous or heavily fouling streams (reducing heat transfer coefficient),
2. corrosive or high-purity streams (requiring exotic metal or alloy exchangers, at high cost),
3. low utility prices (reducing the monetary gain from energy saving),
4. capital shortages (leading to shorter payback times being required on projects),
5. low plant utilisation (one-shift or intermittent operation, rather than continuous three-shift),
6. small flows (energy cost savings are then too small to repay fixed costs of exchangers and project planning, particularly for small-scale batch plants).

Network-based factors can play a major part because of the flat optimum. Often, the need to shift to an extra shell, for size or temperature cross reasons, gives a discontinuity in the cost graph (see Section 3.8). Likewise, topology traps may be the overriding factor in selecting ΔT_{min}. Hence it is rarely worthwhile to try to refine supertargeting calculations to a high degree of precision.

The flat nature of most cost–ΔT_{min} curves has other implications. In many complex problems, there is not a single clear-cut "best" network but a range of "good" ones. A set of networks devised by Sagli *et al.* (1990) illustrate this. They carried out much detailed analysis to compare the merits of networks containing 5, 6 and 7 exchangers in different configurations and showed that the theoretical optimum could shift. In practice, however, the capital costs of the networks were within 1–2% of each other. In any practical design situation, this is immaterial. The choice of network configuration between the half-dozen "best" ones would then be made on the basis of other factors, such as plant layout, location and size of existing exchangers, undesirable matches, etc. All these considerations are best tackled in the detailed design phase. Again, there is no point in prolonging the targeting process to give meaningless precision – indeed it may lead to the premature dismissal of a viable option which is not quite at the optimum. A relatively crude estimate of the energy and cost targets gives virtually all the information that is ever likely to be useful for optimising the plant.

3.7.4 Approximate estimation of ideal ΔT_{min}

An alternative way of estimating ΔT_{min} can be used where problems have a very "sharp" pinch. Here, changes in ΔT_{min} have a much more marked effect on capital cost at the pinch than elsewhere in the problem. Thus if ΔT_{min} is doubled from 10°C to 20°C, then the change in driving force at the pinch is 100%. However, away from the pinch where driving forces are, say, of the order of 100°C, an increase in ΔT_{min}

of 10°C means only a 10% increase in driving force. This suggests a quick method for finding the optimum ΔT_{min}. By just considering the sizes of exchangers in the region of the pinch for the MER design at different values of ΔT_{min}, capital cost can be traded against utilities cost without considering the complete network. However, if the pinch is not sharp, driving forces are significantly squeezed throughout the network and area targeting calculations are often more appropriate.

A further option is to position a design near a concave point on the energy–ΔT_{min} targeting line, where a low energy use can be obtained without squeezing temperature driving forces too much. Examples are points A and B in Figure 3.27. If an initial design has already been sketched out or a current plant exists and is found to be non-optimal, a strategy can be mapped out to move it towards one of these points. Alternative routes are possible; strategy 1 in Figure 3.27 is to reduce utilities at roughly constant capital cost, while strategy 2 is to reduce capital cost at constant utilities usage.

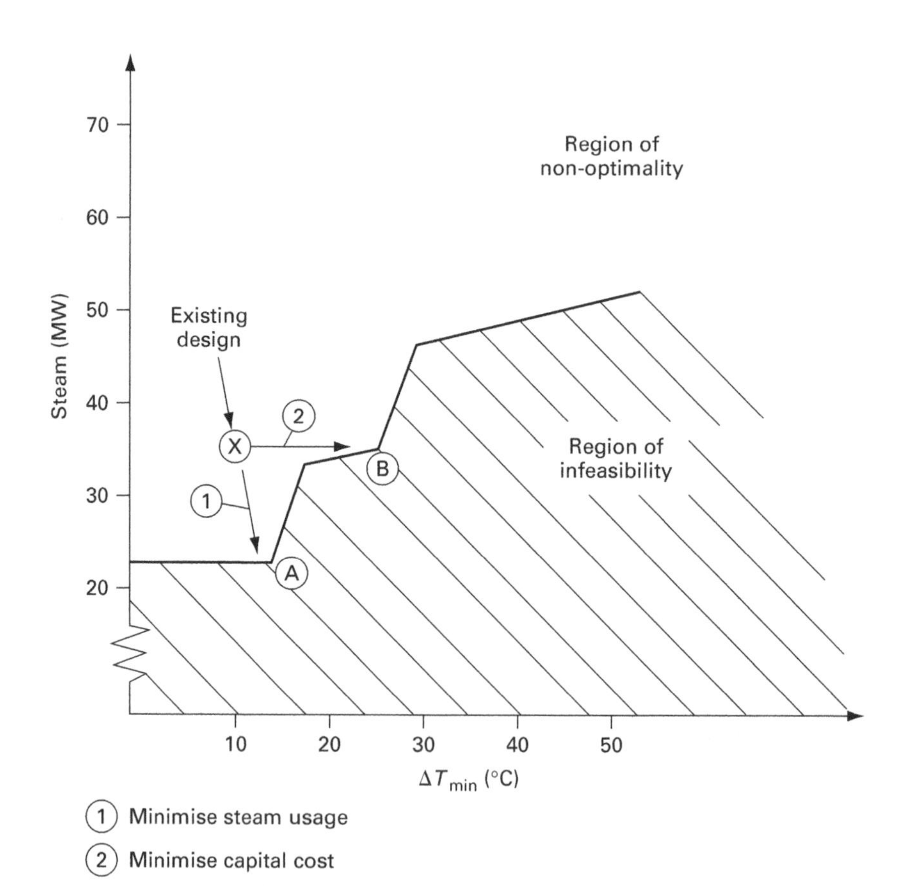

Figure 3.27 Concave points on the energy–ΔT_{min} graph

3.8 Targeting for organics distillation plant case study

3.8.1 Energy targeting

Initially, we can estimate a value of ΔT_{min}; let us choose 20°C, which is similar to the values used in the earlier case studies.

If we have software capable of handling streams with variable *CP*, we can calculate the exact energy targets immediately. However, if we only have simple software (such as the spreadsheet supplied with this book) we initially calculate the targets and pinch temperature using a constant *CP* of 25. This is not "safe-side" linearisation, as the effect is to shift more of the cold stream heat load to lower temperatures and overestimate possible heat recovery. The targets come out as 4,635 kW hot utility and 1,135 kW cold utility, with the pinch at 113°C shifted temperature, corresponding to 123°C for hot streams and 103°C for cold streams. Now, we divide the crude stream into two segments at the pinch and recalculate. The new data for the two segments are illustrated in Table 3.4.

Recalculating, we get energy targets of 4,795 kW hot utility and 1,295 kW cold utility, and a pinch at 113°C. Note that although the targets have changed by 160 kW (the change in heat load on the crude feed segments), the pinch temperature is exactly the same as before. The targets are 2,065 kW lower than the existing utility requirements, so there is considerable potential for heat recovery. The composite curves (Figure 3.28) show a considerable overlap, illustrating the potential for heat recovery. The GCC (Figure 3.29) shows a reasonably sharp pinch and no other near-pinches.

3.8.2 Area targeting

Since we have estimates of film heat transfer coefficients available, we can estimate the required heat exchange area if suitable software is available. For countercurrent exchangers, the calculated area is 558 m^2 with no area increase factor and 669 m^2 with a factor of 20%, while the Ahmad–Smith algorithm gives 605 m^2. These are much higher than the present area of 128.5 m^2, so clearly considerable capital investment will be needed to achieve the heat recovery target.

As an alternative, the ΔT_{min} for the current exchangers may be used. On both exchangers, the ΔT_{min} at one end has a minimum value of 63°C. The targets with

Table 3.4 Stream data for segmented crude feed stream

Stream name	*Stream type*	*Initial temperature (°C)*	*Target temperature (°C)*	*CP (mean) (kW/K)*	*New heat flowrate (kW)*	*Old heat flowrate (kW)*
Crude feed	Cold 1A	20	103	23.075	−1,915	−2,075
Crude feed	Cold 1B	103	180	27.075	−2,095	−1,925

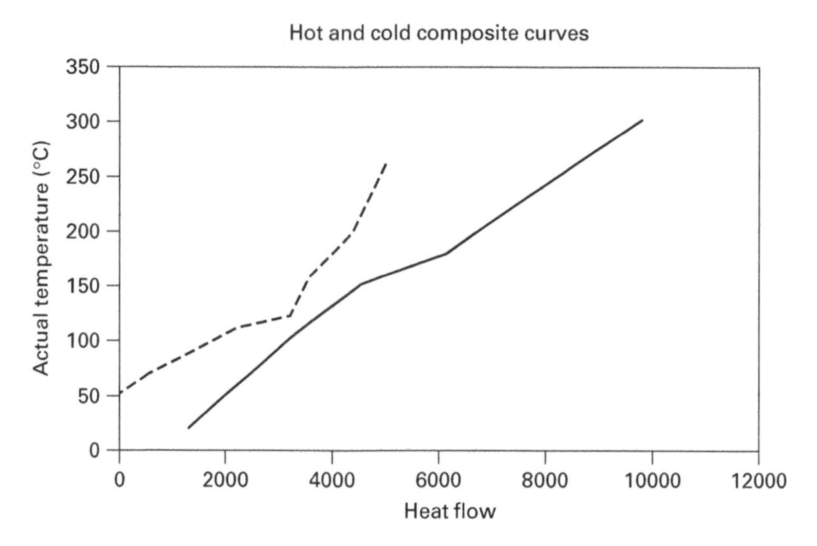

Figure 3.28 Composite curves for organics distillation plant (ΔT_{min} = 20°C)

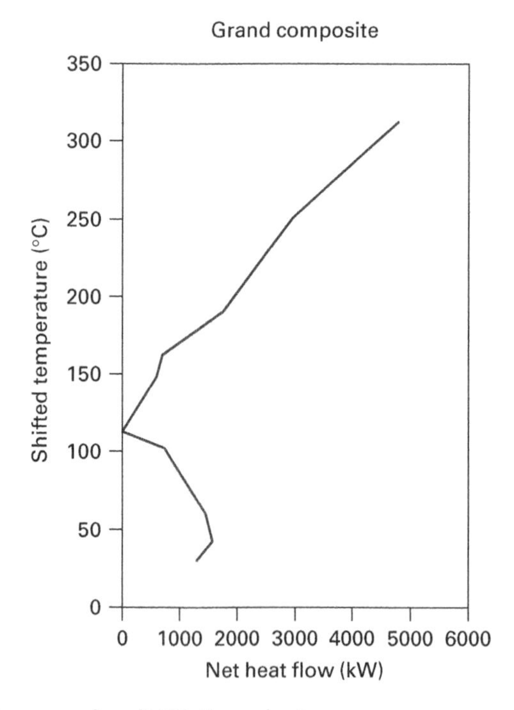

Figure 3.29 GCC for organics distillation plant

Table 3.5 Comparison of targets with current energy use

Situation	*ΔT_{min} value (°C)*	*Hot utility (kW)*	*Cold utility (kW)*	*Heat recovery (kW)*	*Exchanger area (Ahmad–Smith) (m^2)*
Current	63	6,860	3,360	1,640	128.5
Target	63	5,830	2,330	2,670	225.4
Target	20	4,795	1,295	3,705	605

this ΔT_{min} are 5,830 kW hot utility and 2,330 kW cold utility, with the pinch still at 123°C for hot streams but at 60°C for cold streams. These targets are higher than for the first calculation, but still significantly lower than the current energy use. The results are summarised in Table 3.5.

It would be possible at this point to perform calculations with other values of ΔT_{min} and to explore how the targets varied. However, in this case we can also use the economic information to perform supertargeting.

3.8.3 Cost targeting

Data for utility cost, heat exchanger cost and the number of working hours per year was given above and the cost of hot and cold utility can be entered directly from the information in Section 3.2.4. When entering the coefficients for heat exchanger cost, make sure they are in the correct order; $a = 300$, $b = 0.95$, $d = 10{,}000$ (5,000 for coolers). The input and output cooling water temperatures are known and the minimum temperature approach can be taken as 10°C the furnace temperature can be assumed to be 400°C, and a simple payback of 2 years will be assumed. The heat transfer coefficients on heaters and coolers are assumed similar to those for the crude feed/overheads exchanger and a value of 0.2 kW/m^{2}K can be used (these are overall coefficients, not film coefficients).

We now wish to find the optimum value of ΔT_{min}. Since the current value is high (63°C) a large range seems appropriate, and we will consider values up to 70°C. The energy–ΔT_{min} graph (Figure 3.30) shows a steady rise in utility use with ΔT_{min} with no discontinuities. Conversely, the area–ΔT_{min} graph (Figure 3.31) shows a sharp initial fall. Heat recovery initially rises sharply as heat exchanger area is increased from zero, but thereafter a law of diminishing returns clearly applies. The plot for number of units (Figure 3.32) uses the Ahmad–Smith method, which gives large numbers of units at low ΔT_{min} because it correctly allows for temperature crosses within the exchangers.

The graph of most interest, however, is the cost–ΔT_{min} plot, Figure 3.33. Capital cost falls with ΔT_{min} and utility cost rises. The total cost curve is rather flat and this is, in fact, typical of a very large number of process plants. The optimum is in the range 25–30°C but clearly in this case a design could deviate a long way from this value without substantially affecting the total cost over the 2-year period. In particular,

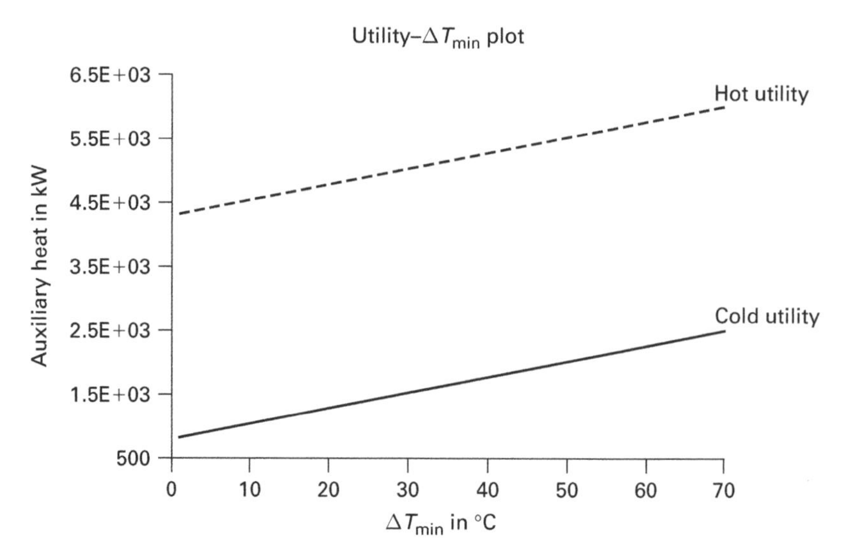

Figure 3.30 Variation of utility use with ΔT_{min}

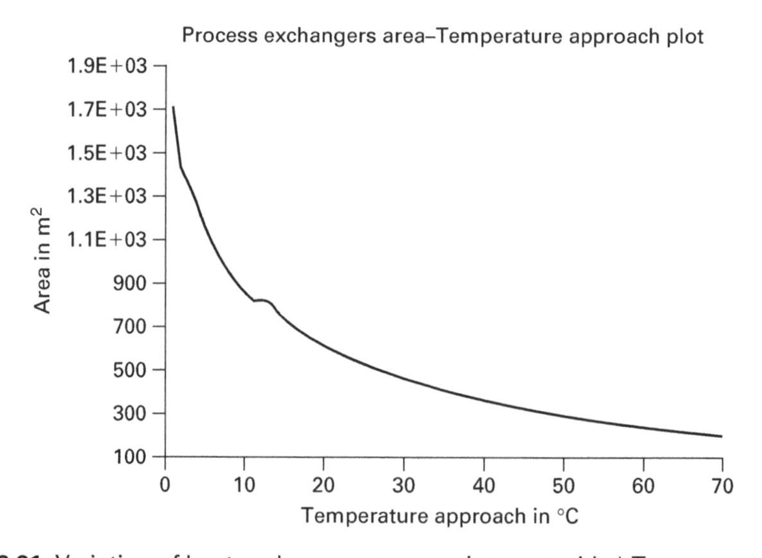

Figure 3.31 Variation of heat exchanger area requirement with ΔT_{min}

higher values of ΔT_{min} give little economic penalty on this short timescale; however, if a longer view is taken, the higher energy costs incurred every year become more apparent. The results are similar for the countercurrent and Ahmad–Smith algorithms, but the graph for the latter (as illustrated) is not so smooth because it allows for the changing number of shells.

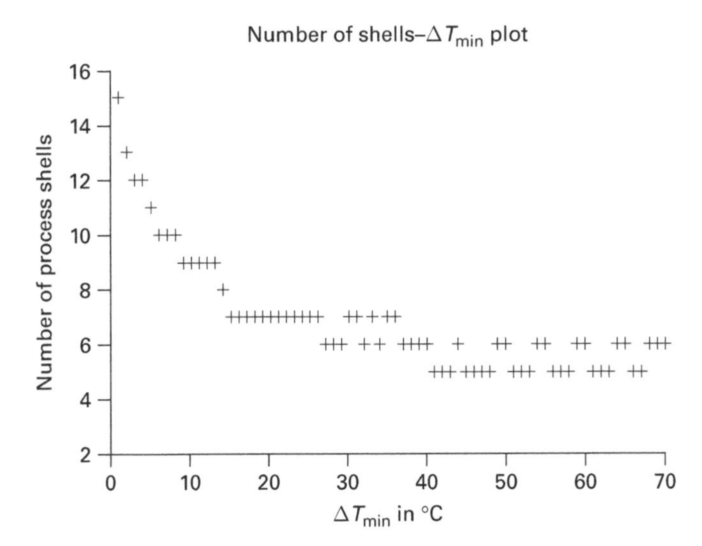

Figure 3.32 Number of heat exchanger shells required at various ΔT_{min} values

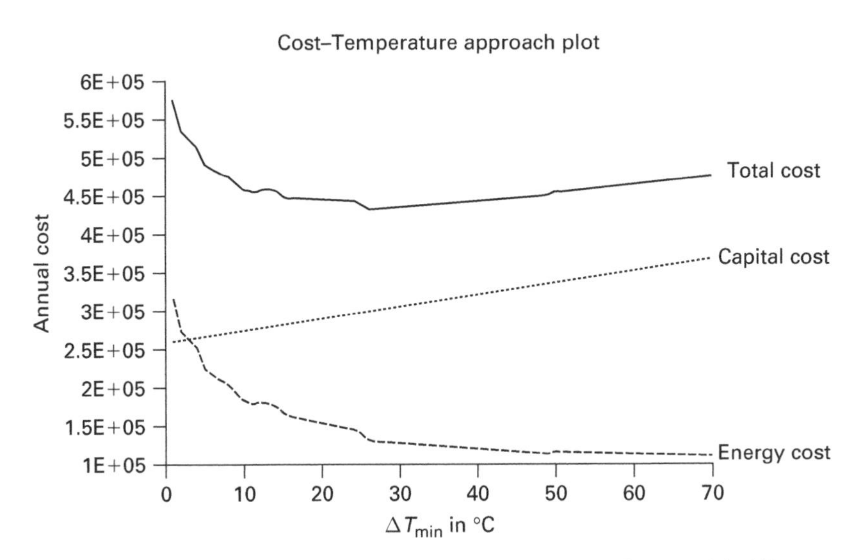

Figure 3.33 Variation of energy, capital and total annualised cost with ΔT_{min}

Overall, the figures of 20°C and 63°C used in the targeting analysis seem reasonable, coming at either end of the acceptable range. Studies could also be made at the optimum ΔT_{min} of 30°, but the cost–ΔT_{min} graph suggests that this will make little difference to the results. Using the economic data, cost estimations could be obtained for capital and energy costs for any selected ΔT_{min} value.

3.8.4 Zonal targeting

So far we have just considered the atmospheric pressure distillation column and its associated streams. However, there is a further section of the plant in series. The residue (bottom product) from the atmospheric pressure distillation column is reheated to become the feed to a second column working under vacuum. It enters via a pressure reducing valve which causes vapour to flash off, and there is no reboiler. Likewise, heavy oil is taken off part way up the column and cooled; some is taken off as product and the remainder is reintroduced to the top of the column, where it condenses the remaining upward vapour flow by direct contact. Finally, a heavy bitumen product is withdrawn from the bottom of the vacuum column, but is not cooled as this would make it too viscous; the same is done with a small wax fraction. The heat and mass balance and stream data are extracted in the same way as for the atmospheric pressure unit and are shown in Table 3.6. The flowrate for the heavy oil draw is back-calculated from the measured heat load on the cooler. Only the two flows with non-zero heat loads become streams for pinch analysis purposes (Figure 3.34).

Is there any incentive for exchanging heat between the vacuum and atmospheric pressure units? To find out, we target them separately and together. The results are shown in Table 3.7.

The vacuum unit shows a pinch region rather than a single pinch temperature, and it currently achieves its energy targets for the simple reason that the hot stream (heavy oil) is always below the temperature of the cold stream (vacuum feed), so no heat recovery is possible! However, the heavy oil is above the pinch of the atmospheric unit, so there is potential for heat recovery between the two plants, shown by the calculations to be 350 kW. This is an additional saving of 4.1% compared with current total heat use (8,500 kW). The overall pinch temperature is the same as for the atmospheric distillation unit alone, and is still caused by the start of the big overheads stream.

The composite and GCC for the combined process are shown in Figure 3.35. We note that an effect of adding in the vacuum distillation unit (VDU) has been to tighten the curves in the region above the pinch, giving a larger area of low net heat flow, so network design will be more challenging.

Table 3.6 Heat and mass balance and stream data for vacuum distillation unit

Flow	*Production rate (te/h)*	*Mass flow (kg/s)*	*Specific heat (kJ/kgK)*	*CP (kW/K)*	*Initial temperature (°C)*	*Final temperature (°C)*	*Heat flow rate (kW)*
Vacuum crude	14.4	4	2.5	10	155	319	−1,640
Heavy oil draw	?	(5)	2.5	12.5	151	67	1,050
Heavy oil product	7.2	2	2.5	5	67	67	0
Top recycle	?	(3)	2.5	12.5	67	67	0
Wax	1.8	0.5	2.5	1.25	204	204	0
Bitumen	5.4	1.5	2.5	3.75	252	252	0

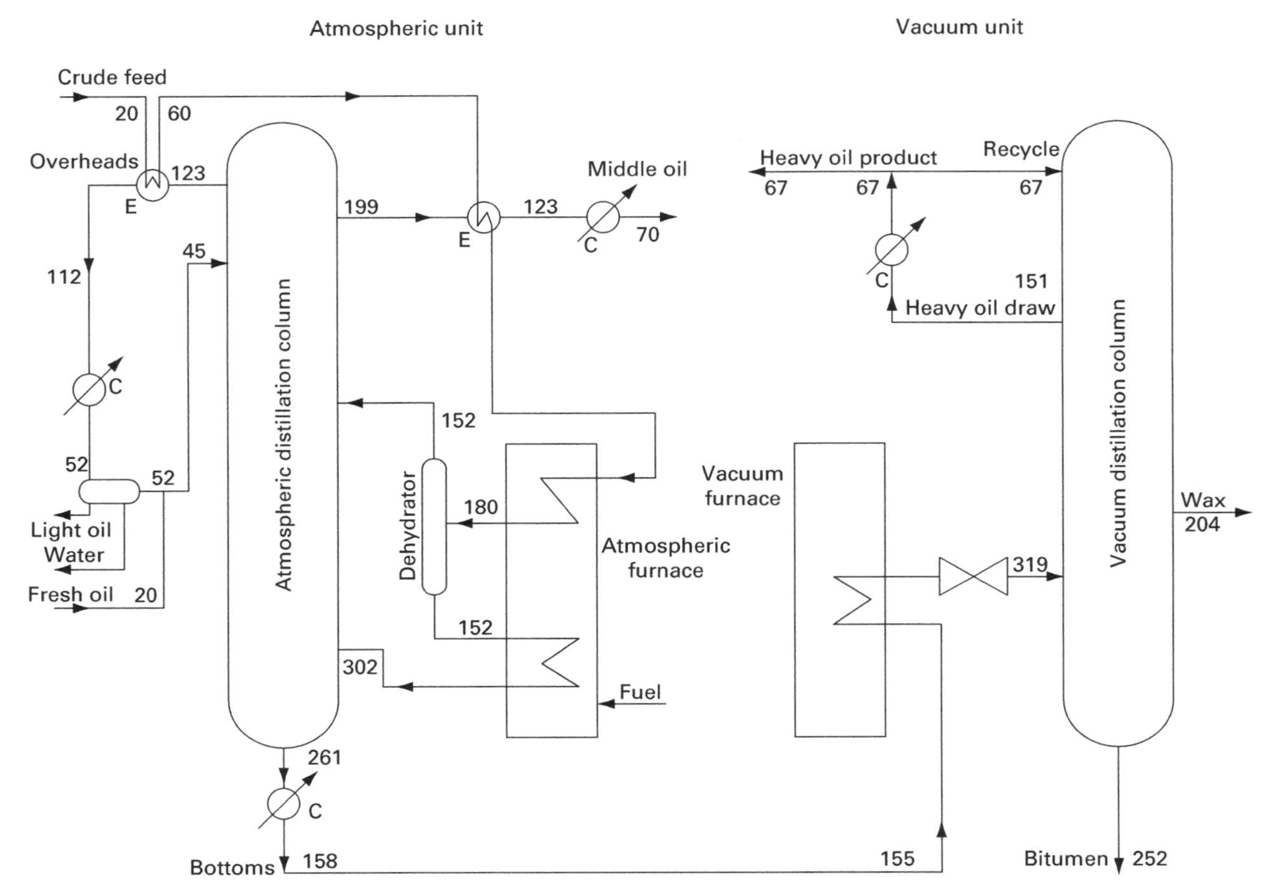

Figure 3.34 Flowsheet of organics distillation process including vacuum distillation unit

Table 3.7 Targets for atmospheric and vacuum distillation units

Situation	*Current hot utility use (kW)*	*Hot utility target (kW)*	*Potential heat saving (%)*	*Cold utility target (kW)*	*Pinch (shifted) temperature (°C)*
Atmospheric unit	6,860	4,795	30.1%	1,295	113
Vacuum unit	1,640	1,640	0.0%	1,050	140–165
Both units separately	8,500	6,435	24.3%	2,345	–
Both units combined	8,500	6,085	28.4%	1,995	113

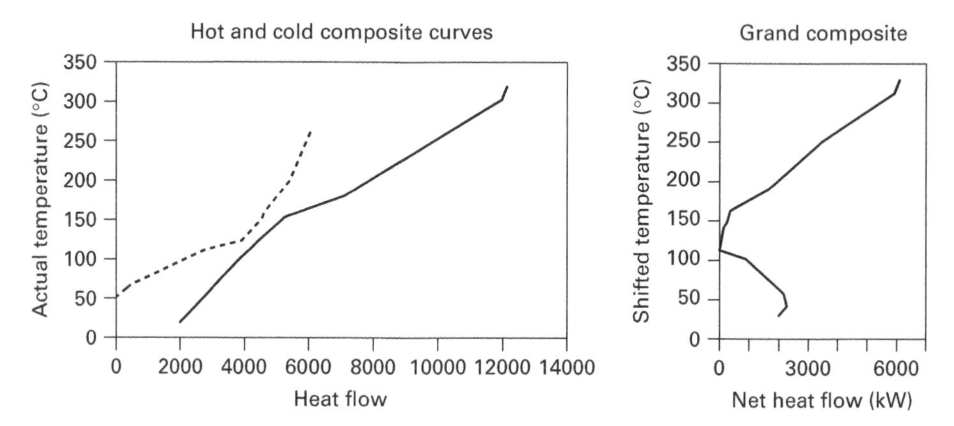

Figure 3.35 Composite and GCC for combined atmospheric and vacuum units

3.8.5 Targeting with utility streams included

Finally, let us consider the associated hot and cold utility systems (for the atmospheric unit alone). The cooling water system simply rejects heat to cooling towers and nothing much can be done there. However, the heating is supplied by the gas-fired furnace. The burners are fed with natural gas and combustion air, giving flame temperatures of well over 1,500°C; since this would damage the furnace tubes, a large quantity of dilution air is mixed in, bringing the furnace bottom temperature down to 400°C. The hot gases pass up the furnace, heating the dehydrate and crude feed, and the flue gas emerges from the top at 200°C. It would be wasteful to discharge this directly, so heat is exchanged with the incoming combustion and dilution air in an air preheater. As gas is a relatively clean fuel, the flue gas can be brought down to 120°C without worrying about acid dewpoint and sulphuric acid corrosion (which can be a problem for oil-fired boilers), but this is considered the practical lower limit to avoid possible condensation in the stack.

Measurements give the total airflow as 25 kg/s. There is no discernible difference between the inlet air and flue gas flows, and this is explained by the heat of combustion for natural gas being 55,000 kJ/kg, whereas the sensible heat change of air

Table 3.8 Heat and mass balance and stream data for furnace and associated streams

Flow	*Mass flow (kg/s)*	*Specific heat (kJ/kgK)*	CP *(kW/K)*	*Initial temperature (°C)*	*Final temperaure t(°C)*	*Heat flow rate (kW)*
Air below 200°C	25	1.0	25	20	200	−4,500
Air above 200°C	25	1.05	26.25	200	400	−5,250
Flue gas	25	1.0	25	200	120	2,000

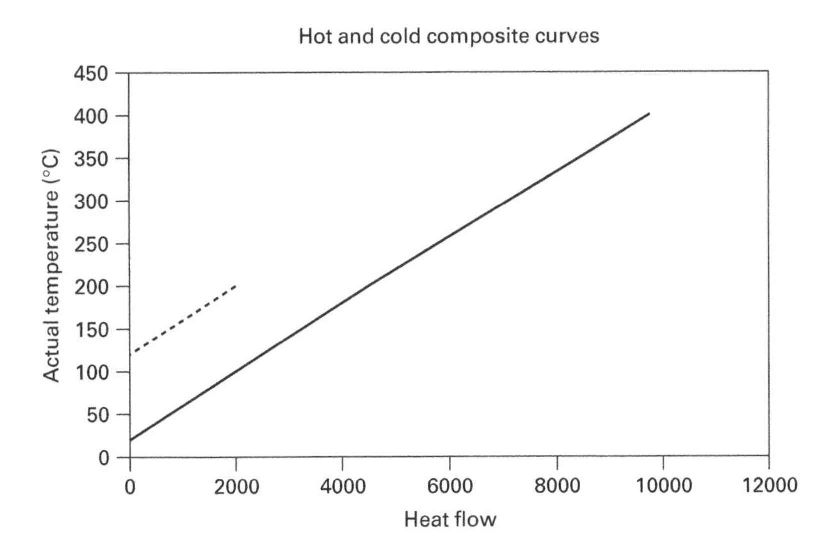

Figure 3.36 Composite curves for ADU furnace utility streams

from 20°C to 400°C is about 400 kJ/kg, so that the fuel flow is less than 1% of the airflow. The specific heat capacity of both air and flue gas will be taken as 1.0 kJ/kgK below 200°C and 1.05 kJ/kgK above 200°C. The stream data for the utility system is then shown in Table 3.8 (we have not distinguished between combustion air and dilution air streams).

The net heat supplied by the fuel, by subtraction, must be 7,750 kW. This compares with 6,860 kW absorbed by the crude feed and dehydrate as they heat up. The difference (roughly 12%) is due to heat losses from the furnace structure.

Targeting the furnace system alone, with a ΔT_{min} of 80°C (since $\Delta T_{cont} = 40$°C for gaseous streams), gives a hot utility target of 7,750 kW and a cold utility target of zero – precisely the same as at present. This is a threshold problem, as can be seen from Figure 3.36, and all the heat available in the flue gas is being used. The $\Delta T_{threshold}$ is in fact 100°C, the difference between the flue gas outlet temperature of 120°C and the air inlet temperature of 20°C.

However, if we combine the utility streams with the process data and retarget, a very different picture emerges. Two alternatives have been given; one for the current airflows, the other with a 30% reduction (to correspond to the 30% less fuel required if the process met its energy targets). If the furnace streams are included, an extra hot stream of 4,795 kW must be added back into the analysis to balance the net heat required by the cold streams in the furnace (otherwise this would be counted twice) (Table 3.9).

The result is that combining the process and utility streams gives a reduction in hot and cold utilities of 500 kW at current airflows or 350 kW at the lower airflows. This corresponds to the portion of the air streams which was below the pinch but could not be matched with the flue gas stream. The pinch temperature does not change; in this region, the additional air and flue gas streams precisely cancel each other out. The composite curves including the utility streams (Figure 3.37) are markedly different from those for the process streams alone, because of the much wider temperature separation for the utility streams and the longer overlap region.

A corresponding exercise could be performed for the VDU if desired (see Section 3.9).

Table 3.9 Targets for atmospheric distillation unit including utility streams

Situation	*Current hot utility use (kW)*	*Hot utility target (kW)*	*Net saving (kW)*	*Cold utility target (kW)*	*Pinch (shifted) temperature (°C)*
Atmospheric unit	6,860	4,795	0	1,295	113
Furnace (current usage)	7,750	7,750	0	0	60–160
Furnace (30% reduction)	7,750	5,425	2,325	0	60–160
Combined (current usage)	7,750	7,250	500	795	113
Combined (30% reduction)	5,425	5,075	350	945	113

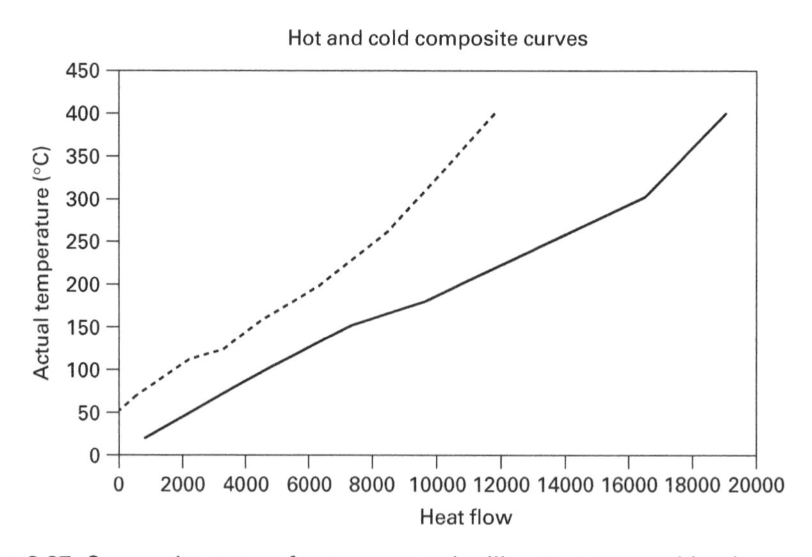

Figure 3.37 Composite curves for process and utility streams combined

3.9 Appendix: Algorithms for Problem Table and composite curves

3.9.1 Problem Table and GCC

1. Select a global ΔT_{min} for the calculation (see also Note (a)).
2. Subtract $\Delta T_{min}/2$ (or ΔT_{cont}, where different – Note (b)) from all hot stream temperatures T_H and add $\Delta T_{min}/2$ (or ΔT_{cont}) to all cold stream temperatures T_C to give the shifted temperatures S for each stream.
3. Make a list of all the shifted temperatures S at which streams (hot or cold) begin, end or change heat capacity flowrate CP.
4. Rank the list of shifted temperatures in descending order (highest temperature at the top).
5. In each temperature interval i between two shifted temperatures, add together the heat capacity flowrates of all the hot streams which exist in that temperature interval and subtract the heat capacity flowrates of all the cold streams, to give a net heat capacity flowrate CP_{net} (see Notes (c) and (d)).
6. Multiply CP_{net} for each interval by the temperature range of the interval ($T_{i+1}-T_i$) to give the net heat released (positive) or required (negative) in the interval (see Notes (c) and (d)).
7. Starting from a zero input at the highest temperature, work down the table, adding on the net heat change in each temperature interval to give a heat cascade (cumulative heat passing through at a given shifted temperature).
8. The cascade in step 7 normally contains negative heat flows and is thermodynamically infeasible. Take the minimum net heat flow in the table $\{-Q_{min}\}$ (largest negative value, or zero) and add this amount of heat Q_{min} as hot utility to the first interval in the cascade. All the net heat flows in the cascade now increase by this amount, and the minimum value becomes zero. This is the feasible heat cascade, or Problem Table.
9. The heat added to the first interval is the hot utility requirement (target) Q_{Hmin}. The heat removed from the final interval is the cold utility target Q_{Cmin}. The point(s) at which there is zero net heat flow in the cascade is the pinch. The plot of the net heat flow (horizontal axis) against the shifted temperature (vertical axis) is the GCC.

Notes:

(a) Targets can be calculated at a range of different ΔT_{min} values and plots can be obtained showing the variation over a chosen range of ΔT_{min} of:
 – hot and cold utility targets,
 – pinch temperature.

(b) Different streams can be allocated a different ΔT_{min} contribution, ΔT_{cont}. When calculating shifted temperatures, ΔT_{cont} for the individual stream should be used instead of $\Delta T_{min}/2$. (This creates some logistical problems if global ΔT_{min} is varied as well.)

(c) In the original Problem Table analysis in the *User Guide* 1st edition, net *CP* was obtained by adding *CP*s for cold streams and subtracting those for hot streams, and the heat change in an interval was negative for a net release and positive for a net demand. This was counter-intuitive and standard practice has been to subtract cold stream *CPs* from hot stream *CPs*, so that a net heat release in an interval is positive, and net energy consumption in an interval is negative. This is the practice adopted in this algorithm.

(d) It is often more convenient or accurate to use enthalpy (heat load) H rather than heat capacity flowrate CP, especially for streams with a temperature-dependent CP. In this case, steps 5 and 6 are replaced by the following new steps:
   5. Check which streams exist in the temperature interval. For each of these, calculate the enthalpy at the upper and lower temperatures of the interval, and subtract the latter from the former to give the change in enthalpy (net heat load) of the stream in that interval.
   6. Add together the net heat loads for all the streams that exist in this interval, giving the combined net heat load change released (positive) or required (negative) over the interval.

(e) Subsets of the streams can be chosen to show the effect of integrating between different zones or groups of streams. (Hand calculation method is to list all the streams, obtain overall target, then delete selected streams and retarget separate groups.)

3.9.2 Composite curves

The calculation and plotting method for the hot and cold composite streams is similar to the Problem Table:

1. Make a list of all the temperatures T at which hot streams begin, end or change heat capacity flowrate CP.
2. Rank the list of temperatures in ascending order (lowest temperature at the top).
3. In each temperature interval i, add together the heat capacity flowrates of all the hot streams which exist in that temperature interval, to give a total heat capacity flowrate for cold streams CP_H.
4. Multiply CP_H for each interval by the temperature range of the interval $(T_{i+1}-T_i)$ to give the net heat required by hot streams in the interval.
5. Starting from a zero input at the lowest temperature, work down the table, adding on the total heat change in each temperature interval to give a cumulative heat load for hot streams at each temperature.
6. The heat load at the end of the final interval gives the total heat load due to cold streams H_H. The plot of the cumulative heat flow (horizontal axis) against the actual temperature (vertical axis) is the hot composite curve.
7. Repeat steps 1–5 for the cold streams to give the cumulative heat loads for cold streams. The heat load at the end of the final interval gives the total heat load due to cold streams H_C. Add the minimum cold utility requirement Q_{Cmin} (calculated using the Problem Table) to all the heat loads at the various temperatures (thus shifting the curve to the right by Q_{Cmin}). This gives the cold composite curve as the plot of the adjusted cumulative heat flow (horizontal axis) against the actual temperature (vertical axis).

The shifted composite curves (SCCs) are generated by a combination of the Problem Table and composite curve plotting methods. Temperatures are adjusted from real temperatures to shifted temperatures (as in step 3 of the Problem Table algorithm, and allowing for differences in ΔT_{min} contribution, ΔT_{cont}, for different streams) but the hot streams and cold streams are considered separately. The hot SCC will again need to be moved to the right by Q_{Cmin}; when this has been done, the hot and cold SCCs should just touch at the pinch. If ΔT_{cont} is the same for all individual streams (and equal to half the global ΔT_{min}), the SCCs are simply the composite curves shifted by $\Delta T_{min}/2$. However, if some streams have different ΔT_{cont} values, the SCCs will need to be generated separately.

Note, if the ΔT_{min} contributions vary between different streams, the GCC and SCC will give a clearer picture than the composite curves.

Exercises

E3.1 For the five-stream problem below (from Section 2.6), find the energy targets and pinch and plot composite curves and GCC if not done previously. What is the significance of the temperature range and *CP* of stream 3? What effect does this have on the pinch?

Stream	*Supply temperature (°C)*	*Target temperature (°C)*	*Heat capacity flow rate* CP *(kW/K)*	*Heat load Q (kW)*
Hot 1	200	50	3	450
Hot 2	240	100	1.5	210
Hot 3	120	119	300	300
Cold 4	30	200	4	−680
Cold 5	50	250	2	−400

E3.2 For the organics distillation unit in Section 3.8, calculate the airflows and fuel use for the vacuum furnace, assuming the same level of heat losses as for the atmospheric unit furnace, and the same temperatures for air and flue gas. Perform the targeting calculation for the vacuum unit with its furnace included, and for the combined units including both furnaces. Evaluate and rank the various options for energy reduction from integrating the utility systems.

References

Ahmad, S. and Smith, R. (1989). Targets and design for minimum number of shells in heat exchanger networks, *Chem Eng Res Des*, 67(5): 481–494.

Ahmad, S. and Hui, D. C. W. (1991). Heat recovery between areas of integrity, *Comp Chem Eng*, 15(12): 809–832.

Ahmad, S., Linnhoff, B. and Smith, R. (1990). Cost optimum heat exchanger networks – 2. Targets and design for detailed capital cost models, *Comp Chem Eng*, 14: 729.

Amidpour, M. and Polley, G. T. (1997). Application of problem decomposition to process integration, *TransIChemE*, Vol. 75, Part A, 53–63, January.

Hohmann, E. C. (1971). *Optimum Networks for Heat Exchangers*, PhD Thesis, University of Southern California, USA.

Kemp, I. C. (1991). Some aspects of the practical application of pinch technology methods, *ChERD (TransIChemE)*, 69:A6, 471–479, November.

ESDU. (1987). Process integration. ESDU Data Item 87030, ESDU International plc, London.

Linnhoff, B. and Ahmad, S. (1990). Cost optimum heat exchanger networks, Part 1: Minimum energy and capital using simple models for capital cost, *Comp Chem Eng*, 14(7): 729–750. Ahmad, S., Linnhoff, B. and Smith, R. Part 2: Targets and design for detailed capital cost models, *Comp Chem Eng*, 14(7): 751–767.

Linnhoff, B., Mason, D. R. and Wardle, I. (1979). Understanding heat exchanger networks, *Comp Chem Eng*, 3: 295.

Linnhoff, B. and Hindmarsh, E. (1983). The pinch design method of heat exchanger networks, *Chem Eng Sci*, 38(5): 745–763.

Linnhoff, B. (1993). Pinch analysis – a state-of-the-art overview, *TransIChemE*, Part A, 71(A5): 503–522.

Polley, G. T., Panjeh Shahi, M. H. and Jegede, F. O. (1990). Pressure drop considerations in the retrofit of heat exchanger networks, *ChERD*, 68(A3): 211–220.

Polley, G. T. and Heggs, P. J. (1999). Don't let the "pinch" pinch you, *Chem Eng Prog, AIChE*, 95(12): 27–36, December.

Saboo, A. K., Morari, M. and Colberg, R. D. (1986). RESHEX — An interactive software package for the synthesis and analysis of resilient heat exchanger networks, Part 1: Program description and application, *Comp Chem Eng*, 10(6): 577–589. Part 2: Discussion of area targeting and network synthesis algorithms, *Comp Chem Eng*, 10(6): 591–599.

Sagli, B., Gundersen, T. and Yee, T. (1990). Topology traps in evolutionary strategies for heat exchanger network synthesis. In Bussemaker, H. and Iedema, P. (eds), *Process Technology Proceedings* 9: 51/58.

Smith, R. (2005). *Chemical Process Design and Integration*. John Wiley & Sons Ltd, Chichester, UK. ISBN 0-471-48680-9/0-471-48681-7.

Tjoe, T. N. and Linnhoff, B. (1986). Using pinch technology for process retrofit, *Chem Eng*, 47–60, April 28.

Townsend, D. W. and Linnhoff, B. (1983). Heat and power networks in process design. Part 1: Criteria for placement of heat engines and heat pumps in process networks, *AIChE J*, 29(5): 742–748. Part 2: Design procedure for equipment selection and process matching, *AIChE J*, 29(5): 748–771.

Townsend, D. W. and Linnhoff, B. (1984). Surface area targets for heat exchanger networks. Paper presented at *The IChemE Annual Research Meeting*, Bath, April.

4 Heat exchanger network design

4.1 Introduction

In Chapter 2 we showed how to develop a simple network achieving energy targets for a given. This is the maximum energy recovery or minimum energy requirement (MER) design. In this section, we cover more advanced concepts, including:

- More complex MER designs involving stream splitting (Section 4.3).
- Network relaxation – eliminating small exchangers with a minor energy penalty (Section 4.4).
- Situations with constraints, multiple pinches, utility pinches and pinch regions (Sections 4.5 and 4.6).
- Revamp and retrofit of existing heat exchanger networks (Section 4.7).
- Operability aspects and multiple base cases (Section 4.8).
- We also briefly look at the main available types of heat exchangers (Section 4.2) and illustrate all the key themes by application to our case study on the organics distillation unit (Section 4.9).

4.2 Heat exchange equipment

4.2.1 Types of heat exchanger

In simple terms, heat exchange equipment can be divided into three families: shell-and-tube, plate and recuperative exchangers.

The shell-and-tube family is generally used for heat exchange between liquids, but may include gases or condensing/boiling streams. Fluid flows through a set of tubes and exchanges heat with another fluid flowing outside the tubes in crossflow, countercurrent, cocurrent or mixed flow. Double-pipe exchangers are a special case of this type where there is just a single central tube with an annular shell around it. Construction is strong and rigid, well suited for high pressures and temperatures as found in many chemicals applications. However, adding additional area requires either major retubing or additional shells.

The plate family is again generally used for liquids, and includes gasketed plate, welded plate and plate-fin units. The basic construction is a large number of pressed or stamped plates held against each other, with the recesses between the plates forming narrow flow channels. These give excellent heat transfer but are also liable to fouling. It is easy but tedious to dismantle the gasketed type for cleaning; this is much more difficult with welded types. The plates are mounted on a frame and there is usually spare space to add more plates; thus, it is easy to increase the heat transfer area if desired. They are frequently used in the food and beverage industries. They are suitable for use as multi-stream exchangers (Section 4.2.6).

Recuperative exchangers cover a variety of types mainly used for heat transfer to and from gas streams. Because of the low-heat transfer coefficients, heating surfaces are frequently extended to provide additional surface area (e.g. with fins). Some types are simple variants of shell-and-tube units; others work on completely different principles, such as rotary regenerators (heat wheels) or Cowper stoves, where the equipment is alternately fed with hot and cold gases and acts as short-term heat storage.

Detailed design of heat exchangers is a huge subject in itself and thoroughly covered elsewhere, so is beyond the scope of this text. Good sources include the Heat Exchanger Design Handbook (confusingly, there are two completely separate books with the same title; Kuppan 2000 and Hewitt 2002) and, of course, heat exchangers are extensively featured in general chemical engineering texts such as Sinnott (2005).

4.2.2 Shell-and-tube exchangers

There are three main types of shell-and-tube exchanger; fixed tubeplate, floating head and U-tube (Figure 4.1). The first two types have straight tubes with the tube side fluid entering at one end and leaving at the other. The fixed tubeplate is cheaper but the shell side is hard to clean and expansion bellows may be needed to deal with thermal stresses. The U-tube type only needs a header at one end and the tubes can easily be withdrawn for external cleaning, but internal cleaning is hard and the flow reversal reduces effective ΔT.

The Tubular Exchanger Manufacturers Association (TEMA) has classified shell-and-tube exchangers by shell type, front end head and rear end head types. These distinctions are important in choice of a suitable exchanger for a given duty, but head types do not affect initial network design.

Double-pipe exchangers are not true shell-and-tube exchangers but show many similarities. In essence, one fluid flows in an annulus around the inner tube, although a convoluted route with multiple flow reversals may be used. Their great advantage is that almost pure countercurrent flow is achieved. However, surface area is considerably less than for a multi-tubular exchanger of the same volume.

4.2.2.1 *Implications for network design*

Temperature crosses will be a problem in "long" matches, especially for U-tube exchangers and other types with multiple tube passes. For these matches, especially near the pinch, it may be best to use multiple shells or countercurrent exchangers. Conversely, if the shell side fluid is boiling or condensing at constant temperature, the U-tube unit is at no disadvantage.

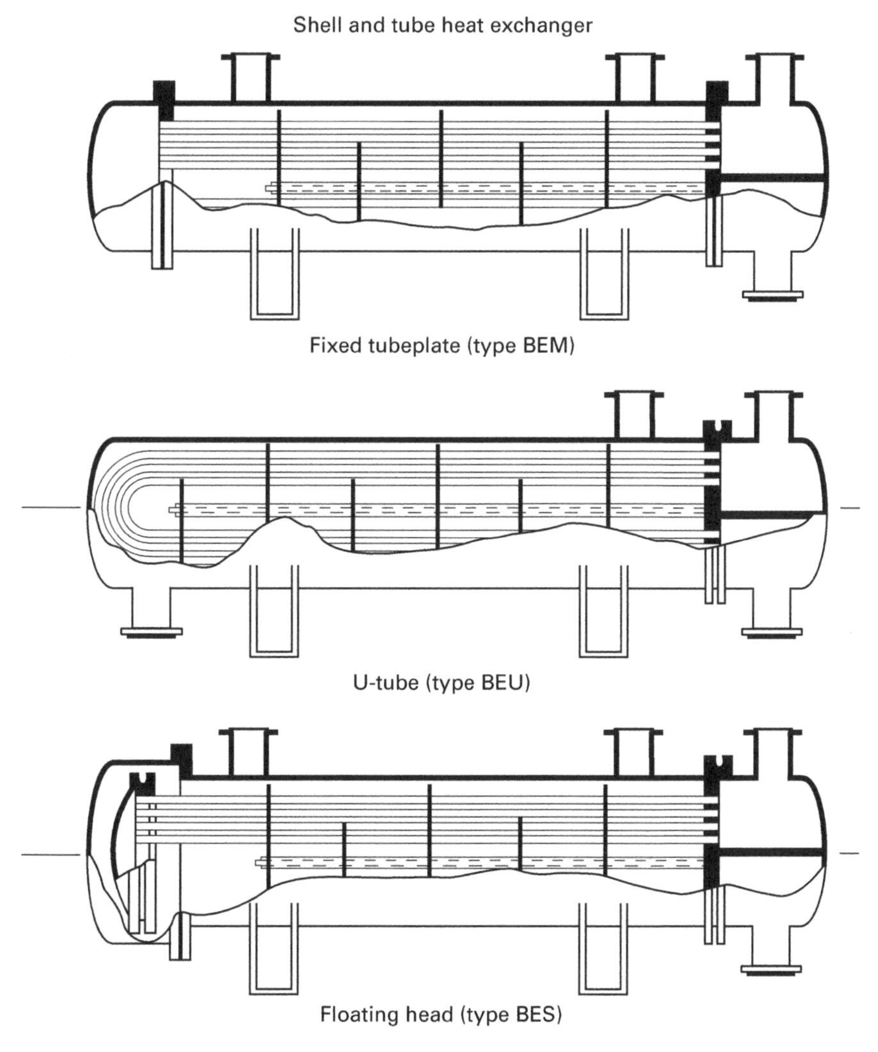

Figure 4.1 Major types of shell-and-tube heat exchanger

Which fluid should go on the tube side and which on the shell side in a match? The following preferences may be applied:

- Put a condensing or boiling stream on the shell side (easier flows and better temperature differences).
- Put the fluid with the lower temperature change (or higher CP) on the shell side (tends to give better temperature differences).
- Put corrosive fluids on the tube side; cheaper to make tubes from exotic alloys than shells, and easier to repair than a shell if corrosion does occur.
- Streams whose pressure drop must be minimised should go on the shell side (ΔP through the exchanger is much lower).

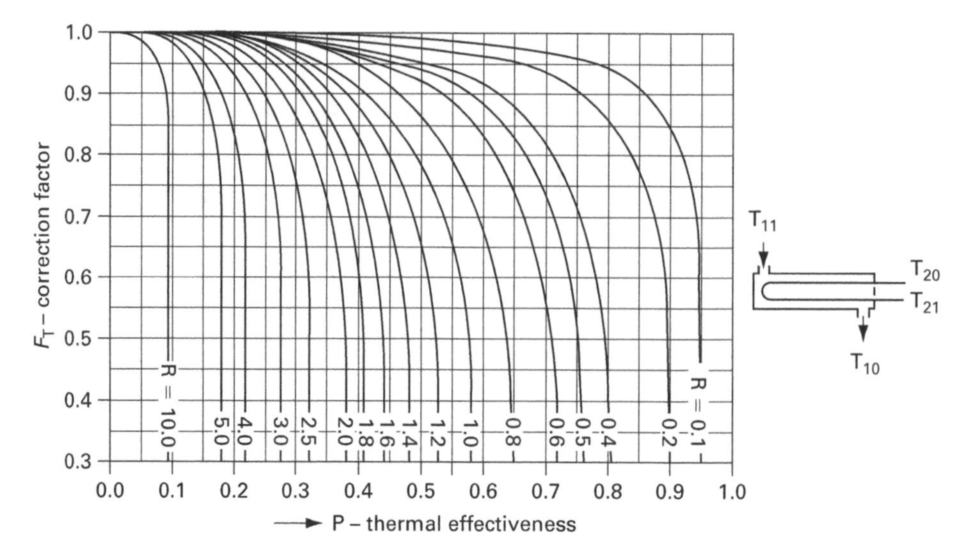

Figure 4.2 F_T correction factors for U-tube exchangers

- In fixed tubeplate units, heavily fouling fluids should go on the tube side; in U-tube units, they should go on the shell side.
- Putting the hot fluid on the tube side minimises structural heat losses.

4.2.2.2 *True temperature driving forces in matches*

In Chapter 2 we stated the formula for heat exchange, $Q = UA\,(\Delta T_{LM})$. However, this only applies for pure countercurrent heat exchange. In shell-and-tube exchangers, the shell side fluid is normally in crossflow. Moreover, in U-tube units and other types with an even number of tube passes, the hottest and coldest tube side fluid is at the same end of the exchanger. Even double-pipe exchangers do not show perfect countercurrent exchange; there is some mixing.

To allow for this, the log mean temperature difference is multiplied by a correction factor F_T (<1). F_T itself is expressed in terms of two other parameters P and R. R is the ratio of the temperature change for the hot stream to that for the cold stream (and therefore also equal to CP_C/CP_H if heat losses are discounted). P is the temperature change on the cold stream divided by the temperature difference between hot and cold streams at inlet. Graphs and formulae are available to give F_T for a wide range of exchanger types; two especially useful ones are U-tube exchangers (Figure 4.2) and crossflow shells (Figure 4.3).

$$Q = UAF_T\Delta T_{LM} \qquad \text{where } \Delta T_{LM} = \frac{T_{hi} - T_{ho} + T_{ci} - T_{co}}{\ln\left(\dfrac{T_{hi} - T_{co}}{T_{ho} - T_{ci}}\right)} \tag{4.1}$$

$$R = \frac{T_{hi} - T_{ho}}{T_{ci} - T_{co}} \tag{4.2}$$

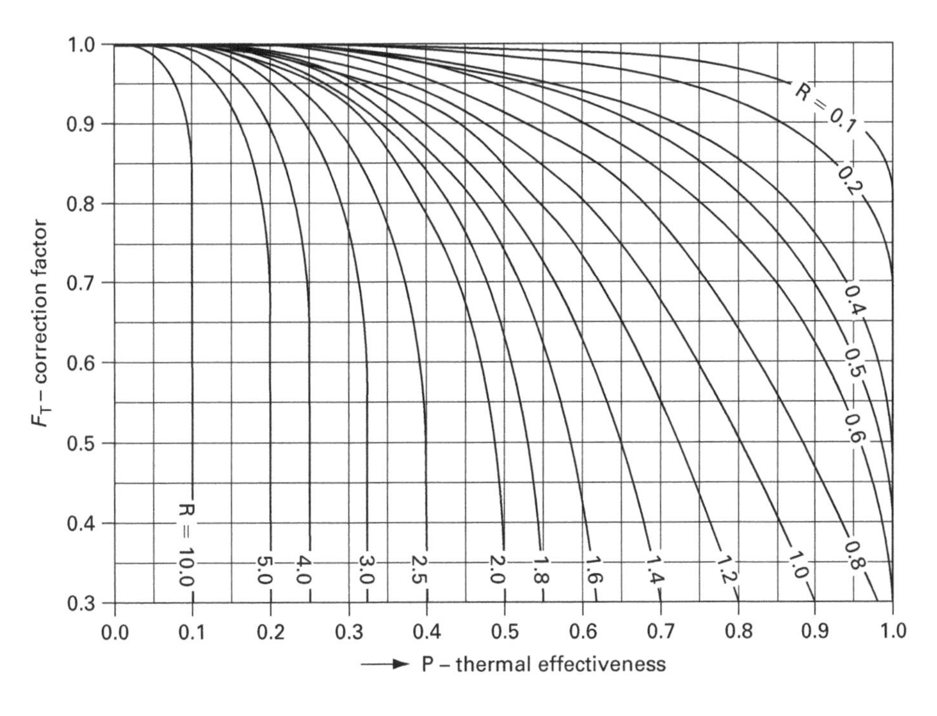

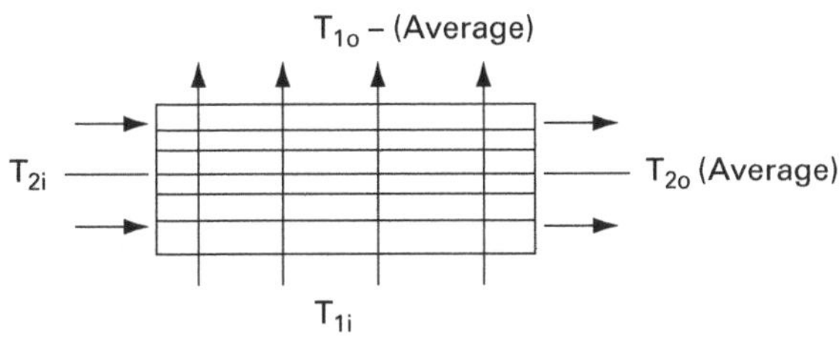

Figure 4.3 F_T correction factors for one-pass exchangers with pure crossflow on shell side

$$P = \frac{T_{ci} - T_{co}}{T_{hi} - T_{ci}} \tag{4.3}$$

Typical film heat transfer coefficients for shell-and-tube exchangers are shown in Table 4.1.

4.2.3 Plate exchangers

Plate exchangers first made an impact in the "non-chemical" process industries, such as food and drink, but are now extensively used in all industries.

The relatively narrow passages mean that the pressure drop tends to be high. However, the flow pattern through the plates (Figure 4.4) means that it is easier to

Table 4.1 Typical film heat transfer coefficients for shell-and-tube heat exchangers

	Cold side			*Hot side*		
	Film coefficient W/m²K (clean)	*Fouling factor Km²/W*	*Overall film coefficient W/m²K*	*Film coefficient W/m²K (clean)*	*Fouling factor Km²/W*	*Overall film coefficient W/m²K*
Low pressure gas ~1 bar	112	0.0002	110	112	0.0002	110
High pressure gas ~20 bar	682	0.0002	600	682	0.0002	600
Process water	–	–	–	6,000	0.0005	1,500
Treated cooling water	5,000	0.0002	2,500	–	–	–
Low viscosity organic liquid	1,667	0.0004	1,000	1,667	0.0004	1,000
High viscosity liquid	210	0.0008	180	170	0.0008	150
Condensing steam	–	–	–	8,182	0.0001	4,500
Condensing hydrocarbon	–	–	–	1,410	0.0002	1,100
Condensing hydrocarbon/ 1 bar	–	–	–	435	0.0002	400
Boiling treated water	5,676	0.0003	2,100	–	–	–
Boiling organic liquid	1,667	0.0004	1,000	–	–	–

achieve a nearly countercurrent flow pattern than in most shell-and-tube exchangers. They are easy to enlarge by adding more plates (there is normally free space between the movable cover and the end mounting) which is very helpful when revamping an existing plant.

Gasketed plate heat exchangers are normally limited to about 150°C and 5 bar by the gasket material, although special materials may be used. However, at higher temperatures and pressures it may be preferable to use a welded plate heat exchanger, which has higher integrity. The drawback is that it is more difficult to dismantle for cleaning, and it is less easy to add extra plates.

Plate-fin heat exchangers are like welded plate exchangers with extended surfaces. They are popular for cryogenic applications, where ΔT must be minimised because of the very high cost of low-temperature refrigeration.

Other exchangers in this group include spiral and lamella types.

Typical film heat transfer coefficients for plate heat exchangers are shown in Table 4.2.

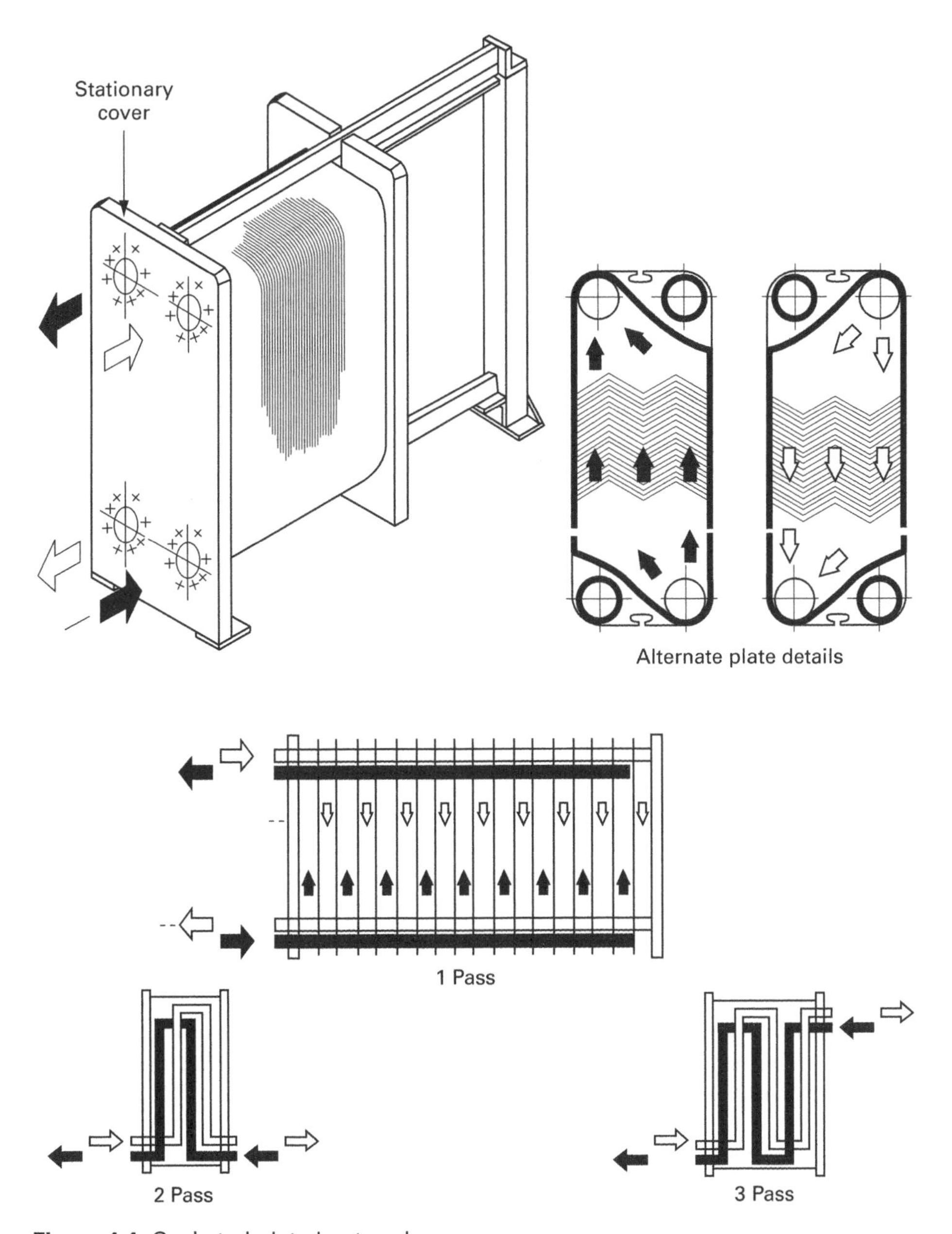

Figure 4.4 Gasketed plate heat exchanger

Table 4.2 Typical film heat transfer coefficients for plate heat exchangers

	Clean coefficient (W/m^2K)	*Fouling factor (Km^2/W)*	*Overall film coefficient ($W/m^2/K$)*
Process water	12,000	0.00003	8,824
Treated water	14,000	0.000015	11,570
Low viscosity organic liquids	7,000	0.00002	6,140
High viscosity liquid	350	0.00004	345
Steam (low pressure)	9,000	0.0000125	8,090

(Basic data courtesy of Johnson Hunt Ltd.) Note also that the wall resistance is relatively high; a figure of 40×10^{-6} $(W/m^2K)^{-1}$ for stainless is used.

4.2.4 Recuperative exchangers

This broad group covers both gas-to-liquid and gas-to-gas duties. Heat transfer coefficients from gases are substantially lower than from liquids, and to achieve a reasonable ΔT without using excessive area, extended heat exchange surfaces (e.g. finned tubes or elements) are often required. In addition, hot gas streams are often wet, dusty and heavily fouling. For these, glass tube exchangers may be used (relatively poor heat transfer but easily cleaned).

Cast iron, stainless steel, plastic or glass tubes may be used, depending on the nature of the process streams and their temperatures. All of these are basically variants on shell-and-tube exchangers. Heat pipe exchangers enhance heat transfer between the hot and cold sides.

Another class of recuperator is based on alternating heat storage using a solid medium; there are both static and rotating types. The static unit is typically a set of chambers made out of firebrick, which are fed first with hot gases and then with cold. These are suitable for very high-temperature dusty gases, such as in the smelting industry, and are used to recover heat from the hot exhaust gases in blast furnaces, where they are known as Cowper stoves. The dynamic unit is a large "heat wheel" with hot gases passing through one side and cold air through the other; this slowly rotates and the heated structure is exposed to the cold air. These are used at moderate temperatures.

Air coolers may also be mentioned here; they provide an alternative to cooling water. Again, fins or other extended surfaces are common on the air side, and air movement and heat transfer coefficients are enhanced by a large fan (whose power consumption must be accounted for).

4.2.5 Heat recovery to and from solids

Process engineering is not just about liquids and gases. A majority of processes involve solids at some point, either as an intermediate or as a product. Quite often, there are

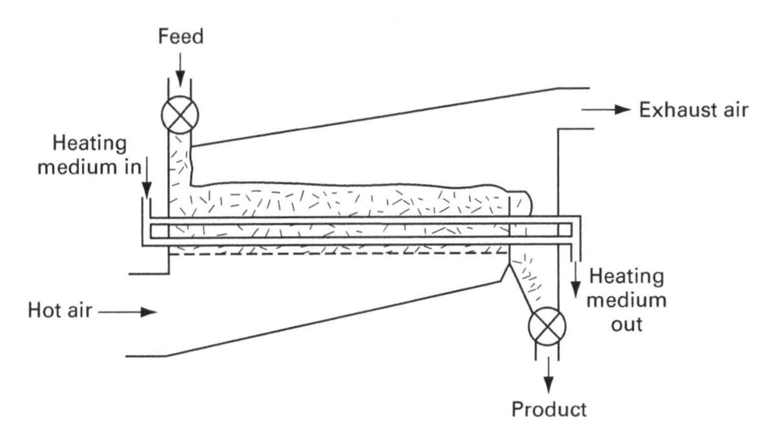

Figure 4.5 Fluidised bed dryer with immersed heating coils

solids streams which include quite significant heat loads. Usually these are sensible heat loads; specific heat capacities of solids (kJ/kg) are generally comparable to those for gases and rather lower than for liquids.

Unfortunately, heat recovery to and from solids streams is very difficult. Heat transfer coefficients between solids and heat exchange surfaces are generally very poor compared with those from liquids, and as a result, any heat exchange from solids requires a high ΔT. One possibility is to pass air through a bed of solids, giving direct contact heat exchange.

An exception to the rule is the fluidised bed, where air is passed through a bed of particles so that they move freely but without becoming elutriated. If plates or tubes are immersed in the bed, the heat transfer from the fluidised solids/gas mixture is an order of magnitude higher than in an unfluidised packed bed of solids. This method is more commonly used to supply heat to solids rather than recover heat from them. Fluidised bed dryers frequently contain immersed heating coils heated by steam, hot water or thermal fluid to supply the large heat demands for latent heat of evaporation (Figure 4.5).

4.2.6 Multi-stream heat exchangers

It will be seen in the following sections that achieving maximum energy recovery (MER) often requires the splitting of streams into parallel branches. As an alternative, a multi-stream heat exchanger could be used. Let us say we want to match two hot streams simultaneously against a single cold stream. We could divide the "hot" side into a section through which hot stream 1 flows and a separate section in which stream 2 flows. In a shell-and-tube exchanger this may be achieved by putting the cold stream on the shell side and streams 1 and 2 through separate tube bundles, but thermal design is difficult if the tubeside fluids are at widely different temperatures. However, plate or plate-fin heat exchangers are very suitable for the task. Multi-stream plate-fin exchangers are used in the cryogenics industry, and have been most reliable and successful. However, they are difficult to clean and therefore are only advisable for non-fouling streams. In dairies and similar industries, this problem has been overcome by using gasketed plate exchangers, and combined exchangers/heaters/coolers are common.

4.3 Stream splitting and cyclic matching

4.3.1 Stream splitting

The principle of design at the pinch was illustrated in Section 2.3 by simple example. However, in practical, more complex cases, a more comprehensive set of rules and guidelines is required, based on the "Pinch Design Method" of Linnhoff and Hindmarsh (1983).

For design at the pinch, we noted that all matches between process streams must fulfil the CP criteria, repeated below:

Above the pinch, $CP_{\mathrm{HOT}} \leqslant CP_{\mathrm{COLD}}$

Below the pinch, $CP_{\mathrm{HOT}} \geqslant CP_{\mathrm{COLD}}$

The CPs for the four-stream example were carefully chosen such that these criteria would be met. In general, however, this will not be the case. For example, consider the organics distillation plant (Section 3.8). Below the pinch we have 3 hot streams and just 1 cold stream, and one hot stream (the overheads) has a very large CP, so it is not difficult to fulfil $CP_{\mathrm{HOT}} \geqslant CP_{\mathrm{COLD}}$. However, above the pinch, we have 2 hot streams and 1 cold stream. So, regardless of stream CPs, one of the hot streams cannot be cooled to pinch temperature by interchange! The *only* way out of this situation is to **split** a cold stream into two parallel branches, as in Figure 4.6. Now, the number of cold streams plus branches is equal to the number of hot streams and so all hot streams can now be interchanged down to pinch temperature. Hence, in addition to the CP feasibility criterion introduced earlier we have a "number count" feasibility criterion, where above the pinch,

$$N_{\mathrm{HOT}} \leqslant N_{\mathrm{COLD}}$$

where N_{HOT} = number of hot stream branches at the pinch (including full as well as split streams).

N_{COLD} = number of cold stream branches at the pinch (including full as well as split streams).

Likewise, below the pinch, we have the additional criterion $N_{\mathrm{HOT}} \geqslant N_{\mathrm{COLD}}$.

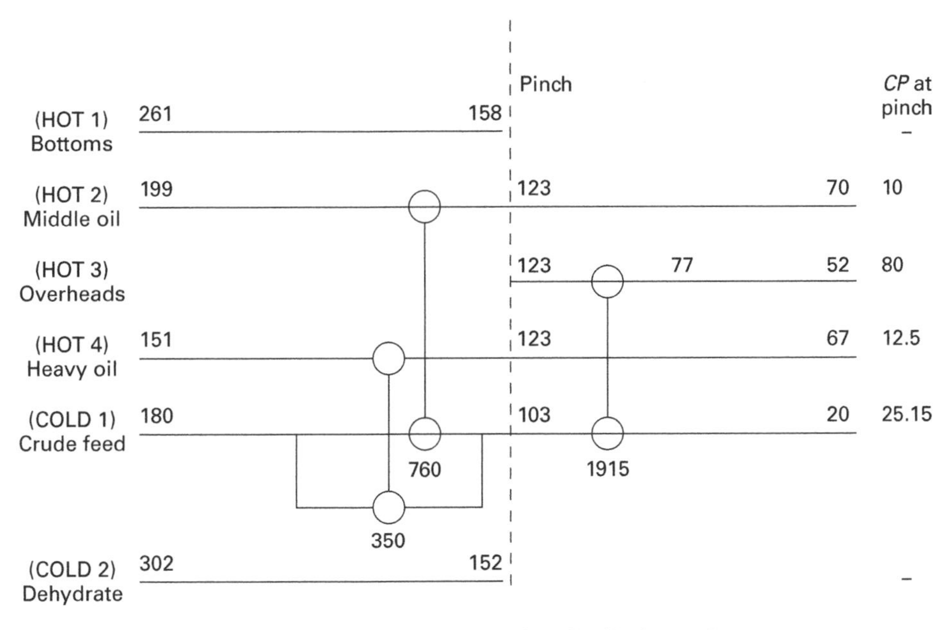

Figure 4.6 Above-pinch stream split for organics distillation unit

Look now at the example shown in Figure 4.7(a). The number count criterion is satisfied (one hot stream against two cold streams) but the CP criterion $CP_{\mathrm{HOT}} \leqslant CP_{\mathrm{COLD}}$ is not met for either of the possible two matches. In this example the solution is to split a hot stream as shown in Figure 4.7(b). Usually in this type of situation the solution is to split a hot stream, but sometimes it is better to split a cold stream as shown in Figure 4.7(c) and (d). In Figure 4.7(c) the number count criterion is met, but after the hot stream of $CP = 7.0$ is matched against the only cold stream large enough ($CP = 12.0$), the remaining hot stream of $CP = 3.0$ cannot be matched against the remaining cold stream of $CP = 2.0$. If a hot stream were now to be split, the number count criterion would not then be satisfied and a cold stream would then have to be split as well! It is better to split the large cold stream from the outset as shown in Figure 4.7(d), producing a solution with only one split. Step-by-step procedures for finding stream splits are given for above and below the pinch in Figure 4.8(a) and (b), respectively. The below-the-pinch criteria are the "mirror image" of those for above the pinch.

The procedure will now be illustrated by example. The stream data above the pinch are shown in Figure 4.9(a), and the CP data are listed in Figure 4.9(b) in what we shall call the "CP-table". Hot-stream CPs are listed in the column on the left and cold-stream CPs in the column on the right, and the relevant CP criterion noted in the box over the table. There are two possible ways of putting in the two required pinch matches, shown at the top of Figure 4.9(c). In both of these, the match with the hot stream of $CP = 5.0$ is infeasible, hence we must split this stream into branches $CP = X$ and $CP = 5.0 - X$ as shown in the bottom table in Figure 4.9(c). Now, $CP_{\mathrm{H}} = X$ or

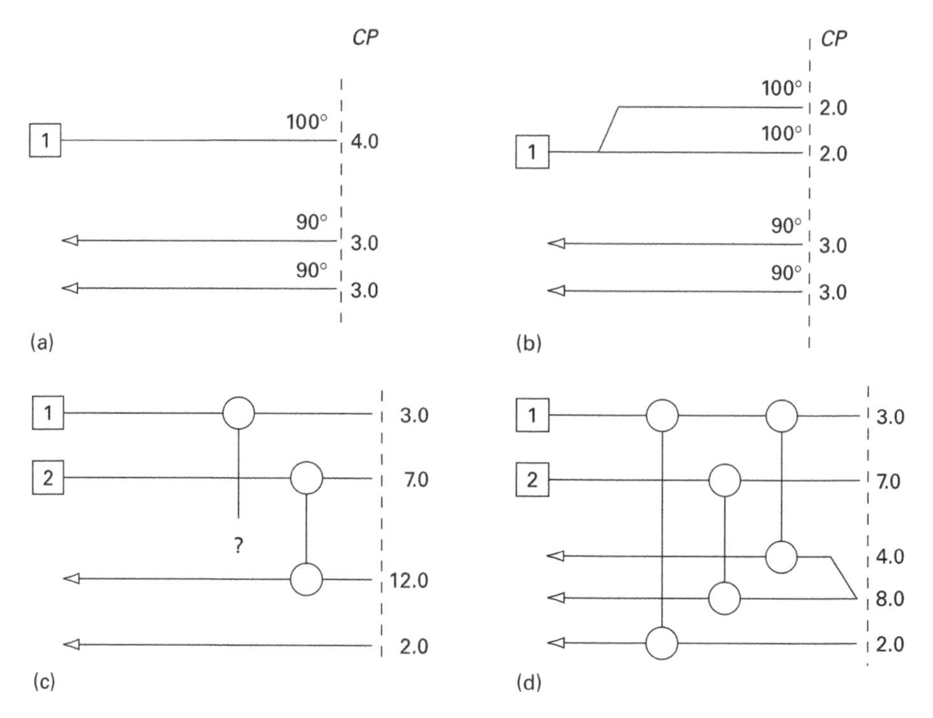

Figure 4.7 Splitting streams to satisfy CP criteria

5.0 − X can be matched with $CP_C = 4.0$, as shown. However, one of the split branches has no partner (i.e. the number count criterion has failed and a cold stream must be split). Either $CP_C = 4.0$ or $CP_C = 3.0$ could be split, and Figure 4.10(a) shows $CP_C = 3.0$ split into branches Y and 3.0 − Y. To find initial values for X and Y it is recommended that all matches except for one are set for CP equality. Thus in Figure 4.10(b), X is set equal to 4.0 and Y set equal to 1.0, leaving all the available net CP difference (i.e. $\Sigma CP_C - \Sigma CP_H$) concentrated in one match. The procedure quickly identifies a set of feasible limiting values. Starting from this set, it is then easy to redistribute the available CP difference amongst the chain of matches, for example as shown in Figure 4.10(c). This design is shown in the grid in Figure 4.10(d). The way in which the branch CPs are distributed is often dictated by the loads required on individual matches by the "ticking-off" rule. Where this is not a constraint, or where choices exist, total exchanger area tends to be minimised if the CPs on the split streams are roughly proportional to those on the hot streams they are matched against, as this gives the most even distribution of temperature driving forces. This is preferable to putting all the slack on one match. Figure 4.10(d) is fairly close to this criterion; exact proportionality would be given by CPs of 3.43/1.57 on stream 1 and 1.83/1.17 on stream 4, giving a ratio of 6:7 for all hot stream:cold stream CPs. However, the temperature range and heat loads on the streams should also be taken into account when splitting.

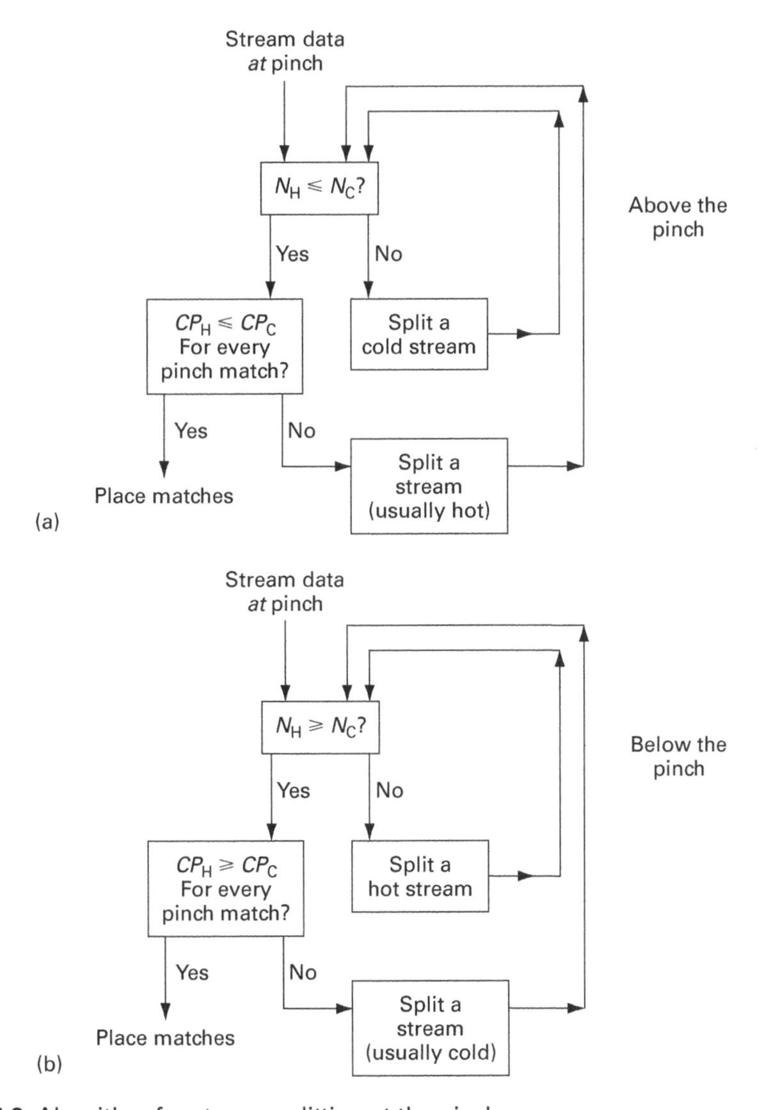

Figure 4.8 Algorithm for stream splitting at the pinch

Figure 4.11 gives another simple example. Here we have the above-pinch network for a process with ΔT_{min} = 20°C. There are two hot streams and one cold stream above the pinch, so the cold stream must be split. We have a range of options on the percentage stream splits and two are shown here. In (a), the CPs are split in proportion to the CPs in the matched streams. Because the supply temperatures of the latter are different, the two halves of the cold stream end up at different temperatures before remixing. Conversely, in (b), the CPs are split in proportion to the matched streams' heat loads, and both branches of the cold stream are raised to 205°C.

The procedure described in this section begs the question "is it always possible to find a solution to the pinch design problem?" The answer to this question is "yes", as

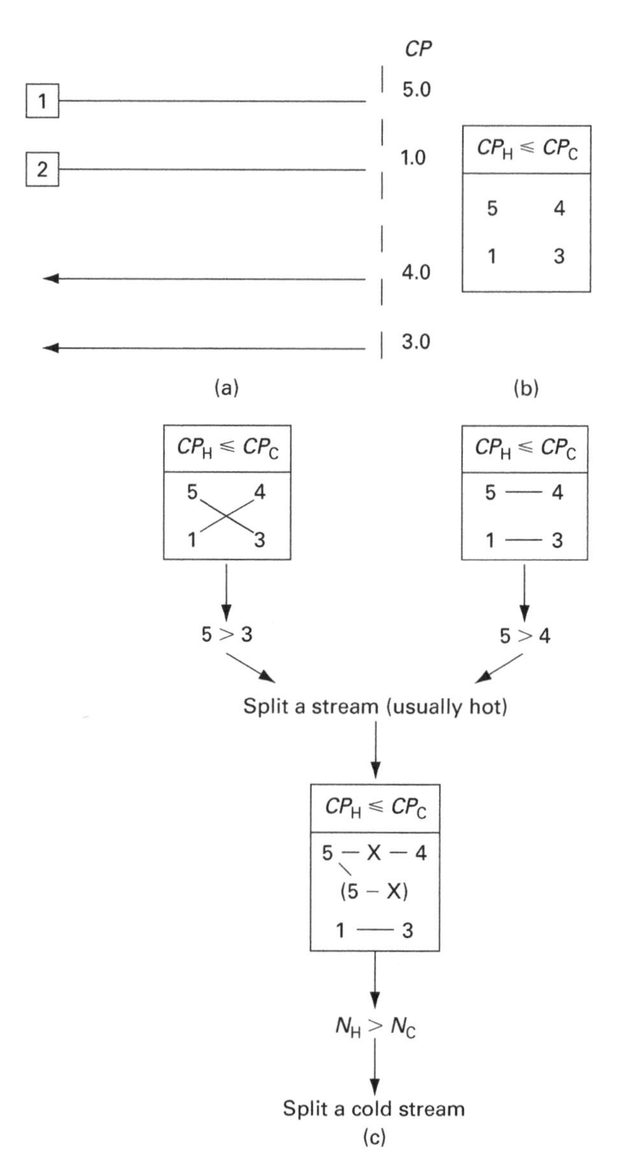

Figure 4.9 Use of the CP-table above the pinch

can be appreciated by remembering the composite curves. Above the pinch $\Sigma CP_{\rm H} < \Sigma CP_{\rm C}$, and below the pinch $\Sigma CP_{\rm H} > \Sigma CP_{\rm C}$, are always true.

Finally, it will be clear to the reader that stream splitting at the pinch will commonly be required to produce an MER design. In some cases this may not be a desirable feature. However, stream splits can be evolved out of the design by energy relaxation, in a manner similar to the energy relaxation for reduction in number of units, which will be described in Section 4.7.

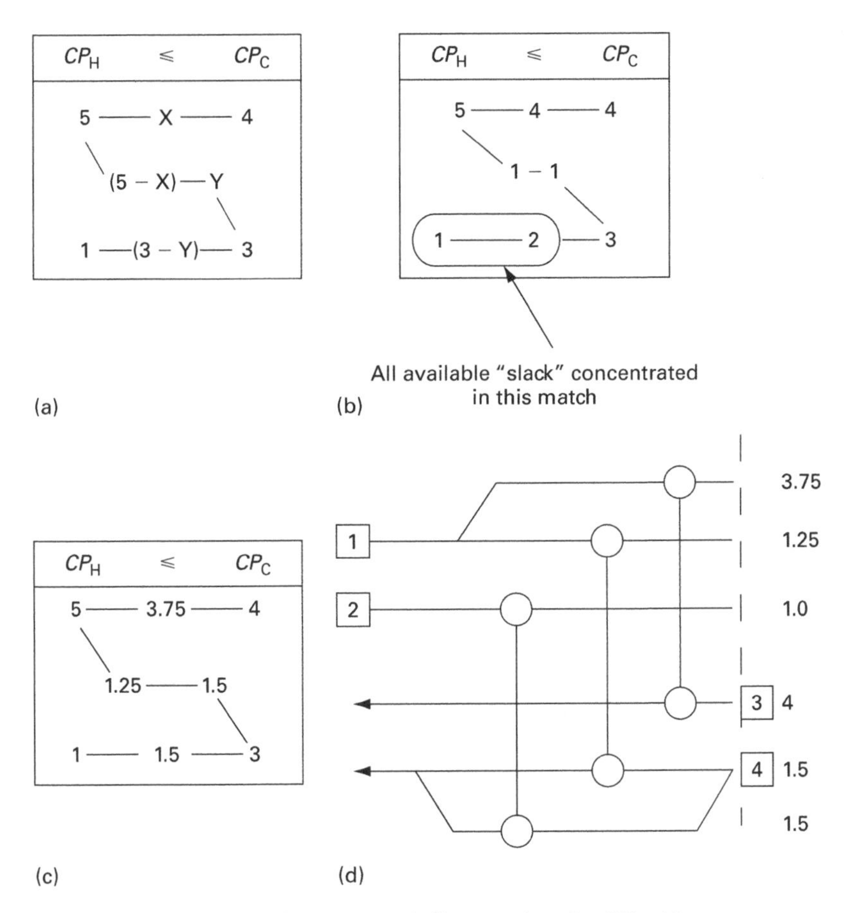

Figure 4.10 Determination of split branch flows using the CP-table

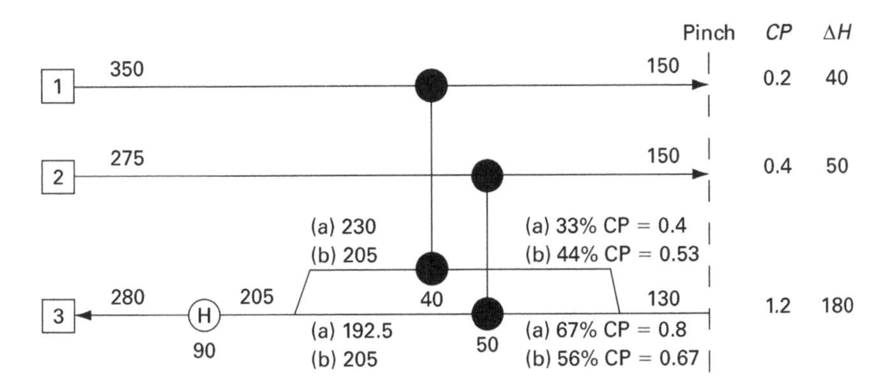

Figure 4.11 Simple above-pinch network with stream splitting

4.3.2 Cyclic matching

So far so good, but stream splitting requires extra pipework and valves, and the flow down each section of the split will need to be controlled. What happens if it is decided that stream splitting is unacceptable for the plant?

Obviously the pinch design criteria cannot be fully met and an energy penalty will be incurred. However, this can be minimised by having streams which should have been split, it is possible to have a series of smaller exchangers. Thus stream C3 can be matched first against H2, then H1, then H2 again, then H1 and so on. This is known as **cyclic matching**. The size of each match will be limited by the appearance of ΔT_{min} violations or temperature crosses. Obviously, the Euler value for minimum number of units will not be met when cyclic matching is used, and the increased cost of the additional exchangers must be set against the energy saved. Theoretically, if an infinite number of infinitely small exchangers were cyclically matched, there would be no energy penalty. However, a much smaller number of cycles may be sufficient to recover most of the energy. Cyclic matching is particularly effective where the pinch region is relatively short, as one quickly moves away into a region where a wider range of other matches is possible.

Taking our simple stream split example in Figure 4.11, if the split is removed and replaced by two matches in series, as in Figure 4.12(a), we still recover 81.7 kW but there is an energy penalty of 8.3 kW. However, for three matches (one loop), Figure 4.12(b), the penalty falls to 3.2 kW and for four matches (two loops) in Figure 4.12(c), the penalty is only 0.8 kW and over 99% of the possible heat is recovered.

What order should cyclic matching be done in? For two simple matches in series, temperature is the key criterion; the stream which extends further from the pinch should be the "outer" match. Thus, in Figure 4.12(a), stream 2 (T_S = 275) should be the match nearer the pinch on stream 3, and stream 1 (T_S = 350) should be the further match, as shown. However, for a larger number of cycles, it becomes more important to match the stream with the highest CP closest to the pinch, as in Figure 4.12(c). Readers may like to try this for themselves (see the Exercises, Section 4.10). The two principles may conflict (see Figure 4.12(b)).

In some cases a physical "stream split" is not really required at all, notably in plate and plate-fin exchangers. The stream has to be divided up anyway to pass through the channels, and two separate streams can easily be run through the same side of an exchanger in parallel. Theoretically this can also be achieved on the tube side of a shell-and-tube unit by having separate groups of tubes for different streams, but in practice this requires too much complexity in the headers, and sealing against cross-contamination is difficult.

4.3.3 Design away from the pinch

It has been shown that if for each design decision at the pinch the designer maximises match loads to tick-off streams or residuals, then a u_{min} solution results. However, in many problems it is not possible to do this in the simple way illustrated in the example in Chapter 2.

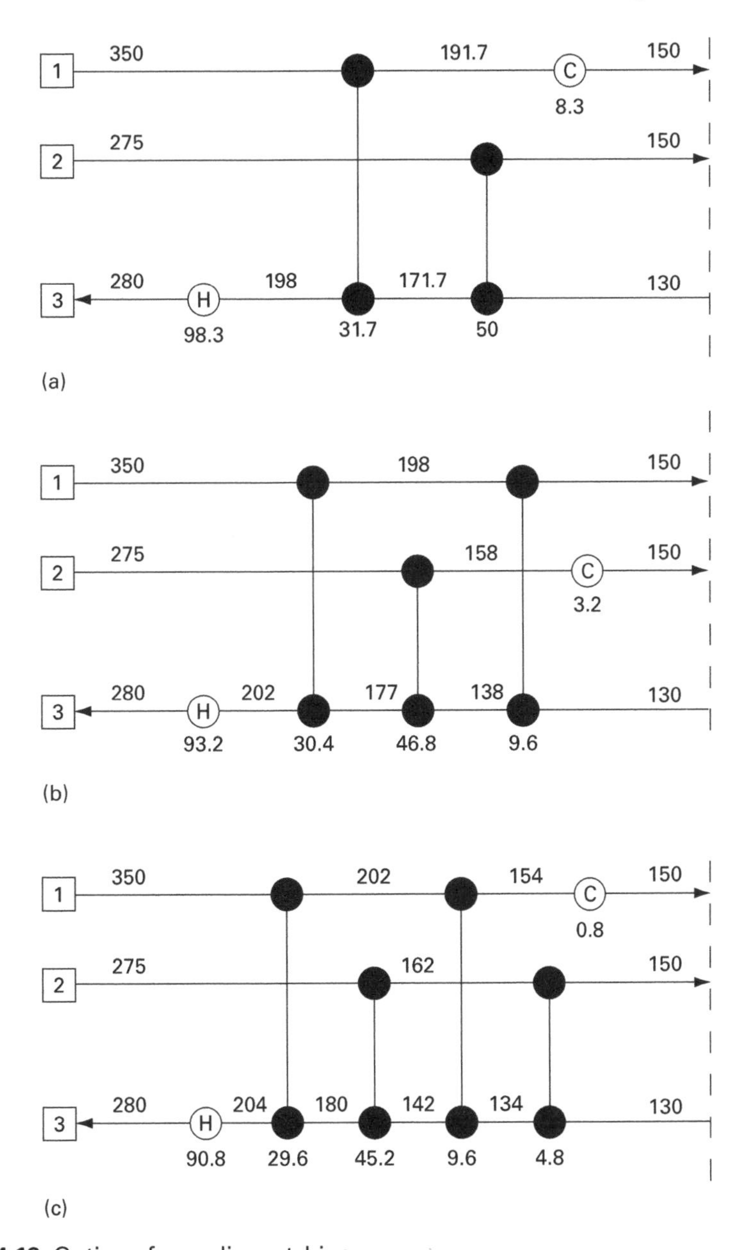

Figure 4.12 Options for cyclic matching

Consider the example shown in Figure 4.13(a). Analysis of the stream data shows a pinch at the supply temperature of stream 1 and the target temperature of stream 2 and hot and cold utility requirements both of zero. The design problem is therefore entirely "below the pinch", with only one pinch match possible (i.e. that between streams 1 and 2).

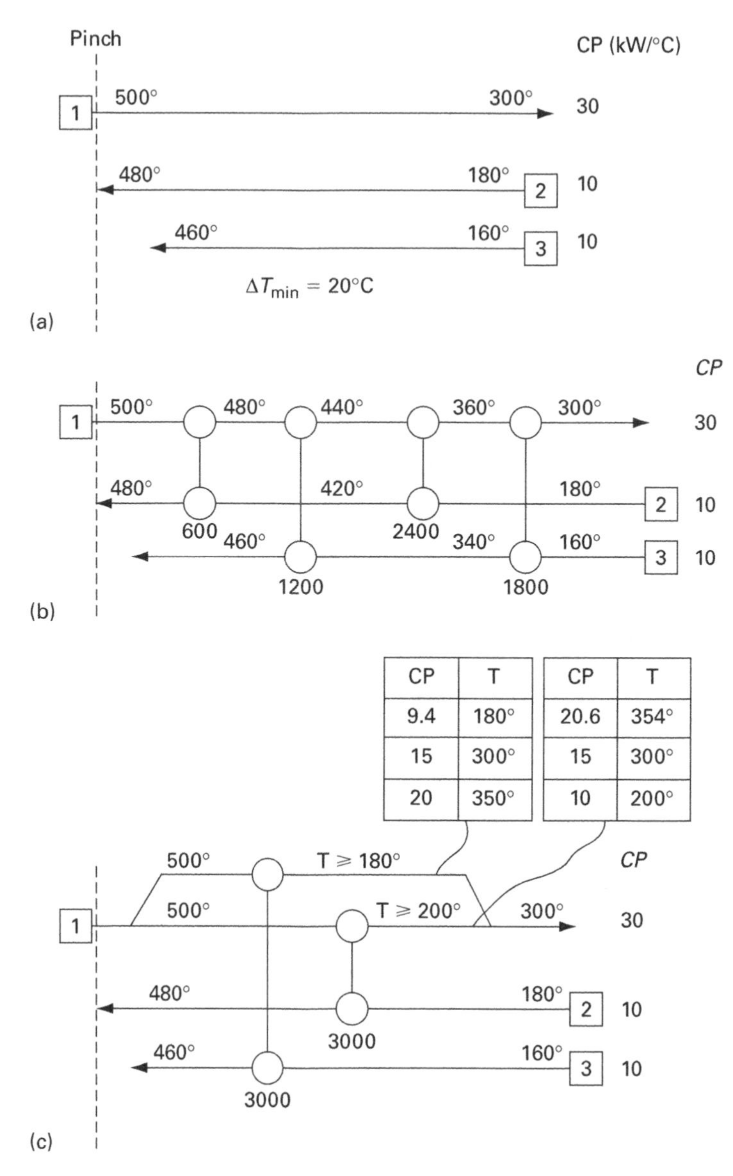

Figure 4.13 Cyclic matching and stream splitting away from the pinch

This is a feasible match ($CP_{HOT} > CP_{COLD}$), but if its load is maximised to tick off stream 2 (a load of 3000 units), stream 1 is cooled to 400°C. This is not then hot enough to bring stream 3 up to its target temperature of 460°C. Since heating below the pinch is not allowed for an MER solution, the design step of ticking off stream 2 would lead to a design that failed to reach the energy target. An alternative strategy is shown in Figure 4.13(b). The load on the pinch match is limited to 600 kW so that stream 1 remains just hot enough (at 480°C) to bring stream 3 up to its target temperature.

However, the next match (between streams 1 and 3) also cannot be maximised in load, because now stream 2 has to be brought up to 420°C by stream 1. The load on the second match between streams 1 and 2 has to be limited, allowing a final match (between streams 1 and 3) to finish the design. This is another example of "cyclic matching".

Cyclic matching always leads to structures containing loops and hence more than the minimum number of units. The only way to avoid cyclic matching is to employ stream splitting away from the pinch. In Figure 4.13(c) the heavy stream, stream 1, is split into two parallel branches, and each branch matched separately to a cold stream. Because this technique "slims down" the heavy hot stream it prevents the phenomenon of repeated pinching of individual matches. Hence the two matches can now be maximised to tick off the two cold streams without running into temperature problems. A u_{min} design results. Notice again that the stream split gives an element of flexibility to the network. The split stream branch flowrates can be chosen within limits dictated by the cold stream supply temperatures. Thus if the branch matched against stream 3 is cooled to 180°C (the minimum allowed) it will have a CP of 9.4 and by mass balance the CP of the other branch will be 20.6. A CP of 20.6 in the branch matched against stream 2 leads to an outlet temperature on this branch of 354°C which is much higher than the minimum allowed (200°C). The same argument can be applied to define the other set of limits based on stream 2 supply temperature. The branch matched against stream 2 then has a CP of 20 and an outlet temperature of 350°C. The CP of the branch matched against stream 3 may therefore vary between 9.4 and 20 with the parallel limits on the other branch being 20.6 and 10. These results are summarised in Figure 4.13(c), along with the results for equal branch flows. This type of flexibility is normally available in stream split designs and can be very useful.

To summarise this section on stream splitting:

- Stream splitting at the pinch is often necessary to achieve an MER design.
- If stream splitting is judged to be undesirable, it can be eliminated by cyclic matching or network relaxation.
- If the designer runs into trouble away from the pinch in applying the ticking-off rule, he can attempt to find a stream split design before resorting to cyclic matching.
- Stream splitting adds complexity to networks as well as flexibility, hence if a non-stream-split, u_{min} solution can be found, it will normally be preferable to a stream-split solution. Note that stream splitting cannot reduce the number of units below the target value.

An example of a safe, operable and flexible stream-split design is given in Section 9.2 of this Guide.

4.4 Network relaxation

4.4.1 Using loops and paths

In Section 3.6 we saw how to obtain targets for the minimum number of units – heat exchangers, heaters and coolers. We also noted that the addition of the pinch constraint, by dividing the problem into two subproblems, increased the number of units required. It is often beneficial to trade-off units against energy by eliminating small exchangers giving little energy benefit, in order to simplify the network, reduce capital costs and improve the overall payback of the project.

Let us illustrate using our four-stream example. As the design in Figure 2.18 has six units, rather than the minimum of five for the total problem ignoring the pinch, there must be a **loop** in the system, as discussed in Section 3.6.1. The loop is shown traced out with a dotted line in Figure 4.14(a) and reproduced in the alternative form in Figure 4.14(b). Since there is a loop in the system, the load on one of the matches in the loop can be chosen. If we choose the load on match 4 to be zero, that is we subtract 30 kW of load from the design value, then match 4 is eliminated and the 30 kW must be carried by match 2, the other match in the loop. This is shown in Figure 4.14(a). Having shifted loads in this way, temperatures in the network can be recomputed as shown in Figure 4.14(c). Now, the value of ΔT at the cold end of match 2 is less than the allowed value ($\Delta T_{min} = 10°C$). The offending temperatures are shown circled. In fact we could have anticipated that a "ΔT_{min} violation" would occur by "breaking" the loop in this way by consideration of Figure 4.14(a). The loop straddles the pinch, where the design is constrained as described in Section 2.3. So changing this design by loop-breaking, if the utilities usages are not changed, must inevitably lead to a ΔT_{min} violation. In some problems, loop-breaking can even cause temperature differences to become thermodynamically infeasible (i.e. negative).

The question is, then, how can ΔT_{min} be restored? The answer is shown in Figure 4.15(a). We exploit a **path** through the network. A path is a connection through streams and exchangers between hot utility and cold utility. The path through the network in Figure 4.15(a) is shown dotted, going from the heater, along stream 1 to match 2, through match 2 to stream 4 and along stream 4 to the cooler. If we add a heat load X to the heater, then by enthalpy balance the load on match 2 must be reduced by X and the load on the cooler increased by X. Effectively we have "pushed" extra heat X through the network, thereby reducing the load on match 2 by X. Now match 3 is not in the path, and so its load is not changed by this operation. Hence the temperature of stream 1 on the hot side of match 3 remains at 65°. However, reducing the load on match 2 must increase T_2, thus opening out the ΔT at its cold end. This is exactly what we need to restore ΔT_{min}! There is clearly a simple relationship between T_2 and X. The temperature fall on stream 4 in match 2 is $(120 - X)$ divided by the CP of stream 4. Hence,

$$T_2 = 150 - \frac{120 - X}{1.5}$$

Alternatively, applying the same logic to the cooler,

$$T_2 = 30 + \frac{60 + X}{1.5}$$

Since $\Delta T_{min} = 10°$ we want to restore T_2 to 75°. Solving either of the above equations with $T_2 = 75°$ yields $X = 7.5\,kW$. Since ΔT_{min} is exactly restored, 7.5 kW must be

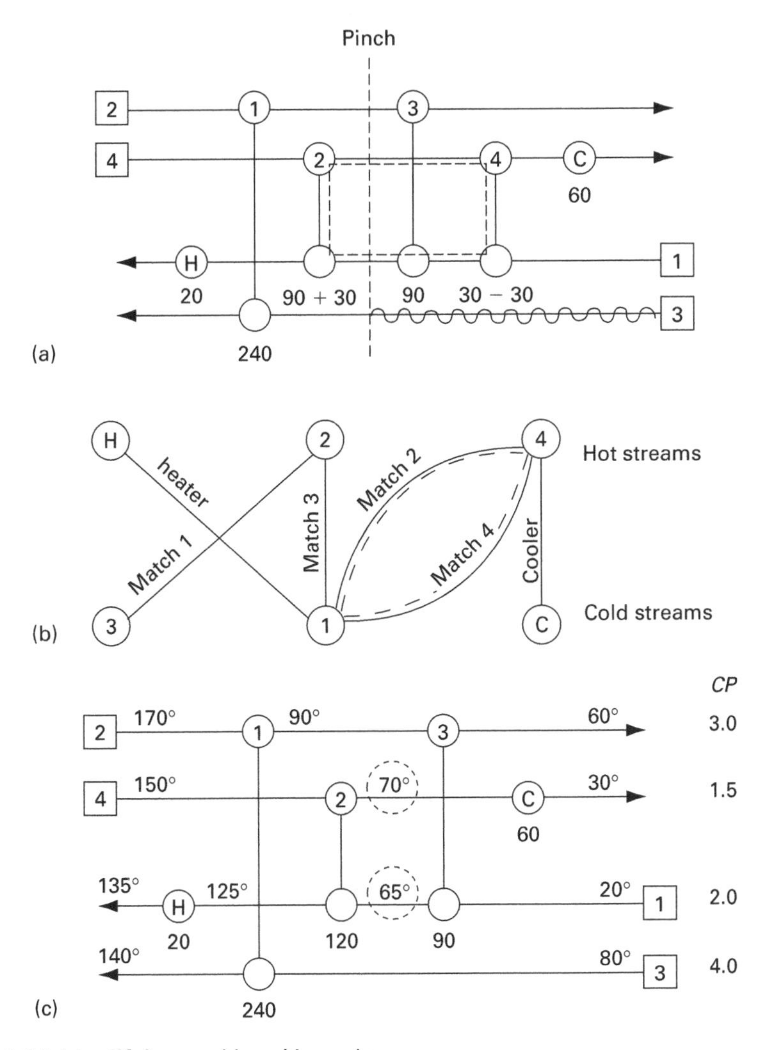

Figure 4.14 Identifying and breaking a loop

the minimum energy sacrifice required to produce a u_{min} solution from the $u_{min\ MER}$ solution. The "relaxed" solution is shown in Figure 4.15(b), with the temperature between the heater and match 2 on stream 1 computed. As expected, we now have 5 units and a hot utility use of 27.5 kW (instead of 20).

A path does not have to include loops. Looking at Figure 4.16, the reader will see that an alternative path exists via exchanger 4 alone. Simply transferring 30 kW of heat down this path, equal to the exchanger duty, will eliminate the exchanger. However, the energy penalty from using this direct path is the full 30 kW, four times as much as by the loop-breaking method! The reason is that we have not exploited the opening out of the temperature driving forces between streams 4 and 1, and the closest approach on the match is a long way from ΔT_{min}. Hence, breaking loops,

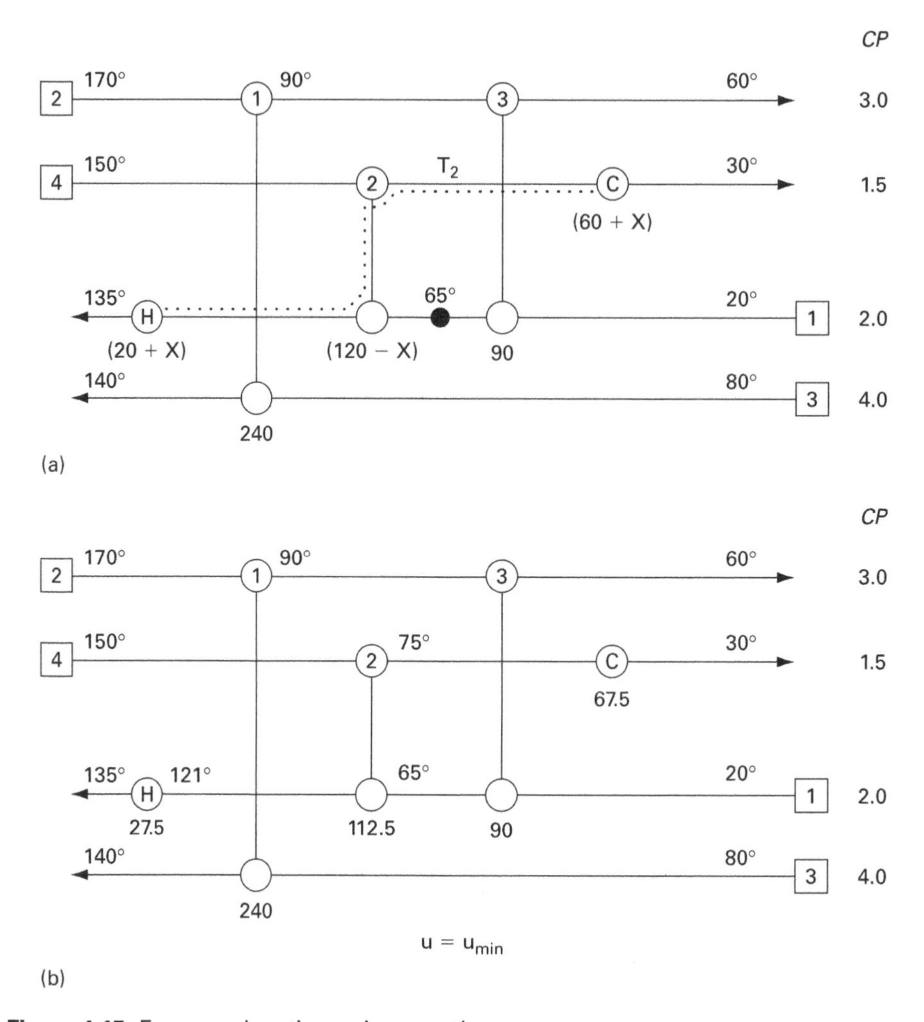

Figure 4.15 Energy relaxation using a path

where they exist, is usually preferable to using simple paths. Nevertheless, there are many situations where there is no alternative to using a direct path.

In summary on this subject of "energy relaxation", the procedure for reducing units at minimum energy sacrifice is:

- Identify a loop (across the pinch), if one exists.
- Break it by subtracting and adding loads.
- Recalculate network temperatures and identify the ΔT_{min} violations.
- Find a relaxation path and formulate $T = f(X)$.
- Restore ΔT_{min}.

The procedure can then be repeated for other loops and paths to give a range of options with different numbers of units and energy usage. Several alternative routes

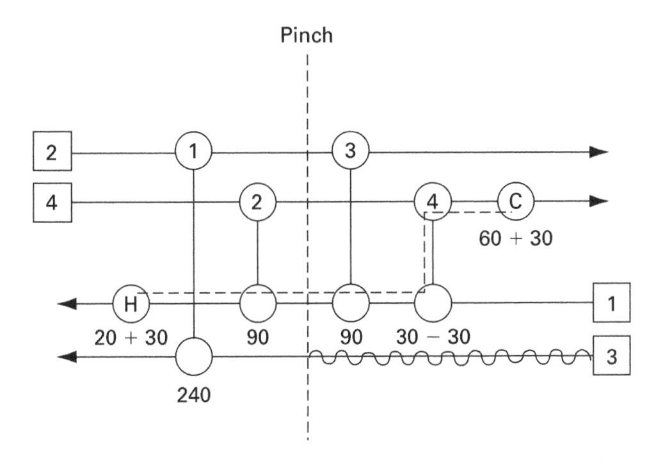

Figure 4.16 Alternative energy relaxation path

for relaxation may exist by eliminating different exchangers, particularly in complex networks.

Loops may be quite complex, or may involve the utility system, as illustrated in Figure 4.17. The loop in Figure 4.17(a) is a simple one involving only two units. In Figure 4.17(a) a more complex one is shown involving four units. However, the loop can be broken in exactly the same way, that is adding and subtracting X on alternative matches round the loop. In Figure 4.17(a) the loop breaks when X equals either L_1 or L_4. Note that the adding and subtracting could have been done in the alternative way, in which case it would break when X equals either L_2 or L_3. In other words, there are two ways of breaking the loop. This is true of the loop in Figure 4.14(a) (90 kW could have been subtracted from match 2 and added to match 4), and in fact is true of all loops. It is not possible *a priori* to say which way will lead to the smallest energy relaxation. However, a good rule of thumb is to go for the way that removes the smallest unit. Note that, when there are, say, two loops in a system, it may be possible to trace out more than two closed routes. This should not cause confusion if it is realised that the number of *independent* loops is always equal to the number of "excess" units ($>N-1$) in the system. Note too that loops can include heaters and coolers, as illustrated in Figure 4.17(b); the "linkage" comes by shifting utility loads from one heater to the other.

A complex path is shown in Figure 4.17(c), and again the alternate addition and subtraction of the load X works in just the same way as for the simple path. Note that although the path goes through match 1 in this example, match 1 is not part of it. Its load is not changed by the energy relaxation, but the temperatures on stream 4 on either side of it *are* changed, and, in fact, temperature driving forces will be increased. When a similar situation occurs within a loop it is possible for the exchanger that does not undergo a load change to become infeasible. Hence the need to recalculate all temperatures after loop-breaking. Finally, paths should not generally double back on themselves; for example, if in Figure 4.17(c), exchanger 2 came to the right of 3 on stream 1, the increased load on exchanger 3 due to the path would be very likely to cause a ΔT_{min} violation at the lower end of the match.

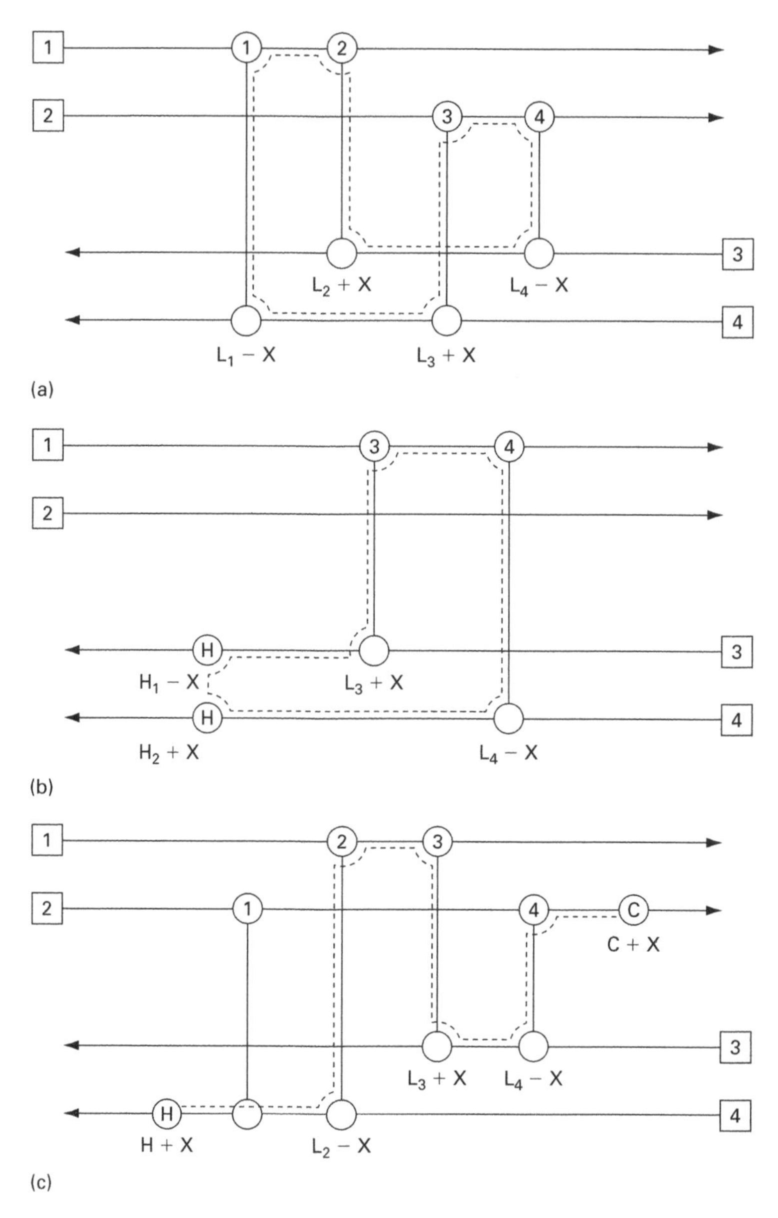

Figure 4.17 Complex loops and paths

In general, network relaxation is a more complicated process than designing the MER network, and there is a wider range of alternative possibilities. Hence the network designer should not omit to try out alternative possibilities. In many cases he will end up with a range of options, with the network progressively relaxed to require less exchangers but use more energy. There may be other families of options (e.g. starting from breaking a loop differently). Examples will be seen in the case studies.

We must also remember that heat exchangers, like all process equipment, come in finite sizes. The surface area of an exchanger in a network may need to be modified to the nearest procurable size, larger or smaller. This is generally more limiting with shell-and-tube exchangers, as with plate units, small increments of size are possible by adding individual plates, allowing fine tuning.

In summary, to design a heat exchanger network on a new plant, the recommended procedure is:

- Use the Pinch Design Method to place matches at the pinch.
- Place utility heaters and coolers for operability, if necessary using remaining problem analysis (RPA) as a check.
- Fill in the rest of the design, if necessary using stream splitting.
- Use relaxation to reduce network capital cost and improve operability by removing small, uneconomic exchangers and inconvenient stream splits.

4.4.2 Network and exchanger temperature differences

In the relaxation process, we have increased the energy consumption away from the original targets. In effect, utility use now corresponds to a new ΔT_{min}, higher than before. However, individual exchangers in the network may still be at the old ΔT_{min}. In effect, we have two values of ΔT_{min}. "Dual approach temperature" methods formalise this, drawing a distinction between HRAT (Heat Recovery Approach Temperature, the ΔT_{min} for the network, giving the spacing between the composite curves) and EMAT (Exchanger Minimum Approach Temperature, the ΔT_{min} for an individual exchanger). In general, HRAT > EMAT. Key papers include those by Trivedi *et al.* (1989) and Suaysompol and Wood (1991), and are comprehensively reviewed by Shenoy (1995). Conversely, a small ΔT_{min} violation on an exchanger may be allowed to avoid adding extra units to the network, so that EMAT has been reduced while HRAT remains the same.

During detailed network design, the temperatures and heat loads on matches often need to be modified to give a desired heat exchanger size, particularly for revamping (retrofit) of existing networks (Section 4.7). Interfacing network synthesis with detailed heat exchanger design is again described in more detail by Shenoy (1995).

4.4.3 Alternative network design and relaxation strategy

In most cases, the best final network is obtained by the method shown here, of beginning with the MER network and relaxing it to eliminate small or inconvenient exchangers. However, there are a small but significant number of cases where a different approach is best. Figure 4.18 shows the MER network for an example based on a real case study on a multi-product plant with several similar parallel processing lines. Eleven heat exchangers are required, because the CP criteria require different streams to be matched on either side of the pinch. In contrast, Figure 4.19 shows a more conventional non-pinch design; only six exchangers are now required, and two coolers have also been eliminated. However, the energy penalty is only 3.2 GJ/h

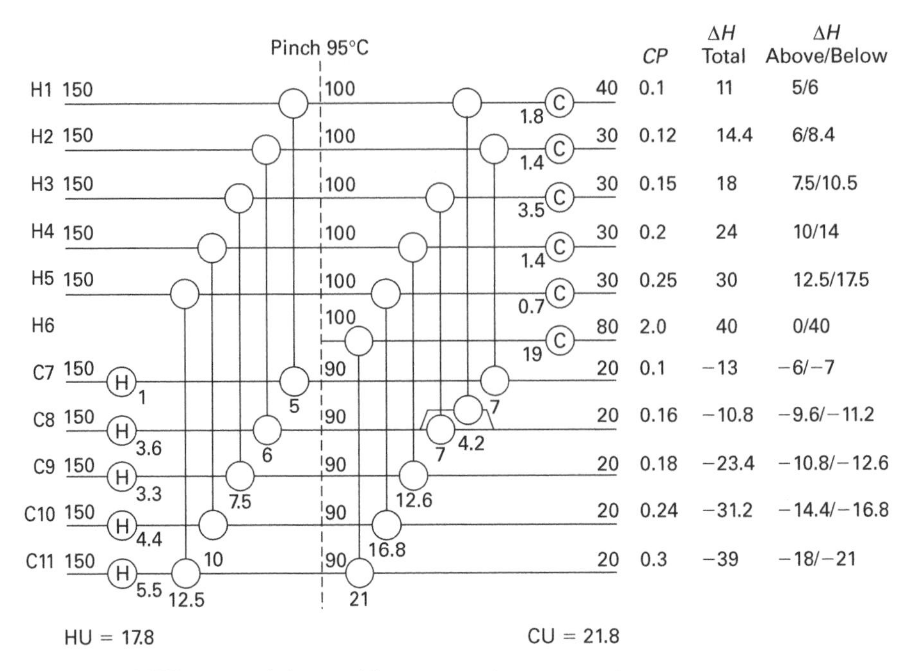

Figure 4.18 MER network for multi-process plant example

(about 18% of the hot utility target of 17.8 GJ/h) and the design achieves 97% of the ideal heat exchange. This network has been developed by matching streams with similar CPs and heat loads over the whole of their temperature range, and is clearly far more cost-effective than the MER design; moreover it would be difficult to relax the latter by the methods given above to reach the design in Figure 4.19. The distinguishing feature of these "anomalous" cases is that a large number or high proportion of streams actually *cross the pinch*. For every cross-pinch stream, the Euler network theorem (Section 3.6.1) shows us that subdividing the problem at the pinch will tend to require one extra exchanger. Nevertheless, the insights of the pinch are still useful; they identify that the vital pinch match is between streams H6 and C11, and it is the presence of this match which allows the alternative "commonsense" design to come so close to the energy target.

	Hot utility	*Cold utility*	*Heat exchange*	*Exchangers used*	*Heaters used*	*Coolers used*	*Total units*	*Stream splits*
MER network	17.8	27.8	109.6	11	5	6	22	1
Alternative	21.0	31.0	106.4	6	5	4	15	0

A similar situation arose in the air-to-air heat exchange of the hospital site case study, Section 9.6.

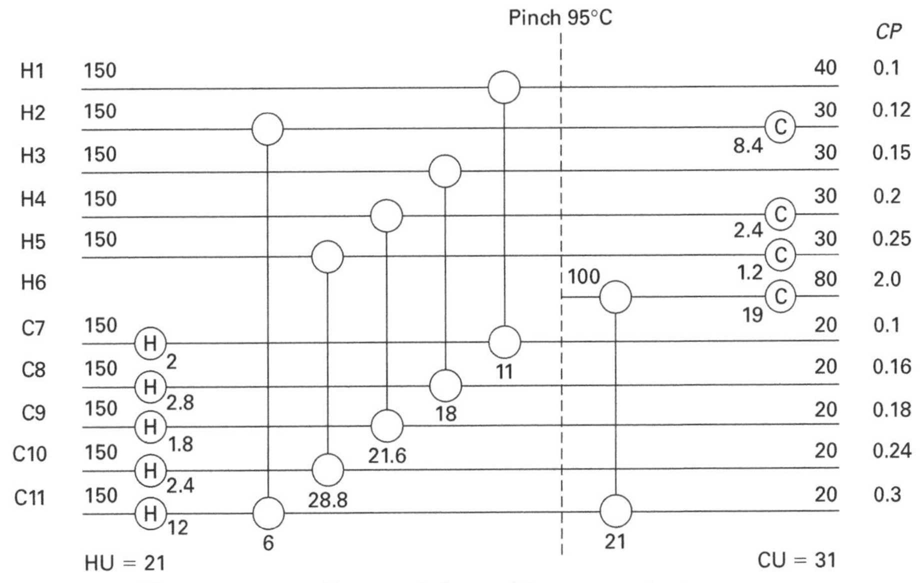

Figure 4.19 "Common-sense" network for multi-process plant

4.5 More complex designs

4.5.1 Threshold problems

As defined in Section 3.3.2, threshold problems are cases where one utility is not required, and fall into two broad categories. In one type, the closest temperature approach between the hot and cold composites is at the "non-utility" end and the curves diverge away from this point. In this case, design can be started from the non-utility end, using the pinch design rules. In the other type, there is an intermediate near-pinch, which can be identified from the composite curves as a region of close temperature approach and from the grand composite as a region of low net heat flow. Here it is often advisable to treat the problem like a "double pinch" and design away from both the near-pinch and the non-utility end.

In both cases, a typical value of ΔT_{min} can be chosen, just as for a pinched problem. If this value of ΔT_{min} is much less than the ΔT at the non-utility end and the problem is of the first type, the network design will be relatively "slack", and a great many designs are possible as the thermodynamic constraint of the pinch does not apply. The design will generally be determined by placing heaters or coolers for good control, applying the ticking-off rule, and by identifying essential matches at the "non-utility" end.

In contrast, the four-stream example with ΔT_{min} = 5°C is of the second type, as shown by the composite curves (Figure 4.20). The composite curve and grand composite curve (GCC) show a near-pinch at 82.5°C shifted temperature (85°C for hot streams, 80°C for cold streams) with a heat flow of only 2.5 kW. We therefore design away from both the non-utility end and the near-pinch, noting which location is

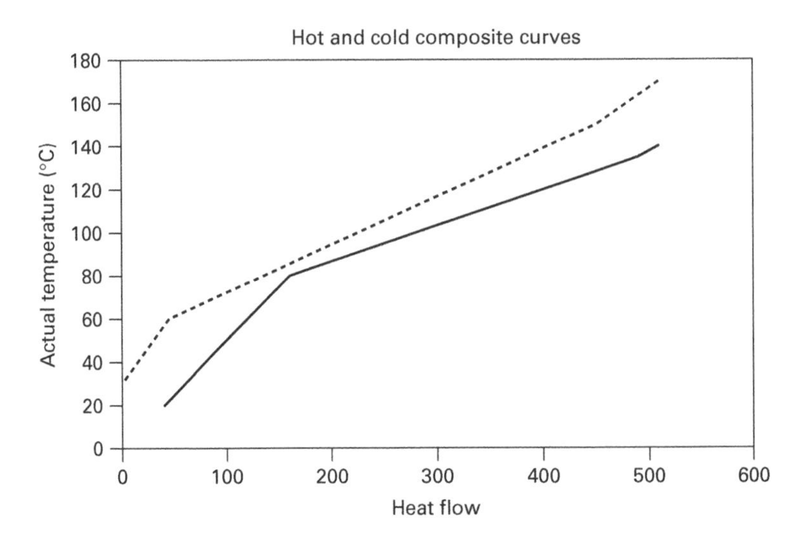

Figure 4.20 Composite curves for threshold problem (ΔT_{min} = 5°C)

more constrained. At the non-utility end there seems no pressing reason to prefer either cold stream 1 or 3 to match against 2, the hottest hot stream. However, at the near-pinch, we will need to obey the Pinch Design Rules, otherwise a violation would quickly occur in one of the matches. We construct the network grid diagram and it is helpful to include the heat loads on each stream immediately above and below the near-pinch, noting that the total heat loads on hot streams exceed those on cold streams by 2.5 kW. We match stream 2 against stream 3, ticking off the 240 kW on the latter, and stream 4 against stream 1, ticking off the 97.5 kW on the former. This leaves 12.5 kW on stream 1 unsatisfied above the pinch, and since we are not using hot utility, this must be provided from stream 2. Below the pinch, we develop the network as for a pinched problem, and end up with the network in Figure 4.21. It is notable that, in this case, the network geometry is almost identical to that for the pinched problem with ΔT_{min} = 10°C (Figure 2.18). The only differences are small changes in loads on individual exchangers and the replacement of the heater on stream 1 by a heat exchanger with stream 2.

The other type of threshold problem can be generated by halving the CP on stream 3, to 2.0 kW/K. The composite curves are shown in Figure 4.22. Net cooling requirement is 160 kW. We start network design at the non-utility (hot) end. The CPs of the hot streams are 3 and 1.5; those for the cold streams are 2 and 2. If this was a pinch, we would have to split one of the cold streams to satisfy the CP criterion below the pinch; we would then in turn have to split a hot stream to satisfy the number of streams criterion. However, because this is a threshold problem, we have more leeway, although we must be careful not to get a temperature cross on stream 4. We match the hottest hot stream (2) with the hottest cold stream (3), and can tick off the latter while bringing stream 2 down to 130°C. Likewise, stream 4 can be matched against stream 1, and if we maintain the ΔT_{min} = 10°C criterion, can bring it down to 130°C. After this, stream 2, with its higher CP, can take over. Adding two coolers gives the network in

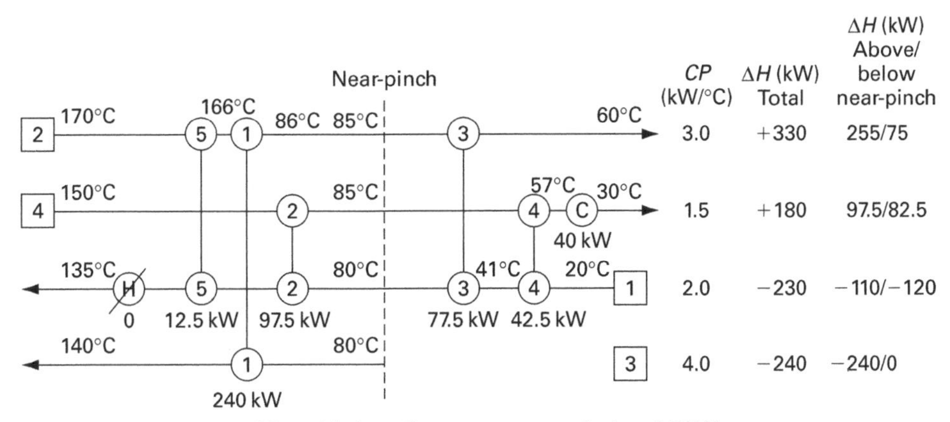

Figure 4.21 Network for threshold problem ($\Delta T_{min} = 5$°C)

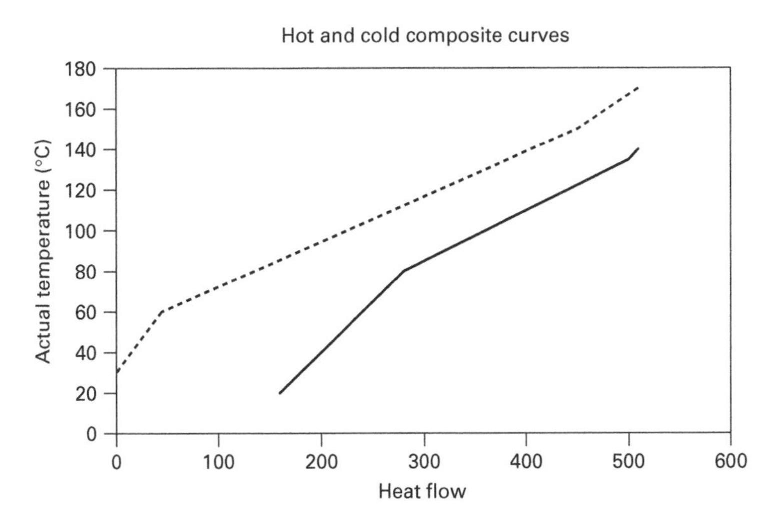

Figure 4.22 Composite curves for alternative threshold problem

Figure 4.23. No stream splits have been needed. Note that one of the coolers is very small and could be eliminated by transferring 10 kW around a path through exchangers 2 and 3, but the ΔT_{min} at the hot end of match 3 would be only 5°C.

4.5.2 Constraints

Designers are always faced with many more constraints than purely thermodynamic ones when designing heat exchanger networks. Two important ones are considered in this section; forbidden and imposed matches.

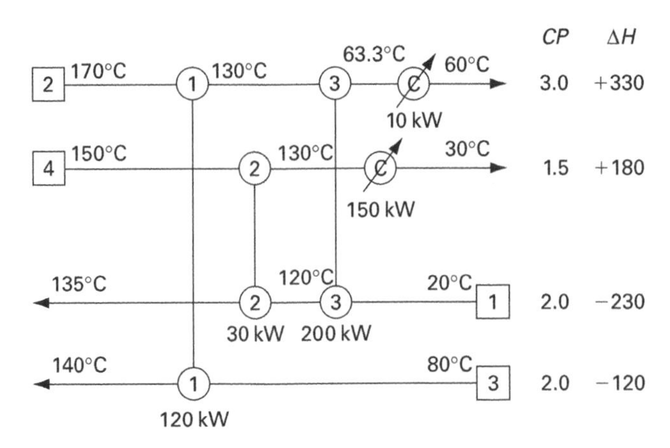

Figure 4.23 Network for alternative threshold problem

4.5.2.1 *Forbidden matches*

There are many reasons why a designer might want to forbid a match between any given pair of streams, for example corrosion and safety problems, long pipe runs required or controllability. Imposing a forbidden match on a design might or might not affect the possible energy recovery of the network. At the top of Figures 4.24(a) and (b) are shown four streams, two hot (A and B) and two cold (C and D). In Figure 4.24(a) it is clear that because the relative temperatures allow, either of A or B may interchange with either of C or D. Forbidding a match between, say, A and C does not impair the chances of producing an MER design. However, in Figure 4.24(b) it can be seen that B is not hot enough to exchange with C, and so a match between A and C is essential if an MER design is to be produced. The consequence of forbidding the A–C match is therefore an increase in utilities as shown at the bottom of Figure 4.24(b). Basic targeting methods do not show whether or not a forbidden match constraint will affect the energy target, and if so by how much. However, the linear programming (LP) method of Cerda *et al.* (1983) rigorously solves this problem. Some advanced software includes this, allowing the rigorous energy targeting element to be retained even with the constrained problem. This allows the designer to define precisely what energy penalty he is paying for the constraint, and so the cost incentive for overcoming it (e.g. by the use of a different, possibly more expensive, mechanical design). In some cases, zonal targeting may provide a simpler alternative. For network design, the simplest approach is to produce an "unconstrained" MER design by the pinch design method, avoiding forbidden matches, and then to modify it in the light of the constraint and the modified energy target. If an essential pinch match is forbidden, then an energy penalty will result. More detailed analysis is given by O'Young *et al.* (1988), O'Young (1989), and Cerda and Westerberg (1983).

4.5.2.2 *Imposed matches and RPA*

Secondly we look at the constraint of imposed matches. For reasons of operability (e.g. start-up and control), layout, and in order to re-use existing units in "revamps", the

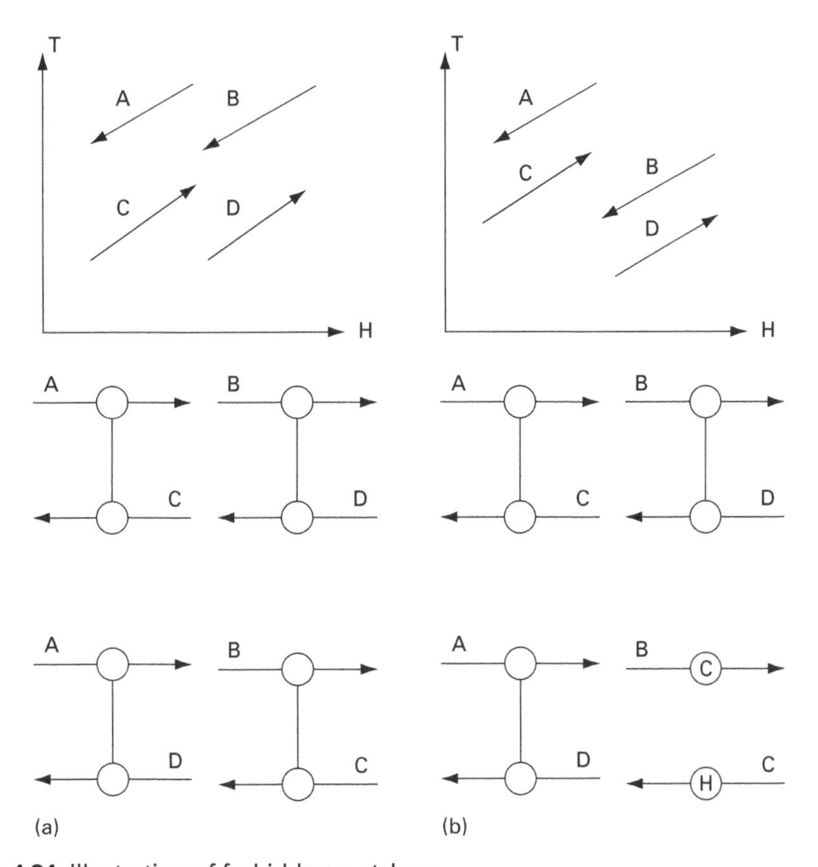

Figure 4.24 Illustration of forbidden matches

designer may want to include a certain match in his design. Suppose in the example problem shown in Figure 4.25 the designer requires a heater on stream 4 for start-up and control reasons. Analysis of the data shows a total utility heating requirement of 302 units and no cooling requirement. In order to meet simultaneously the control objective and the requirement for minimum number of units, the designer would like to place the whole heating duty on stream 4. The question is, does this design step prejudice his chances of achieving an MER design? The way to test for this is to analyse the "remaining problem" indicated by the dotted line in Figure 4.25. That is, apply the energy targeting procedure to streams 1, 2, 3 and 5, and the remainder of stream 4 after placement of the heater. Applying the procedure, two results are possible. Either the remaining problem will require no utility heating, in which case the heater placement does not prejudice MER design, or the remaining problem will require heating X and cooling X, in which case the full heater load cannot be placed on stream 4 for MER design to be achievable.

Let us suppose that the energy target for the "remaining problem" is 60 units. This means that the heater on stream 4 is well located for start-up but not for effective thermal integration; if all 302 units of steady state heat load are placed on stream 4, an overall penalty will be incurred of 60 units. Retargeting with different loads on

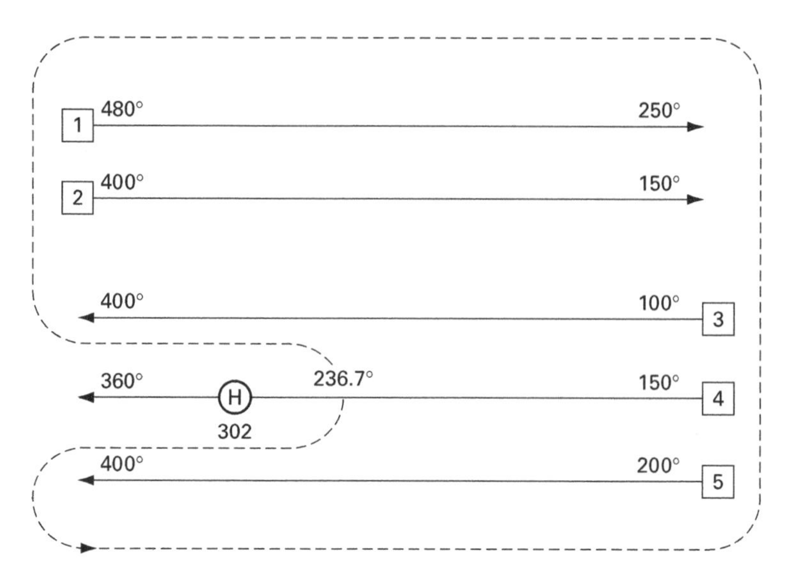

Figure 4.25 Remaining problem analysis

stream 4 may reveal a possible compromise, for example it may be possible to place up to 180 units of utility on stream 4 (for start-up *and* steady-state load) without penalty. At least one more heater, handling 122 units, will be needed on another stream; however, by judicious choice of network structure and heat exchanger loads, an extra unit may not be necessary if an exchanger elsewhere is eliminated.

RPA can be carried out with respect to both energy and capital cost targets, and can be used after the placement of exchangers as well as heaters and coolers. It is a useful tool allowing the designer to evaluate the impact of key design decisions during targeting. Design features such as "a start-up heater on stream 4, no larger than 180 units" are easily agreed at an early stage and form the basis of subsequent design work.

4.6 Multiple pinches and near-pinches

4.6.1 Definition

So far, we have assumed a single, reasonably sharp pinch, and have designed away from it. However, many processes have other areas of *low net heat flow*, which can be called **near-pinches** or, where they extend over a wide temperature range, **pinch regions**. There may even be **multiple pinches**, each with zero net heat flow. All of these are easily identified by looking at the GCC. Network design may have to be modified to take account of this. Multiple and near-pinches represent an additional point where network design is highly constrained, and the pinch rules may have to be re-applied at this point. A network is then designed by working away from both pinches, meeting in the middle.

For a near-pinch, the heat flow Q_n can be calculated from the Problem Table. The network can be designed as if this were a true pinch, but can then be relaxed to allow transfer of up to Q_n across the near-pinch without damaging the overall energy target. Larger violations will mean that this becomes, in effect, the new pinch. This is called a **network pinch**, as it is caused by the choice of network structure rather than being inherent from the stream data.

A frequent cause of additional pinches is the use of multiple utilities. If the use of a low-grade utility is maximised, this gives one or more *utility pinches*, as discussed in Section 3.4.4. When designing such networks, a **balanced grid** diagram should be used, including the utility streams and especially those causing the utility pinches.

4.6.2 Network design with multiple pinches

Let us return to our four-stream example. In Section 3.4.6, we found that if steam was supplied at the lowest feasible shifted temperature of 98.3°C, two new utility pinches would be created at S = 165°C and 98.3°C in addition to the process pinch at 85°C. What effect does this have on the network?

Firstly, we will construct our balanced grid diagram, showing the three pinches and including the lower-temperature steam explicitly as an additional stream.

The philosophy of the Pinch Design Method is to start the design at the pinch and move away. However, where there are two or more pinches, designing away from each into the region in between them can clearly lead to a "clash". The recommendation is, design away from the most constrained pinch first. Here, above the process pinch at 85°C we are forced by the CP criteria to match stream 2 with stream 3, and stream 4 with stream 1, as before, whereas below the utility pinch at 98.3°C we have the additional choice of using the steam, whose CP is infinity. So we work upwards from the process pinch, and find that we can tick off the hot streams 2 and 4 in this interval, leaving residual loads on cold streams 1 and 3 which must be satisfied by two separate steam heaters. Likewise, above the 98.3°C pinch the same constraints on matches apply, whereas we only have one stream (2) actually present at the 165°C pinch – the matches are not at ΔT_{min} and we have flexibility. So, designing upwards, we tick off streams 4 and 3, and find ourselves with a residual load of 13.3 kW on streams 2 and 1, which are therefore matched at the top end. This gives the above-pinch network shown in Figure 4.26.

It is immediately apparent that we have lost the elegant simplicity of Figure 2.18, where we only required three units above the pinch – one heater and two exchangers. Now, we have seven, some of which are very small. We can try to eliminate some of these by shifting heat loads around loops, such as the one shown as a dotted line in Figure 4.26, which passes through both the matches with the LP steam. Shifting loads in one direction or other around the loop will eliminate either of the two heaters, but will also cause a ΔT_{min} violation. Rather than restoring this by transferring heat across the process pinch, we could shift the steam temperature upwards until the violation is removed.

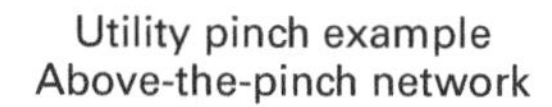

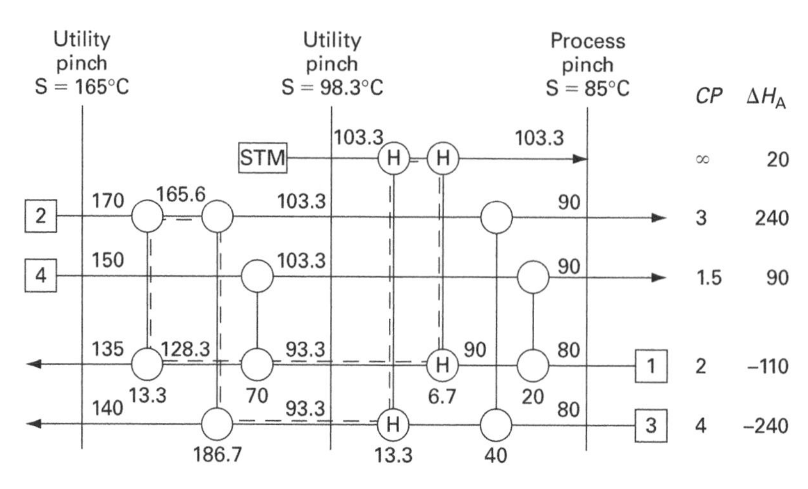

Figure 4.26 Balanced grid for four-stream problem above process pinch with minimum steam temperature

We also remember that we have a stream subset above the pinch – streams 2 and 3 have equal heat loads. It would be useful to preserve this, and have the whole heater load on stream 1. Have we an easy way of evaluating this situation? Yes, we have – RPA! (Section 4.5.2.2). We force the match between streams 2 and 3, remove their above-pinch heat loads from the stream data and re-target. Now we have a new GCC for the remaining problem, and the net CP above the pinch is 0.5. Since we have 20 kW of hot utility, we see that to satisfy Appropriate Placement, we must supply this heat 40°C above the pinch, that is at S = 125°C or an actual temperature of 130°C. Now we have a much simpler network with just four above-pinch units, as shown in Figure 4.27, although we still have one cyclic match on stream 1.

We recall that our original above-pinch network with 3 units required stream 1 to be heated up to 135°C by hot utility, which therefore had to be at 145°C. Conversely, we can also choose to put all the heater load on stream 3. By retargeting and network design, we find that we now need 5 units (as no subset equality is possible in this case) but the steam only needs to be supplied at S = 105°C. Overall, then, we have a clear trade-off between number of units and steam supply temperature, as shown in Table 4.3.

4.7 Retrofit design

4.7.1 Alternative strategies for process revamp

The ideal situation for heat exchanger network design is where one is designing a new plant and can start with a clean sheet of paper. Sadly, this is a relatively rare situation.

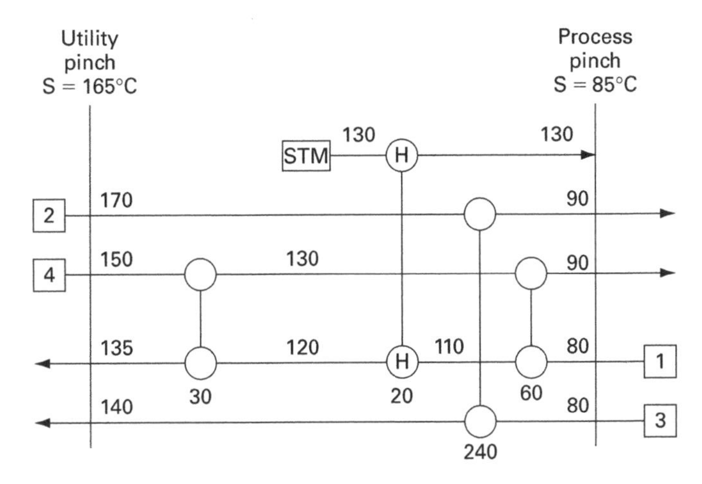

Figure 4.27 Balanced grid above process pinch with steam level increased to simplify network

Table 4.3 Summary of alternative networks with different steam supply temperatures

Heater position	*Above-pinch units*	*Steam temperature (shifted) (°C)*	*Steam temperature (actual) (°C)*
Stream 1 only	3	140	145
Stream 1 only	4	125	130
Stream 3 only	5	105	110
Streams 1 and 3	7	98.3	103.3

Far more common for most of us is to be faced with the task of analysing an existing plant and seeing whether we can make improvements, to reduce energy and emissions and increase profitability. This is known as a **retrofit** or **revamp** situation.

The strategy for retrofit problems needs to be somewhat different from that for new design. In fact, at least three different approaches are possible:

1. Develop an MER design as for a new plant, but where a choice exists, favour matches which already exist in the current network. This may help us to choose between alternative pinch matches, and is certainly a key criterion when choosing matches in the less constrained regions away from the pinch. In terms of Figure 4.28, we note the configuration of the current design and work towards it, both in the MER design and during relaxation. This was the approach used in the earliest pinch studies, including those described in Sections 9.2 and 9.3.
2. Start with the existing network and work towards an MER design. We note the current ΔT_{min} and calculate the targets and the pinch temperature. Now we plot the existing exchangers, heaters and coolers on the grid diagram and look to see which ones are the pinch violators. We can then identify ways to add new matches which correct these problems. This approach was described by Tjoe and Linnhoff

(1986); Section 9.3.6 develops a network using this approach and compares it with the one produced by starting at the MER network.

3. Start with the existing network and identify the most critical changes required in the network structure to give a substantial energy reduction. This method will be appropriate if the MER design is so different in configuration from the existing layout that they are virtually incompatible, as in Section 4.4.3. However, Asante and Zhu (1997) showed that it is also highly effective in other situations. A key insight is the **network pinch**, which shows directly how the network structure affects energy targets.

Depending on the situation, any combination of these three approaches may be valuable for retrofit. For example, one can work from the MER network by relaxation (1), and from the existing network by identifying pinch violations (2) and meet in the middle. Ideally, we find ourselves with a multi-step strategy, progressively adding new matches to the existing network to approach the MER design. We can then evaluate the energy and cost saving for each new match and its likely capital cost, and apply economic criteria to see how far to go. An excellent example of such a strategy is given in the aromatics plant case study in Section 9.3.

Figure 4.28 shows a rather schematic representation of the population of feasible network designs against energy recovery. The population is sparse at maximum energy recovery, but increases, sometimes greatly, as driving forces are increased and energy recovery is relaxed. The existing design will be one of many towards the base of the "pyramid". To work upwards from this design and attempt to find the MER design can be problematic, as the best design will normally not be within easy range of evolutionary steps by the obvious routes (Figure 4.28(a)). However, starting with the MER design at the top of the pyramid can give an overview of the solution space (Figure 4.28(b)) and give obvious evolutionary routes towards the current network. Again, the two approaches can "meet in the middle".

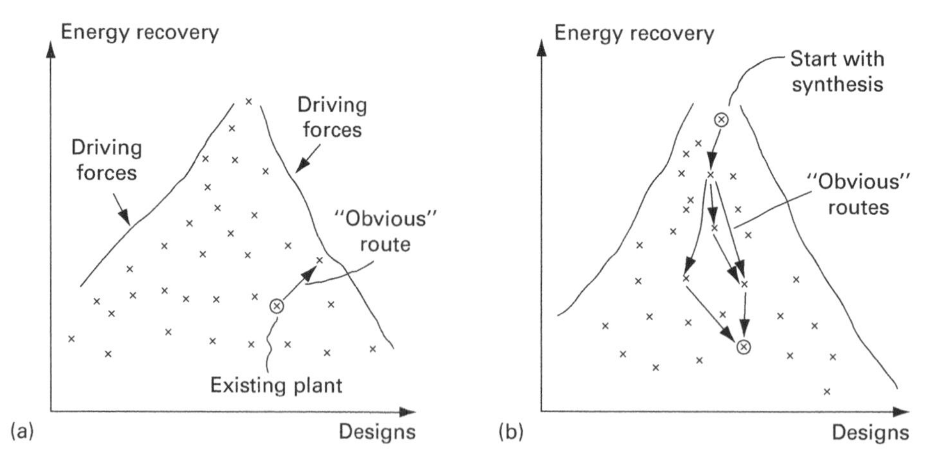

Figure 4.28 Design strategy for revamps

4.7.2 Network optimisation

Network design presents two separate challenges: finding the basic *structure*, and optimising the exchanger *sizes*. Pinch techniques are particularly helpful in the first area, eliminating structures that inherently cannot reach high levels of heat recovery because they violate the pinch. However, the basic methods described so far do over-simplify the situation. In both new design and network relaxation, for example, we must remember that the imposition of ΔT_{min} on all matches throughout the network is an arbitrary constraint put in to simplify the problem. However, "in the country of the blind, the one-eyed man is king". The techniques allowed the designer to arrive at a workable network relatively easily, and to understand how he had got there.

In reality, the presence and size of existing matches will dictate the most cost-effective network modifications. It is often sensible to shift loads between exchangers (even accepting a small violation) so as to maintain some at their existing area requirement. This can be seen in practice in the organics distillation case study in Section 4.9. It will also be easier and cheaper to enlarge one match considerably (especially a new match where one has complete freedom on sizing) rather than making piecemeal area additions to two or more exchangers. To achieve this, we must be able to calculate the relationships between exchanger areas, heat loads and temperatures throughout the network. Hand calculations are tedious, and the calculation can best be done using network design software, where available, or by setting up a spreadsheet with equations for stream heat loads and exchanger duties; iteration is often necessary to achieve convergence.

4.7.3 The network pinch

Let us return to our four-stream example and the network format in Figure 4.29, which achieves the minimum number of units (five – one heater, one cooler and three exchangers, for six process and utility streams). The ΔT_{min} is 10°C and the targets are 20 kW hot utility and 60 kW cold utility. However, with this network we are unable to achieve less than 27.5 kW heating and 67.5 kW cooling without violating ΔT_{min}. Clearly, the chosen network configuration is imposing a constraint which stops us achieving our targets. One approach is to identify the pinch violator; with the pinch at S = 85°C, it can be shown that match 2 is partly across the pinch. However, an alternative is to find the match that limits the heat recovery – the **pinching match** – and the point at which this occurs – the **network pinch**. These can be identified as the point where the existing exchangers reach ΔT_{min}, and in Figure 4.29 it can be seen that this occurs at the cold end of match 2.

A similar conclusion is reached if we reduce ΔT_{min} to zero. The loads on exchangers 1 and 3 are unchanged, but stream 2 can be brought down to 65°C, the load on exchanger 2 increases to 127.5 kW and the hot and cold utility requirements fall to 12.5 and 52.5 kW, respectively. However, this is even further from the targets, which are 0 and 40 kW for $\Delta T_{min} = 0$ (a threshold problem).

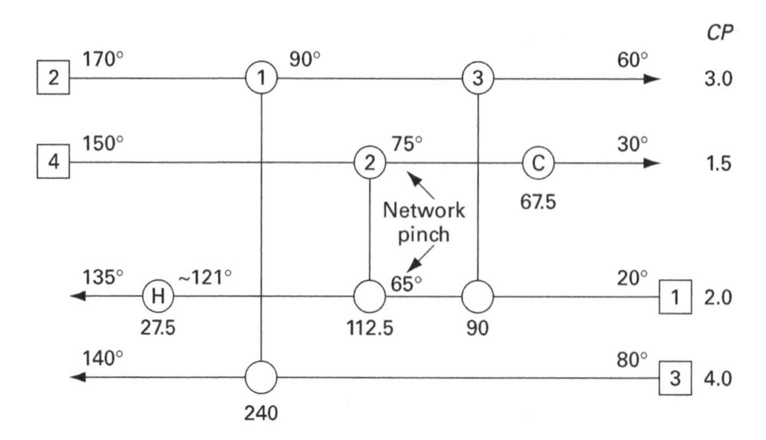

Figure 4.29 U_{min} network for four-stream example

Having identified the network pinch, what can we do about it? Asante and Zhu identified four possible approaches:

1. *Resequencing*: The order of two exchangers can be reversed, and this sometimes allows better heat recovery. For example, exchanger 3 could come at the hot end of stream 1, before stream 2. It still exchanges heat between streams 2 and 1 as before, but in a different network location. In this network, however, it will be found that any resequencing will worsen rather than improve heat recovery.
2. *Repiping*: This is similar to resequencing, but one or both of the matched streams can be different to the current situation. Thus, for example, exchanger 3 could be used to match streams 4 and 1, or 2 and 3, or 4 and 3. Again, a little experimentation will show that this brings no benefits in this case.
3. *Adding a new match*: This can be used to change the load on one of the streams in the pinching match. In this case, a new exchanger 4 could be added between streams 4 and 1, below exchanger 2, shifting the cold end temperatures on match 2 upwards. The targets can then be achieved; in fact, we have regained our MER network (Figure 4.14(a)).
4. *Splitting*: Split a stream, again reducing the load on a stream involved in the pinching match. Here, stream 1 could be split so that stream 2 can be run down to a lower temperature without incurring a ΔT_{min} violation at the bottom end. In practice, the stream split will be very asymmetric in both flow and temperature and a special configuration would be needed (e.g. two shells on match 2, one in parallel with match 3 on the split stream, and one above the stream split).

In general, at least one of these four options will be available. Of course, this may move the network pinch to a different pinching match, and the technique may have to be re-applied to reach the final target for that ΔT_{min}.

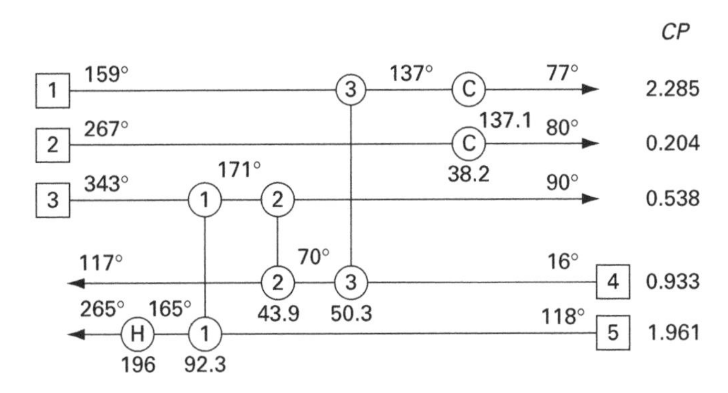

Figure 4.30 Revamp example; existing network

4.7.4 Example retrofit network design

We will now apply the revamp methodologies to the network shown in Figure 4.30, starting with the route based on the MER network. At $\Delta T_{\min} = 10°C$, the calculated utility heating target is 106.4×10^2 kW and the utility cooling target is 85.7×10^2 kW. Current hot utility consumption is 196×10^2 kW, so there is a 46% scope for saving energy. Producing an MER design, there is only one option above the pinch and this is shown in Figure 4.31(a). The heater and match 1 are both present in the base case design, but the match between streams 2 and 5 represents a "new" match. Below the pinch (Figure 4.31(b)), the pinch design method requires one pinch match (i.e. between streams 1 and 5), which is not present in the base case. After placing this match, there are several options for completing the design. The philosophy of the approach here is, where there are options, choose those options which maximise compatibility with the existing design. This philosophy dictates below-the-pinch design shown in Figure 4.31(b), reusing exchangers 2 and 3 and requiring no further new matches. Putting the above and below-the-pinch designs together gives the MER design shown in Figure 4.32(a). It requires one stream split not present in the base case, and two new matches.

To evolve this design at minimum energy sacrifice back towards the base case design, the first target is the "new" match carrying 22.1×10^2 kW of load. Eliminating this match by breaking the loop picked out by dotted line in Figure 4.32(a), the network shown in Figure 4.32(b) is obtained. This network now has two infeasible matches, requiring energy relaxation along the path shown. If $\Delta T_{\min}$ is restored, the design shown in Figure 4.32(c) is obtained. Notice that it was necessary to relax by the full 22.1×10^2 kW lost in the eliminated match. Notice too that energy relaxation led to the elimination of the stream split. Further loop-breaking and energy relaxation with $\Delta T_{\min} = 10°C$ leads to the design in Figure 4.32(d). Notice that this is the same topological (units arrangement) design as the base case. Compared to the base case, energy has been "tightened up" by transferring 9.4×10^2 kW along the path shown, with increase in load on matches 1 and 3 and decrease in load on match 2. This has reduced the minimum ΔT in the network (at the cold end of exchanger 2) from 20°C to 10°C.

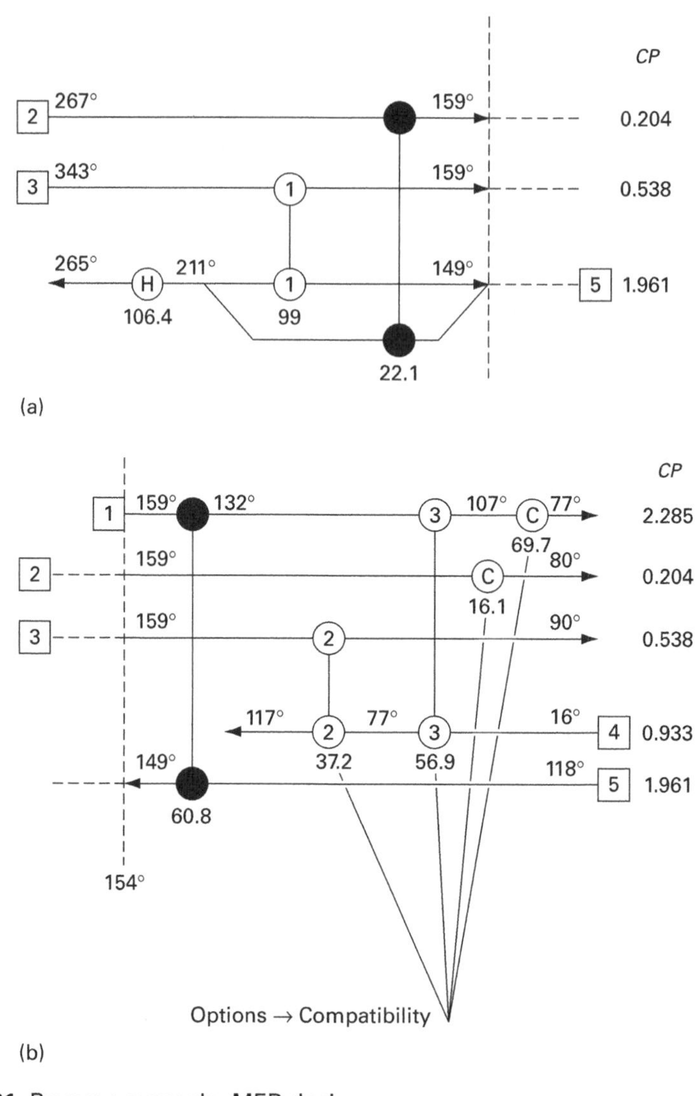

Figure 4.31 Revamp example; MER design

As pointed out earlier, there can be several alternative revamp options. Here, one alternative way of eliminating the stream split is by cyclic matching. Above the pinch, the new match between streams 2 and 5 could be placed in series with the existing match 1, rather than in parallel. Which match should come at the pinch? The criteria in Section 4.3.2 suggest that it should be the new match, because the supply temperature on stream 2 is lower (closer to the pinch) than on stream 3. Moreover, any unused heat above the pinch in stream 3 will at least be recovered below the pinch in exchanger 2, rather than being thrown away in a cooler. The result is shown in Figure 4.33; we have eliminated the stream split with a modest energy penalty of

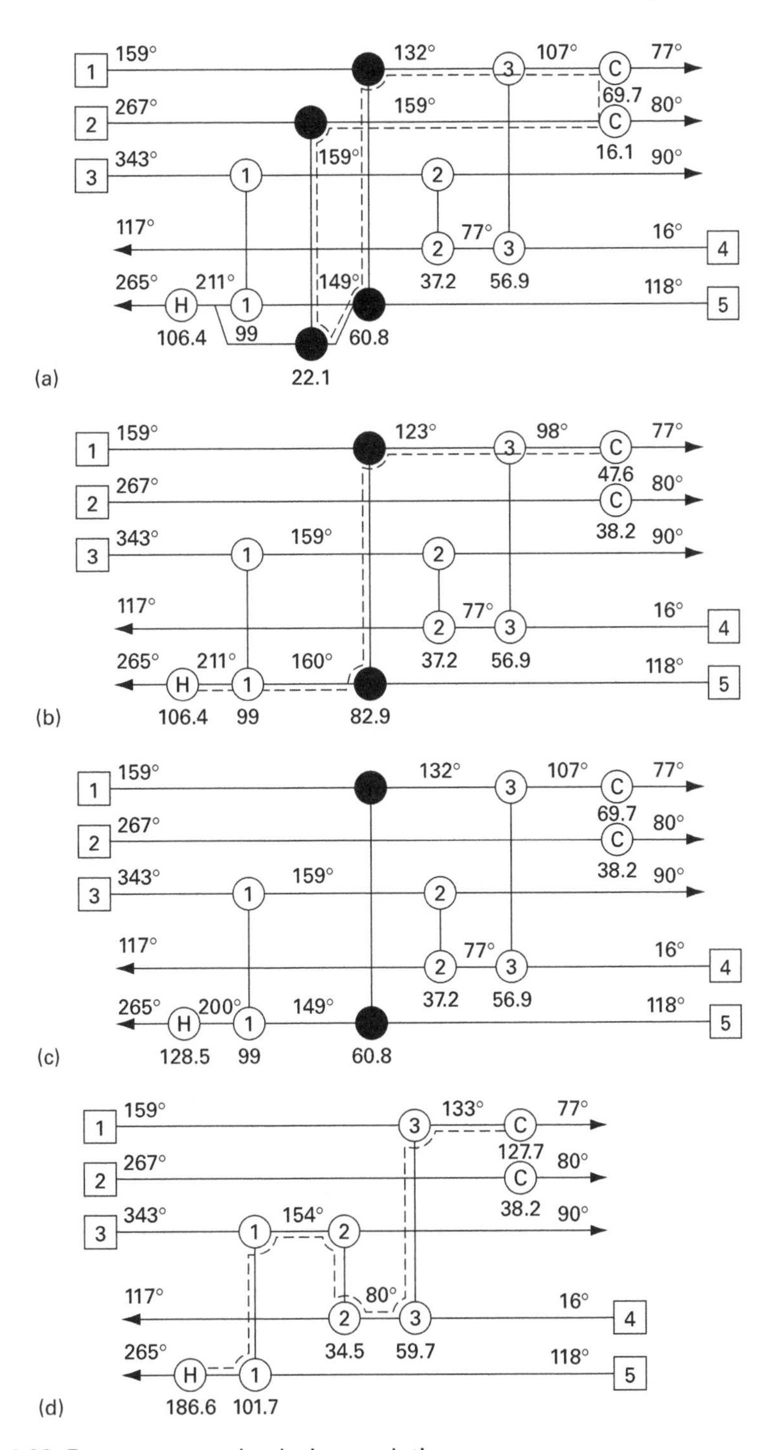

Figure 4.32 Revamp example; design evolution

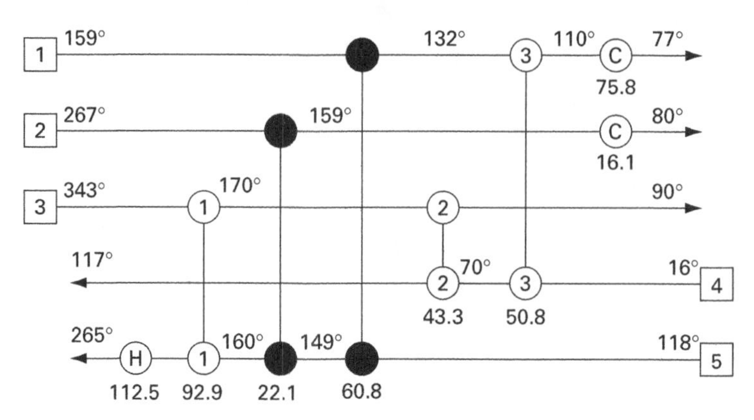

Figure 4.33 Revamp example; alternative design with cyclic matching

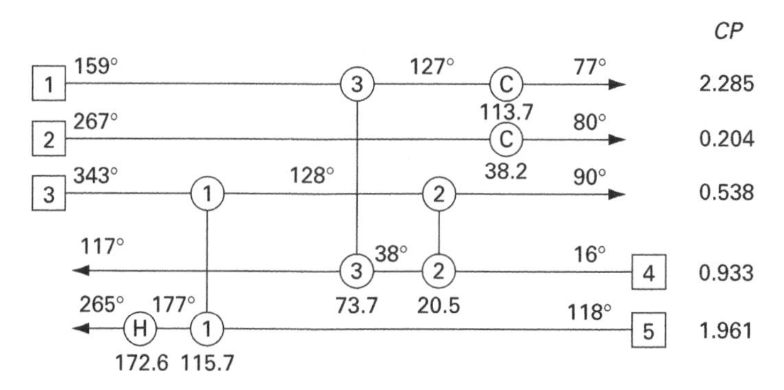

Figure 4.34 Revamp example; resequenced design identified by network pinch

6.1×10^2kW. Relaxation could continue by breaking the same loop as before, to arrive back at the network of Figure 4.32(b).

An alternative approach is to identify the network pinch. For the existing network, although none of the exchangers actually reach ΔT_{min}, the tightest constraint is clearly at the cold end of match 2. Resequencing exchangers 2 and 3 then becomes one obvious possibility, creating the further alternative shown in Figure 4.34. This gives lower energy use than the design in Figure 4.32 with the same ΔT_{min} of 10°C. Further possibilities include repiping exchanger 2; to provide an additional shell in series with either exchanger 1 or exchanger 3, or to match streams 2 and 4. Figure 4.35 shows the first of these options. For all the repiping options, an additional cooler will be needed on stream 3.

The next step is to make a crude evaluation of all the designs produced, comparing them to the base case design. At this stage, ΔT_{min} is abandoned and the effect of the network changes on the individual units is assessed. This is most simply done by "*UA* analysis". By applying the standard equation $UA = Q/\Delta T_{LM}$ to each unit, the

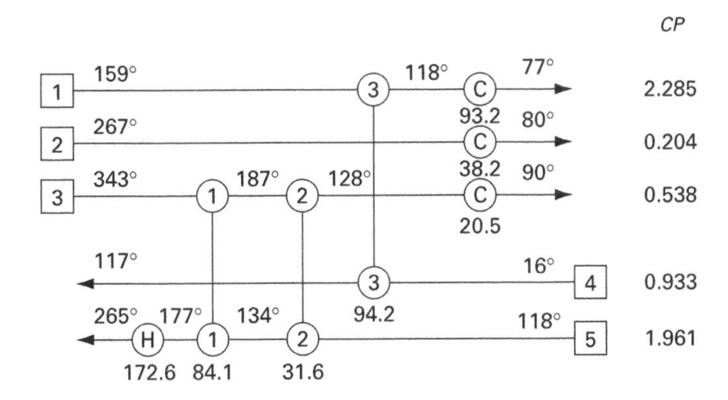

Figure 4.35 Revamp example; exchanger 2 repiped

Table 4.4 Comparison of exchanger *UA* values for alternative revamp schemes

	MER design	*Cyclic matching*	*Design 1*	*Design 1A*	*Design 2*	*Resequenced*	*Repiped*	*Existing network*
Heat recovery	276.0	269.9	253.9	247.8	195.8	210.0	210.0	186.4
Hot utility	106.4	112.5	128.5	134.6	186.6	172.6	172.6	196.0
Energy saving	45.7%	42.6%	34.4%	31.3%	4.8%	11.9%	11.9%	0.0%
UA values:								
E1	2.10	1.93	1.98	1.43	1.17	2.08	0.85	0.89
E2	1.50	1.28	1.50	1.28	1.67	0.25	1.23	1.28
E3	0.80	0.66	0.80	0.66	0.62	1.18	1.39	0.48
N4	0.78	0.54	0.00	0.00	0.00	0.00	0.00	0.00
N5	5.11	5.11	5.11	5.11	0.00	0.00	0.00	0.00
Total *UA*	10.29	9.52	9.39	8.48	3.44	3.51	3.47	2.66
Additional *UA*	7.63	6.86	6.73	5.82	0.78	0.85	0.81	0.0

effect of network changes on the total area of each unit is assessed, on the assumption that heat transfer coefficient U remains constant. A spreadsheet is convenient for this, and Table 4.4 shows the results. Not surprisingly, we see that all the networks with greater energy recovery require considerably more area than the existing case. For example, in the MER design shown in Figure 4.32(a), match 1 is 2.35 times its base case size, partly due to increased load and partly due to reduced driving force. In general, all the existing heat exchangers need to be enlarged, as even where heat loads have not increased, the temperature driving forces have been squeezed.

Having generated a *UA* table, loads should then be shifted around loops or along paths in the networks to restore the *UA* values as far as possible to their values in the base case, *but without eliminating units*. Note that full use should be made of any spare capacity a unit may have available. This network is quite constrained, with heaters and coolers on only 3 of the 5 streams, and it is therefore difficult to shift loads around to keep some exchangers at their existing size. (The organics distillation case study in Section 4.9 gives a situation where load shifting between exchangers is more easily accomplished.) One particular frustration is that resequencing exchangers 2 and 3, which gives a very area-efficient network, matches very badly with the current sizes of the exchangers; 2 is now vastly oversized, whereas 3 would need a lot of extra area. In fact it could be better to repipe 2 and 3 to swap their hot streams, rather than physically changing their sequence on stream 4! However, repiping exchanger 2 in parallel with exchanger 1 allows both of them to be retained at their existing size, and all the new area is on exchanger 3.

It is however noteworthy that the cyclic matching design coincidentally requires no extra area on exchanger E2. Of course a new exchanger N4 is required, but this may well be more convenient than an extra shell for E2, especially as the additional energy recovery is quite substantial. The *UA* values for the cyclic matching design are very similar to Design 1 but the energy recovery is much greater, suggesting that better use is being made of temperature driving forces through the network (by recovering heat from stream 2). However, Design 1 can also be modified by shifting 6.1×10^2 kW on a path through exchangers 1, 2 and 3 to open out driving forces and restore the original size of E2, giving Design 1A.

Where a modest amount of additional area is required in a shell-and-tube exchanger, it may be possible to enhance the heat transfer coefficient instead, by changing the shell-side baffle arrangement or adding tube inserts. Both of these have the disadvantage that they increase pressure drop, which may limit flowrates. For plate exchangers, it is normally easy to increase the area by adding further plates.

Having reviewed all the options, a table as shown in Table 4.5 can be produced, ranking the possible improvement schemes in terms of energy performance and listing the equipment modifications necessary for each. From this table, the "best bets" are identified for further evaluation, involving detailed simulation of the network's performance. In this case, the MER, cyclic matching and "design 1A" options all look more promising than "design 1" and "design 2", with the repiping and resequencing options in between. Clearly, the key match which gives the biggest improvement is the new one between streams 1 and 5 below the pinch.

Table 4.5 Comparison of evolved revamp schemes

Design	*Illustration*	*Energy saving*	*Capital cost implications*	*Total* UA *value*
MER	Figure 4.32(a)	46%	2 new units, 2 new shells, 1 stream split	10.29
Cyclic matching	Figure 4.33	43%	2 new units, 1 new shell	9.53
Design 1	Figure 4.32(c)	34%	1 new unit, 2 new shells	9.39
Design 1A	–	31%	1 new unit, 1 new shell	8.57
Design 2	Figure 4.32(d)	5%	3 new shells	3.45
Resequencing	Figure 4.34	12%	1 new shell	3.47
Repiping	Figure 4.35	12%	1 new cooler, 1 new shell	3.53
Existing case	Figure 4.30	0	–	2.66

It can be seen that there are two distinct "families" of revamp options. The first group are relatively modest changes to the current network, with low capital cost but also fairly low savings. The best of these networks are those with the resequencing or repiping identified by the network pinch method. The second group achieve large savings but also require major investment in additional exchanger area; the key match is between streams 1 and 5, which was identified from the process pinch and MER design but is not obvious from the network pinch. Thus, both approaches have merits, and the best way to identify the full range of possible retrofit options is to use all three of the possible strategies in Section 4.7.1 in parallel. As the number of streams and existing exchangers increase, the network pinch approach based on the existing network becomes relatively more attractive, as the extent of changes required to get at all near the MER design will usually be prohibitive.

Summarising, to find the best potential energy improvement schemes for an existing plant, the designer should:

- Check the existing network and identify pinch violators.
- Obtain an MER design, having as great a compatibility with the base case as possible.
- Where a choice exists on matches, especially away from the pinch region, favour matches which already exist.
- Identify the network pinch and pinching match for the existing network configuration.
- Consider working in two directions: from the MER network by loop breaking and energy relaxation, and from the existing network by eliminating violations of either the process or network pinch.
- Perform a crude evaluation of all the alternative topologies by *UA* analysis, restoring *UA* values of existing units as far as possible.
- Perform detailed simulation and optimisation of the "best bets".

Finally, a word of warning (or encouragement): don't give up on the basis of one route only from MER design to base case! Figure 4.36 illustrates that where there are options there will be more than one route.

4.7.5 Automated network design

Even for a relatively simple problem like this, calculating the exchanger sizes for the various options involves significant work. As with targeting calculations, spreadsheets can help with some of the donkey work of calculation. Network simulators can be even more useful, particularly if they can generate and compare a wide range of alternative designs. Several commercial software packages are now available with well-developed retrofit tools.

As the previous examples show, it is difficult to predict in advance which revamp strategies will prove best out of the many alternative possibilities; a promising route may turn into a dead end, and vice versa. One attempt to overcome this is the superstructure approach, proposed by Floudas, Ciric and Grossmann (1986). A conceptual network is constructed including all possible matches and a computer search then

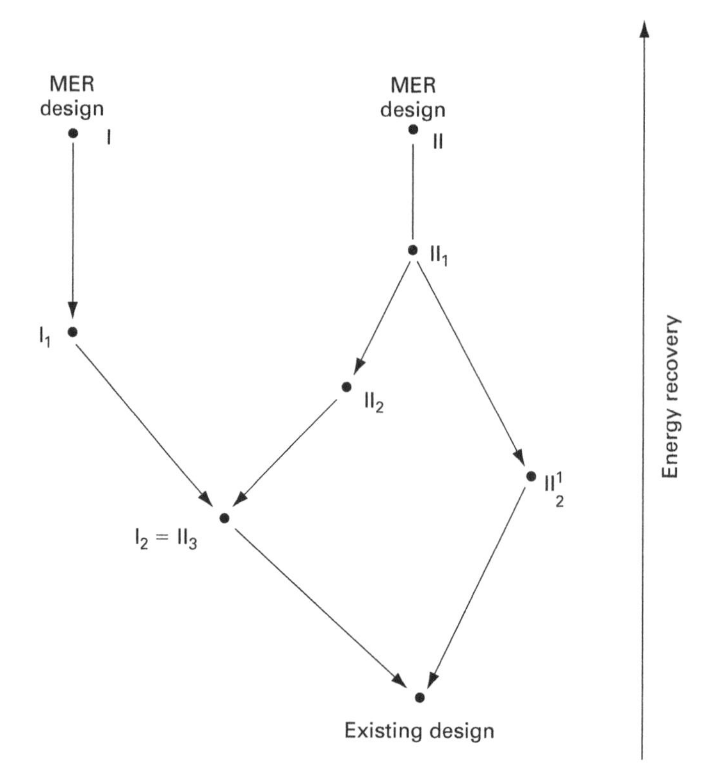

Figure 4.36 Alternative routes from MER designs to existing network

finds the most successful options. However, this is an MINLP (mixed-integer non-linear programming) problem requiring considerable computing power for all except the simplest networks. Various methods have been adopted to simplify the problem by removing some of the less likely options, but this also gives the possibility of missing the optimum.

Network optimisation in fact predates pinch techniques. Synthesis of heat exchanger networks, in particular using advanced computer techniques, was a major subject of study during the 1970s. However, the problem is extremely complex when starting from a blank sheet of paper. The wide range of possible network structures and the even larger range of possible exchanger sizes mean that the overall problem is not amenable to analytical solution; again, it is a MINLP and there are many local minima in the total cost function. Hence, trying to find the overall optimum network is an immensely challenging task, even when substantial computing power is available. In this situation, pinch technology came as something of a bombshell. Here were simple techniques which could be performed by hand calculation (although computers were certainly more convenient) and yet gave better results in many cases.

In fact, even for fairly complex plants with large numbers of streams, pinch analysis tends to home in on a small "family" of networks with similar structures (as can be seen from the examples in this and other sections). The differences between them are in number and position of units, and stream splits. The global optimum will normally be one of these structures with a particular combination of heat exchanger sizes. The optimisation of a network for which the structure is known is a much simpler task than synthesising an unknown structure; it is a standard NLP (non-linear programming) problem, though even here there is great danger of becoming trapped in one of the many local optima.

Further constraints can be added to reduce the "solution space" from a continuum to a large but finite number of options. Typical heuristics are restricting heat exchangers to standard manufacturers' sizes in new design, and in revamps, keeping current exchangers at their existing size as far as possible.

In retrofit the task is to work from the existing network to a better one, which means changing the structure. Again, this is a MINLP, not easily automated. Instead, we can find improvements to the structure by identifying the key constraints on the network. The first methods used the process pinch, as for grassroots design, but more recent techniques are based on the network pinch, the most constrained point in the network. In fact, both approaches are of value, as shown above.

Considerable research is still continuing on the overall network problem, and many attempts have been made to simplify the MINLP problem to one which can be tackled by LP, NLP, MILP or a simplified MINLP. However, as yet, these have not yielded a technique or program which can be used reliably in practice. The overall network optimisation challenge is well explained by Smith (2005).

4.8 Operability; multiple base case design

So far, we have assumed a steady-state flowsheet with all flows and temperatures constant. However, few designs, if any, are always operated as per base case data. Processes need to operate efficiently, reliably and safely for different capacities, different product specifications, different feedstocks, fresh or spent catalyst, varying ambient temperatures, clean and fouled equipment, etc. Multiple base cases may therefore need to be considered, in three categories:

- Intentionally different operating cases (e.g. using a different feedstock composition in an oil refinery).
- Unintentional long-term variation (e.g. "clean" and "fouled" plants with different heat transfer coefficients in the exchangers).
- Short-term and random fluctuations (e.g. in temperature or mass flowrate due to variations in the upstream process).

The first industrial users of pinch analysis were sceptical as to the flexibility of integrated designs. The common thinking was that integration would lead to operability problems.

Initially, this hurdle was overcome only by hard work. Integrated structures had to be evolved for the base case and operability had to be checked in the traditional way (i.e. through simulation, modifications to the design, more simulation, etc.). Experience showed quickly that while there was a relationship between integration and operability, there was not necessarily a conflict. In some cases, a given integration feature

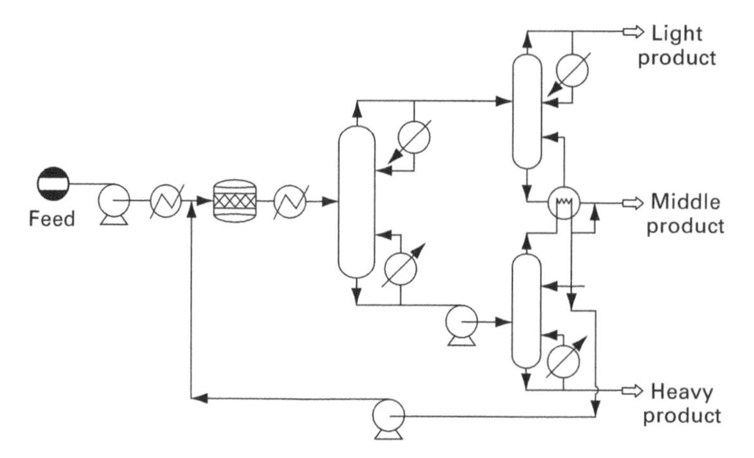

Figure 4.37 Operability effects of integrating two distillation columns

could prove beneficial for operability. In other cases, the same feature could be detrimental.

Consider Figure 4.37. A product separation/recycling system consists of three distillation columns. The condenser of the heavy products column is integrated with the reboiler of the light products column. Now consider two potential changes to operating conditions: (1) change of catalyst performance and therefore reactor product composition, and (2) change of feed flowrate. If the catalyst deteriorates and the reactor product composition changes the proposed integration may prove detrimental to operability. The load on one column may increase while that on the other column may decrease and the condenser/reboiler integration is likely to bottleneck capacity. If, by contrast, the overall feed flowrate changes the proposed integration may prove beneficial. A given variation in overall process capacity would result in a *smaller* variation in utility loads as a result of integration and in a case where the utility system is limiting this may prove beneficial to debottlenecking the overall process.

The overall experience today is that integrated systems can be more operable than their less integrated counterparts *provided* operability is taken into account early during design. The approach taken in pinch analysis is to include operability objectives in targeting and during the development of the integrated structure.

An example of this approach has already been discussed in the context of RPA. In Figure 4.25, RPA was used to settle the issue of a start-up heater *prior to design*.

Another example is referred to in Figure 4.38 (Tjoe and Linnhoff 1986). A given network design has its performance simulated in terms of their predicted energy consumption for three different cases, and is compared with the target curves (for overall surface area vs. energy) for each case. It is found that the design in question suits operating cases (A) and (B) reasonably well but is well above the target energy consumption for case (C). Additional costs will be necessary to make the design flexible with respect to case (C). Alternatively, the mismatch could be accepted. Case (C) may not represent a frequent case and poor efficiency might be acceptable.

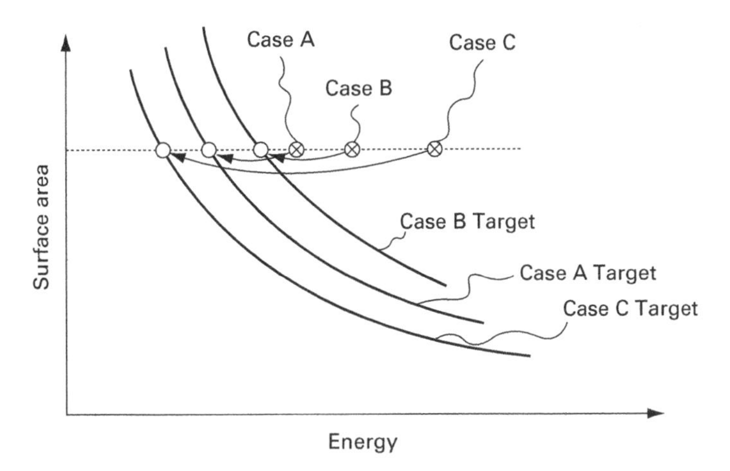

Figure 4.38 Energy targets for three different operating cases

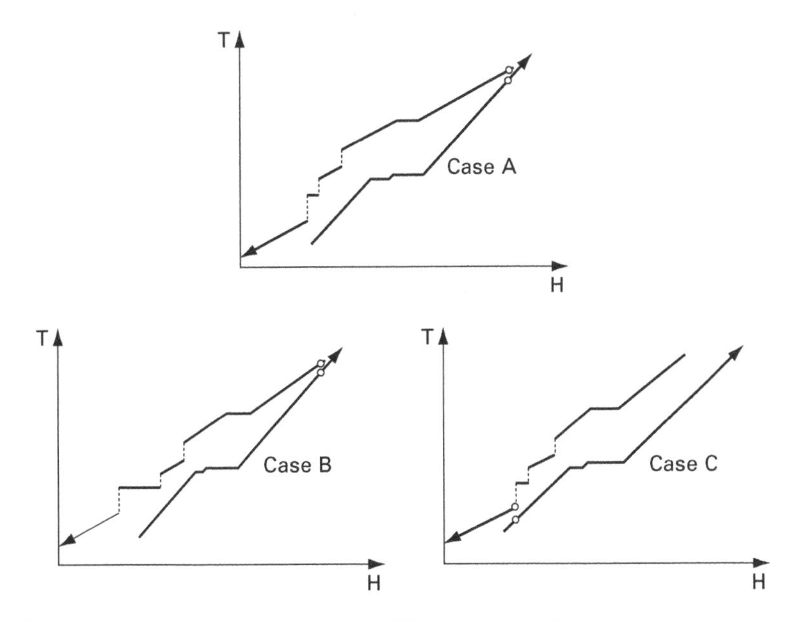

Figure 4.39 Composite curves for three different operating cases

Figure 4.39 shows why certain designs may suit certain operating cases but not others. Simple study of the composite curves reveals that a design which suits case (A) is likely to also suit case (B) with few additional features, as the pinch is the same, but that more significant changes will be required to operate cases (A) and (C) in one design or cases (B) and (C) in one design.

Sensitivity analysis of a network is another useful approach in some cases. The interrelationships of the various network temperatures and areas can be listed, and the effect on output variables of fluctuations in temperatures, flowrates or heat transfer coefficients (due to fouling) can be analysed. This generally requires dedicated software or a spreadsheet. The approach is described by Kotjabasakis and Linnhoff (1986, 1987).

The consideration of operability at the targeting stage has proved a very successful approach. In addition, there has been work addressing the design of flexible structures. Much of this work has stood the test of industrial application. Key references are papers by Calandranis and Stephanopoulos (1986), Colberg *et al.* (1989), Floudas and Grossmann (1987), Linnhoff and Kotjabasakis (1986) and Saboo *et al.* (1985, 1986).

4.9 Network design for organics distillation case study

We will now show how to develop the heat exchanger networks for the case study presented in Sections 3.2 and 3.8. There are two options: one for the atmospheric unit on its own, the other when it is integrated with the vacuum distillation unit. In many complex processes, such as those described in the case studies in Sections 9.2 and 9.3, to meet the targets precisely requires a large number of small exchangers which are not cost-effective, and a compromise has to be made. However, for a simple plant such as this one, with a small number of streams and a sharp pinch, we may hope to achieve the target exactly.

4.9.1 Units separate

This is a retrofit situation so it is valuable to start by looking at the existing network and plotting the pinch temperature on it, as in Figure 4.40. We can then see that there are three pinch violations, for a ΔT_{min} of 20°:

- The bottoms are being cooled above the pinch (1030 kW).
- The middle oil–crude feed exchanger is across the pinch (760 kW).
- The crude feed is being heated below the pinch (275 kW).

These three violations add up to 2065 kW, which is the difference between the current hot utility use (6860 kW) and the target (4795 kW).

Developing the correct MER network structure for the atmospheric distillation unit is quite straightforward (refer to Table 3.3 for the stream data). Below the pinch, the CP inequality dictates that the overheads should be matched against the crude feed. Immediately above the pinch, the middle oil is the only available hot stream and must be matched against the crude feed. All the heat in the middle oil should be used so that this stream is "ticked off". Finally, a new match between the bottoms residue and the crude feed is added in series. This broad network structure, shown in Figure 4.41, is the same whether a ΔT_{min} of 20°, 30° or 63° is

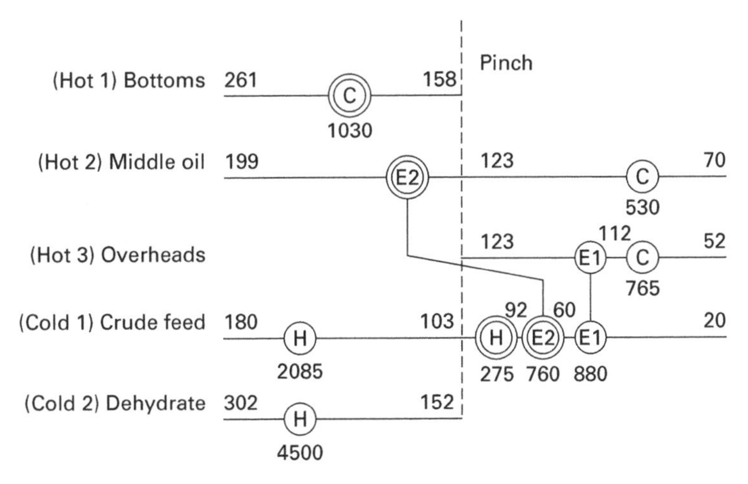

Figure 4.40 Network grid diagram for existing process showing pinch violations

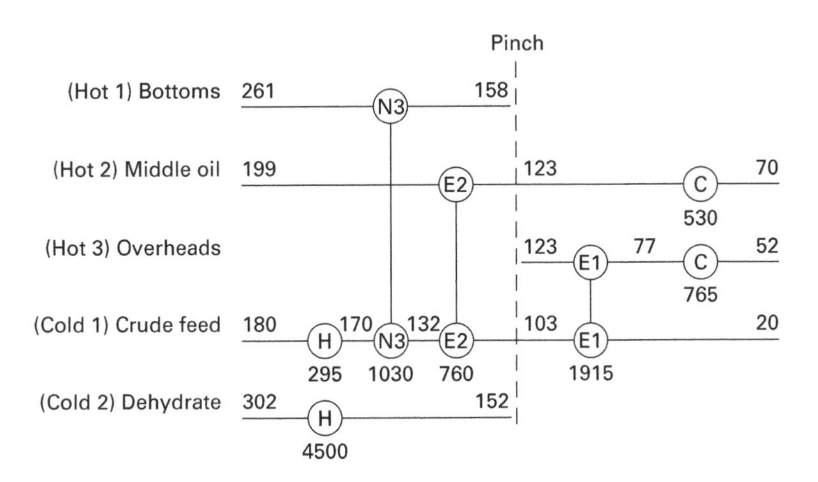

Figure 4.41 Grid diagram for MER network with $\Delta T_{min} = 20°$

used. The only things that change are the sizes of the heat exchangers. It can be seen that the network structure is the same as the current one with the addition of the residue–crude feed exchanger, removing one pinch violation. However, for $\Delta T_{min} = 20°$, both the two existing exchangers must be greatly enlarged. This recovers enough additional heat from the overheads to shift the middle oil–crude feed exchanger completely above the pinch and also eliminate below-pinch heating. In contrast, for $\Delta T_{min} = 63°$, the two existing exchangers need not be enlarged; only the new match is needed.

The choice between the various options for retrofit will depend on the payback time chosen and the amount of capital readily available. Table 4.6 is based on the

Table 4.6 Economic evaluation of three possible projects

ΔT_{min} (°C)	New/ Mod units	Added area (m^2)	Total area (m^2)	Capital cost (£)	Energy saved (kW)	Annual saving (£)	Payback time (yr)	NPV after 2 years (£)
20	1, 2	483	612	140 K	2,065	124 K	1.1	112 K
30	1, 2	330	459	105 K	1,810	109 K	0.9	117 K
63	1, 0	99	228	32 K	1,030	62 K	0.5	93 K

results obtained from cost targeting calculations and shows the trade-off between capital and operating costs. The required heat exchanger area has been calculated using the actual temperatures and heat loads obtained from the network and assuming 1–2 exchangers. It can be seen that the total area required in each case matches well with the estimate obtained from targeting are given in Table 3.5, and the target energy saving has been exactly achieved. Since this is a retrofit situation, the capital costs are lower than the targets because some of the area already exists. (In the table, K stands for thousands of pounds.) It has been assumed that the cost of adding further area is the same as that of a new exchanger of the same size and that none of the "spare area" which exists on coolers or furnaces (because their heat loads have decreased) can be re-used; this is a "worst case". Nevertheless, the payback for all three schemes is under 2 years. The minimum cost option with ΔT_{min} = 63°C and just one new exchanger gives the shortest payback time; but over a 2-year period, the net benefit from the other two schemes (expressed as a net present value) is significantly greater. As expected, the NPV for a ΔT_{min} of 30° is only slightly better than for a ΔT_{min} of 20° over the 2-year period chosen. It may be noted that the cost of 1000 kW for a year is £60 K (£12/MWh, 5000 h/year).

Finally, we will look at the network and consider the practical implementation. In particular, we will compare new exchanger sizes with existing ones, bearing in mind that adding modest amounts of new area to shell-and-tube exchangers is expensive compared to the energy saved. (In plate exchangers, by contrast, it is usually easy to add a few new plates.) For simplicity, we will assume ideal countercurrent exchange in these calculations. It will be seen that the results are close to the targets obtained above for 1–2 exchangers.

We will start by evaluating the network for ΔT_{min} = 63°, as two of the exchangers (E1 and E2) are unchanged and we only have one new exchanger N3. The calculations for the three exchangers are shown in Table 4.7.

However, as a starting point for energy relaxation, we will begin with ΔT_{min} = 20°, on the assumption that we may well have to back off from the "ideal" network and sacrifice some energy savings. Table 4.8 shows the new exchanger sizes.

Exchanger N3 is new so we have complete freedom on sizing. Exchanger E1 needs to be more than quadrupled in size and the obvious way to do this is to install a second shell in series. E2 also needs to have its area more than doubled, but saves no more energy (because of the lower driving forces). One might ask, what if we leave this exchanger as it is? The consequences are shown in Table 4.9. The energy

Table 4.7 Sizes of heat exchangers in MER network with $\Delta T_{min} = 63°C$

Number	Streams	Hot stream temperatures (°C)	Cold stream temperatures (°C)	Log mean temperature difference (°C)	Heat load (kW)	Overall HTC (kW/m²K)	Required area (m²)	Existing area (m²)	Additional cost (£)
E1	CF/Ohds	123–112	20–60	77	880	0.20	57.5	57.5	0
E2	CF/MO	199–123	60–92	85	760	0.125	73.2	73.2	0
N3	CF/Bott	261–158	92–132	94	1,030	0.125	87.7	0	31.0 K
Total					2,670		218.4	130.7	31.0 K

Table 4.8 Sizes of heat exchangers in MER network with $\Delta T_{min} = 20°C$

Number	Streams	Hot stream temperatures (°C)	Cold stream temperatures (°C)	Log mean temperature difference (°C)	Heat load (kW)	Overall HTC (kW/m²K)	Required area (m²)	Existing area (m²)	Additional cost (£)
E1	CF/Ohds	123–77	20–103	36	1,915	0.20	269.6	57.5	58.7 K
E2	CF/MO	199–123	103–132	39	760	0.125	157.2	73.2	30.2 K
N3	CF/Bott	261–158	132–170	52	1,030	0.125	159.6	0	47.1 K
Total					3,705		586.4	130.7	136.0 K

Table 4.9 Sizes of heat exchangers with E2 left unchanged

Number	Streams	Hot stream temperatures (°C)	Cold stream temperatures (°C)	Log mean temperature difference (°C)	Heat load (kW)	Overall HTC (kW/m²K)	Required area (m²)	Existing area (m²)	Additional cost (£)
E1	CF/Ohds	123–77	20–103	36	1,915	0.20	269.6	57.5	58.7 K
E2	CF/MO	199–146	103–124	58	530	0.125	73.2	73.2	0
N3	CF/Bott	261–158	124–162	61	1,030	0.125	134.5	0	41.6 K
Total					3,475		477.3	130.7	100.3 K

Table 4.10 Sizes of existing utility heaters and coolers

Number	*Streams*	*Hot stream temperatures (°C)*	*Cold stream temperatures (°C)*	*Log mean temperature difference (°C)*	*Heat load (kW)*	*Overall HTC (kW/m²K)*	*Existing area (m²)*
H	Crude Feed	400–400	92–180	261	2,360	0.20	45.1
C1	Overheads	112–77	25–35	48	1,800	0.50	75.5
C2	Middle Oil	123–70	25–35	64	530	0.20	41.3
C3	Bottoms	261–158	25–35	175	1,030	0.20	29.4
Total					3,475		191.3

penalty is only 230 kW (worth £14 K/year). The capital cost of enlarging E2 is saved and there is also a slight reduction in the required area for N3 (because temperature driving forces are improved), so that the capital cost falls by £35.7 K. Hence, the *marginal* payback on enlarging E2 is 2.5 years. This is clearly less attractive than the overall project, and could well be dropped. The benefits from enlarging E1 and adding N3 are that 1835 kW of energy is saved, worth £110 K, for a capital expenditure of £100 K, again giving a payback of less than a year.

One other possibility in retrofits is to re-use existing heaters and coolers whose load has been reduced or eliminated as heat exchangers. However, both the temperature driving forces and the heat transfer coefficients tend to be better from utility streams, so the additional area available can be disappointingly low. This is the case here, as shown in Table 4.10. Moreover the materials on the old "utility" side of the exchanger may not be suitable for a process fluid flow, and the heater for the crude feed is a furnace (gas-to-liquid, with finned tubes on the gas side to increase surface area and heat transfer) and is therefore less suitable for a liquid–liquid duty. Therefore, re-use of heaters and coolers in this way will not be considered further in this case. This also gives operability benefits at start-up and shutdown, as the old heaters and coolers are available for use when necessary.

4.9.2 Units integrated

The pinch region is "tighter" in this case, as pointed out in Section 3.8.4. Immediately above the pinch, we have two hot streams (middle and heavy oil) and only one cold, so a stream split is required if the pinch matches criteria are to be met. The natural next step is to tick off both hot streams and then match the crude feed against the residue. However, calculation shows that this will give a ΔT_{min} violation. The middle oil and residue matches must therefore be placed in parallel, either by a new stream split or by placing the bottoms exchanger on the same branch as the heavy oil one; the latter is more convenient. Finally, we find that the crude feed no longer has enough heat to tick off the bottoms residue stream, and to avoid above-pinch cooling, 55 kW must be matched against the dehydrate stream. The resulting network

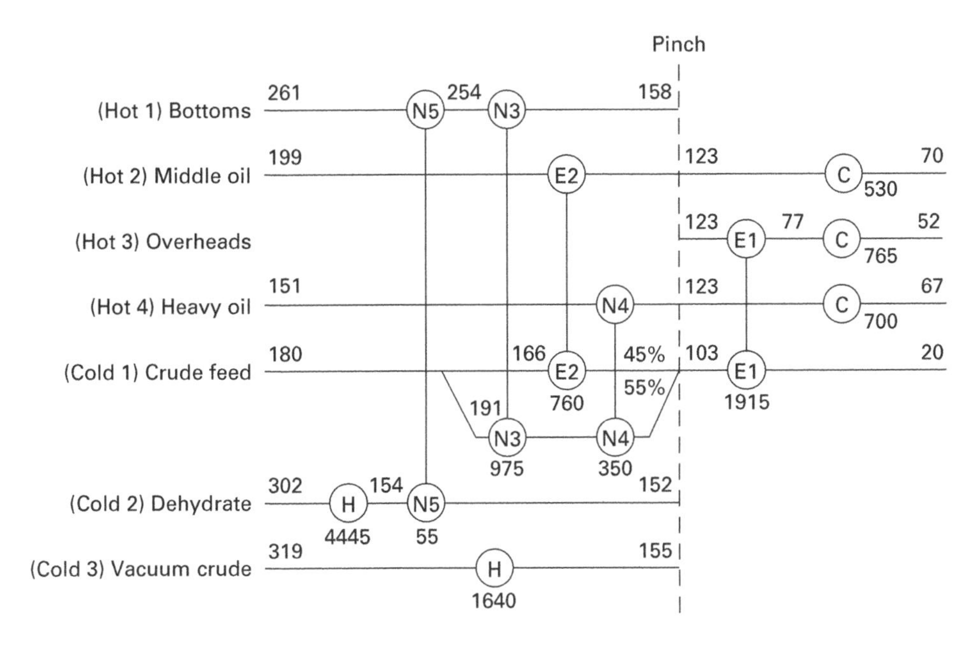

Figure 4.42 Network with atmospheric and vacuum units integrated, $\Delta T_{min} = 20°$

is shown in Figure 4.42; there is some flexibility on the flows in the branches of the crude feed stream, and here it has been assumed that they are divided proportionately to the hot stream CPs above the pinch. As temperature driving forces have been squeezed, the heat exchangers on the matches between the crude feed and the middle oil and bottoms will be larger than those in Table 4.6 and Figure 4.41.

An above-pinch network could also be generated by cyclic matching, with some reduction in energy recovery. However, for operability reasons, the stream split is an attractive option in this case. When the two plants are not operating simultaneously, the branch of the crude feed which goes to the heavy oil exchanger can simply be closed off by valves, and all the crude feed goes through the existing route. Also, the pipework on this new branch can be smaller as it only has to carry half the flow, and the pressure drop of the parallel arrangement is less than for putting the exchangers in series. Finally, only half the flow suffers the heat losses on the long pipe run between the atmospheric and vacuum distillation units (and these are eliminated entirely when the two units are not running together).

The annual saving from a 350 kW energy reduction is about £21K. However, we now have 5 exchangers – two more than if the units are not linked. The small 55 kW exchanger will clearly be uneconomic; one option is to abandon it (putting the load on heaters and coolers via a path) and accept the energy penalty, but this reduces the savings still further. However, the preflash temperature is not set in tablets of stone. Increasing it by just 2°C would allow this 55 kW to be taken up by the crude feed stream, with a new final temperature of 182°C, while the dehydrate will now start at 154°C instead of 152 and its heat load will fall accordingly. This change is easily acceptable – indeed it is less than the typical temperature variation during the operating cycle due to exchanger fouling. Even after this, the £21 K/year saving must pay not only for the cost of the new exchanger but also additional pipework. The need to operate the two units simultaneously to achieve the saving is another drawback. Hence, this project looks very marginal.

A quick calculation suggests that the exchanger will need to be of about 130 m^2 (assuming the normal overall HTC of 0.125 kW/m^2K; the mean ΔT is only just over 20°C) and the estimated capital cost is £40.6 K. This gives a 2-year payback on this item alone, but moreover the temperature driving forces in the rest of the network are squeezed and the other exchangers E2 and N3 will need to be enlarged.

A further network option is identified in the Process Change Section (6.6.2).

4.9.3 Including utility streams

In Section 3.4.6 it was pointed out that by including the utility streams in the analysis, further heat recovery could be achieved. Design of a network in this case should use the balanced grid, which includes the furnace heating and flue gas as additional hot streams, and the air heating requirement as an additional cold stream. Since our network achieves the energy targets, the furnace, flue gas and air streams are all 30% lower than their current values, corresponding to the energy saving achieved.

Figure 4.43 shows a network achieving the targets, considering the atmospheric unit alone. The crude feed and dehydrate are matched against the furnace heating stream as at present. Theoretically there is a small ΔT_{min} violation at the cold end of the dehydrate and furnace heat streams (as the ΔT_{min} on gas–liquid matches is taken as 50°C); in practice, this is immaterial, especially as something similar must also happen in the existing furnace. The flue gas is matched against the furnace air as at present; the heat exchanger straddles the pinch but does not violate it, as the above-pinch and below-pinch loads are carefully matched. The ΔT on the match has fallen from 100°C to 80°C; the initial air preheating from 20°C to 40°C must be done with a below-pinch process stream, and the only one hot enough is the middle oil stream. So one new air-to-liquid exchanger is required between the middle oil and furnace air.

The driving force on the flue gas–air exchanger and furnace heating coil for the crude feed have fallen, but so have the heat loads, because of the 30% hot utility reduction, so on balance the $(Q/\Delta T)$ value and required area should be similar to before. Detailed calculations would show the exact change and a small resizing of the new middle oil exchanger might then be desirable. The network recovers an additional 350 kW of heat compared to the situation where process and utility streams are not integrated. Similar calculations could be performed on the vacuum furnace streams.

4.9.4 Multiple utilities

In Section 3.4, two options were noted for multiple utilities. The first was to use steam at a shifted temperature of 200°C with furnace heating above that; the second

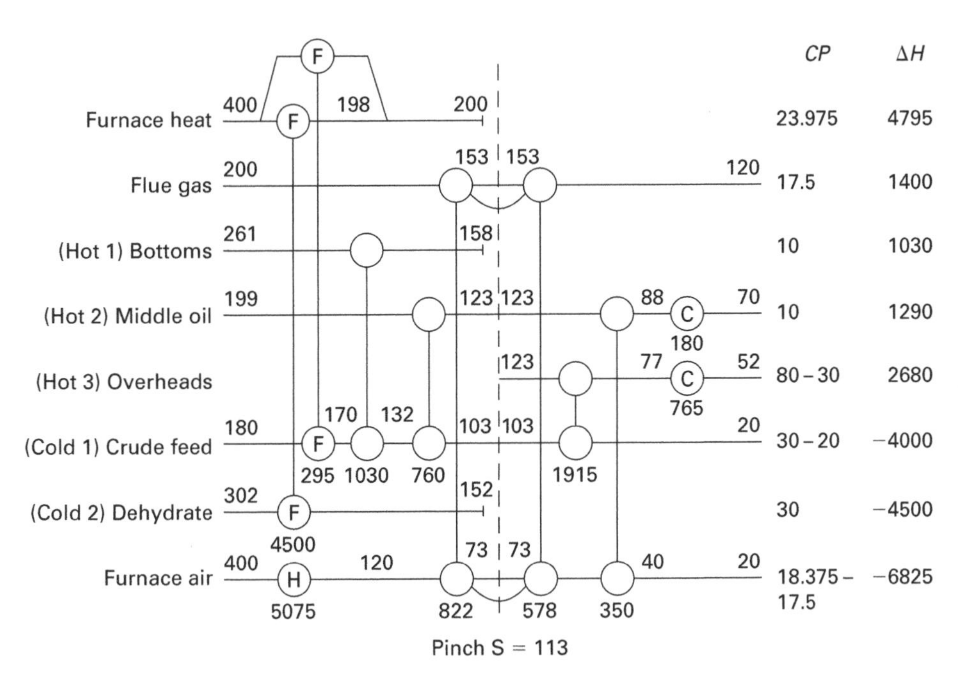

Figure 4.43 Balanced network grid for atmospheric unit integrated with utility streams

was to use hot gas as a variable temperature utility. What effect does this have on network design compared to Figure 4.41?

For the intermediate steam level, there is little change; the dehydrate duty below 200°C shifted temperature (190°C actual temperature) and all the crude feed heating duty will be fulfilled by the steam, and the dehydrate above this temperature will be heated by the furnace, as before. So the only real change is that there are two separate heaters on the dehydrate stream instead of one. This is consistent with our units targeting, as we have added an extra utility stream, so we expect one extra unit.

For the flue gas stream, things become more complicated. A new utility pinch has appeared at S = 162°C and the process pinch at 113°C has disappeared completely! Therefore, a completely new network needs to be designed, using the balanced grid and starting from the new pinch. Above the pinch, there are three hot streams and only two cold streams, so the crude feed stream has to be split. Conversely, below the pinch, the CP of the crude feed stream is higher than any of the three hot streams, so again it must be split. The flue gas stream above the pinch does not have enough heat load to satisfy the dehydrate stream, so an additional match is needed between the bottoms and the dehydrate. Below the pinch, at least 697 kW must be extracted from the flue gas to satisfy the cold stream loads above the process pinch, but it can also be run down further if desired. In this case, the (slightly arbitrary) decision has been taken to bring it down to 153°C, corresponding to the old process pinch. The extra 317 kW of heat recovered does not save any energy overall, but it does reduce the heat load and open out the driving forces on the big overheads–crude exchanger.

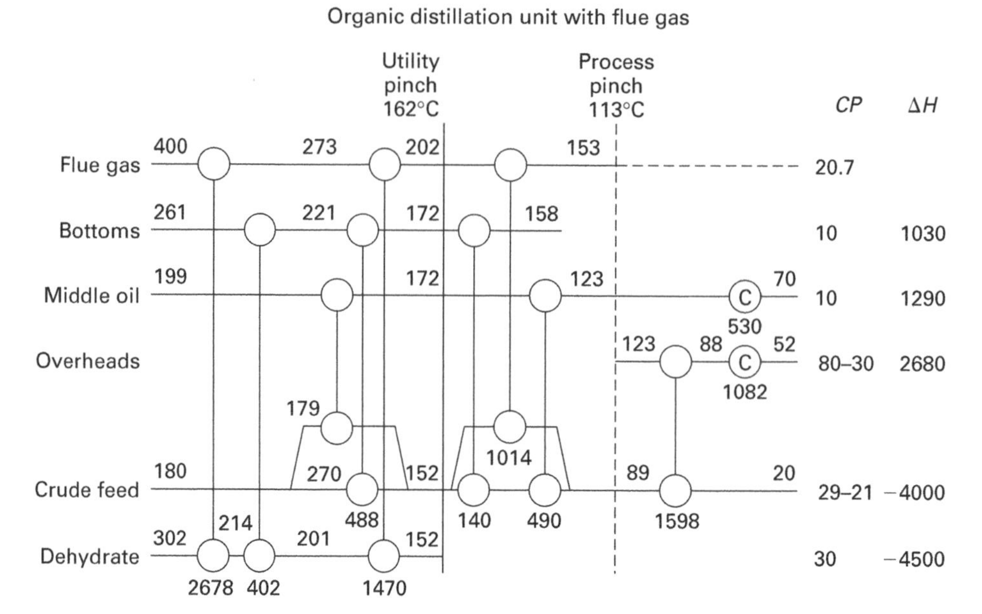

Figure 4.44 Balanced network grid for atmospheric unit with flue gas heating

The resulting network is shown in Figure 4.44. It requires 11 units and two stream splits, which is unlikely to be economic, so we want to relax the network. However, we cannot simply do so by passing in additional unspecified hot utility. Instead, we must increase the flowrate and CP of the flue gas stream to provide the extra heat.

One obvious option is to eliminate the bottoms–dehydrate exchanger by shifting its load on to the flue gas stream. This requires an extra 402 kW and the CP of the flue gas stream will increase from 20.7 to 22.7 kW/K – roughly 10% more. Next comes the bottoms–crude feed exchanger; to eliminate stream splits and cyclic matching, we will accept a small ΔT_{min} violation (to 15°C) at the cold end of this. The middle oil is matched against the crude feed and brought down to the point where the cold end is at the ΔT_{min} of 20°C. A match against the flue gas stream is now needed, and again we can bring this down as far as we like. We have chosen to leave the heat load on the overheads–crude exchanger at 1500 kW, as previously.

The relaxed network is Figure 4.45. We now have only 7 units, so we have saved 4 exchangers; we have also eliminated the stream splits. The extra heat load on the flue gas is 400 kW between 202°C and 400°C – but we must also remember that there is a corresponding penalty below 202°C for the heat lost in the flue gas because of its higher flowrate.

If a furnace is being used and air preheating is possible, it should be included in the stream data, targeting, network analysis and the balanced grid, as described in Section 4.9.3. However, not all flue gas streams have associated inlet air flows that can be preheated. For combined heat and power (CHP) systems based on gas turbines or reciprocating engines, air preheat is often infeasible, as it can adversely affect engine efficiency (see Chapter 5).

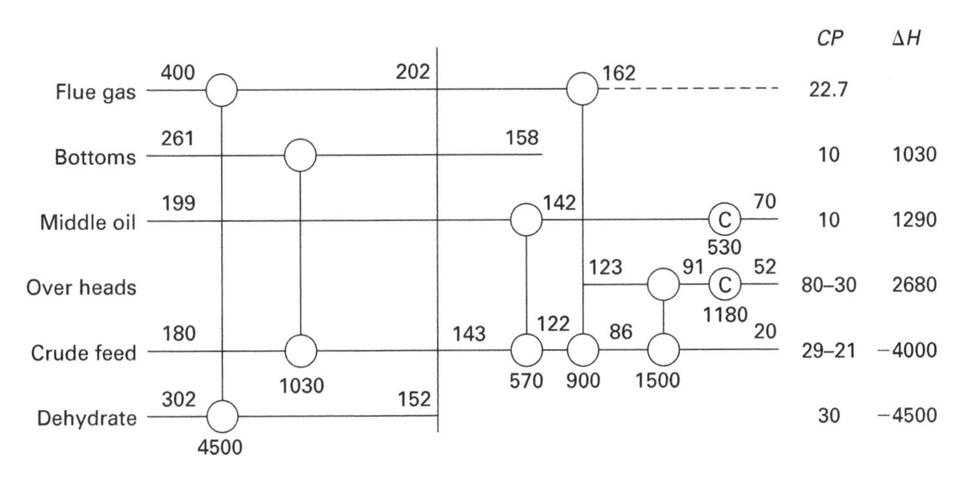

Figure 4.45 Relaxed network for atmospheric unit with flue gas heating

4.10 Conclusions

Heat exchanger networks can be effectively designed to reach the energy targets obtained by the Problem Table analysis, using the pinch design method. The method can be adapted effectively to allow trade-offs between energy, number of units and capital, and to cover retrofit of existing plants.

A great deal of additional research has been done on heat exchanger network synthesis over the last 20 years, looking at the design and optimisation of complex networks in much more depth. A detailed account is given in the book by Shenoy (1995). This material is particularly worth studying where complex plants with many streams are involved (e.g. oil refineries and bulk chemicals plants). Mathematical and computer optimisation methods have also been extensively studied for these cases, and specialised network design software is available.

Network design strategy has also evolved from the basic pinch techniques. In particular, revamps have increasingly tended to start from the existing network rather than the MER design, and network optimisation to fine-tune the heat loads on exchangers is standard practice. However, some of this change in emphasis may be because pinch analysis has been applied mainly to large and complex plants. For smaller and simpler processes, there is still much merit in the older techniques, which happily coexist alongside the newer developments.

Exercises

E4.1 For the main four-stream example, change the CP value to 2.5 and the supply temperature to 120°C. Design a new above-pinch network, using stream splitting at the pinch.

E4.2 For the stream split example in Sections 4.3.1 and 4.3.2:
- Try cyclic matching with 2 exchangers on (a) example as given, with exchangers in reverse order; (b) reversing CPs on streams 1 and 2, with both orders of exchangers; (c) equal supply temperature on streams 1 and 2, with both orders of exchangers. What implications do you draw from the respective energy penalties?
- Try to construct a series of four cyclic matches (in either order). What degrees of freedom are there? What are the best configurations to minimise the energy usage? Can you develop a set of equations to predict the configuration which minimises the energy penalty?

E4.3 Generate the UA-value table for the revamp example in Section 4.7 and compare with the values given in the text. Construct the network diagram for Design 1A, with temperatures.

E4.4 For the organics distillation unit in Section 4.9, construct the set of six simultaneous equations for the three heat exchangers in the basic structure for the atmospheric unit only (Figure 4.41), with six variable temperatures (between exchanger and cooler on the 3 hot streams, and between exchangers on the cold crude feed) and variable areas A1, A2 and A3 for the exchangers. As there are six equations and nine unknowns, how many items must be specified to obtain a solution? Obtain solutions for the equations with (a) sufficient known areas, (b) sufficient known temperatures, (c) a mix of known areas and temperatures, using the values in the MER network. A spreadsheet is recommended for these calculations.

References

Asante, N. D. K. and Zhu, X. X. (1997). An automated and interactive approach for heat exchanger network retrofit, *Trans IChemE*, 75(A): 349.

Calandranis, J. and Stephanopoulos, G. (1986). Structural operability analysis of heat exchanger networks, *Chem Eng Res Des*, 64(5): 347–364.

Cerda, J. and Westerberg, A. W. (1983). Synthesizing heat exchanger networks having restricted stream/stream matches using transportation problem formulations, *Chem Eng Sci*, 38(10): 1723–1740.

Cerda, J., Westerberg, A. W., Mason, D. and Linnhoff, B. (1983). Minimum utility usage in constrained heat exchanger networks – A transportation problem, *Chem Eng Sci*, 38(3): 373–387.

Colberg, R. D., Morari, M. and Townsend, D. W. (1989). A resilience target for heat exchanger network synthesis. *Comp Chem Eng*, 13(7): 821–837.

Floudas, C. A. and Grossmann, I. E. (1987). Synthesis of flexible heat exchanger networks for multiperiod operation, *Comp Chem Eng*, 11(2): 123–142.

Floudas, C. A., Ciric, A. R. and Grossmann, I. E. (1986). Automatic synthesis of optimum heat exchanger network configurations, *AIChE J*, 32(2): 276–290.

Gundersen, T. and Naess, L. (1988). The synthesis of cost optimal heat exchanger networks – An industrial review of the state of the art, *Comp Chem Eng*, 12(6): 503–530.

Hewitt, G. F. (ed.) (2002). *HEDH: Heat Exchanger Design Handbook*, 4 volumes, ISBN 1-56700-181-5 (also available separately). Begell House Inc, Redding, CT, USA.

Kotjabasakis, E. and Linnhoff, B. (1986). Sensitivity tables for the design of flexible processes (1) – How much contingency in heat exchanger networks is cost-effective, *Chem Eng Res Des*, 64(3): 197–211.

Kotjabasakis, E. and Linnhoff, B. (1987). Better system design reduces heat-exchanger fouling costs, *Oil Gas J*, 49–56, September.

Kuppan, T. (2000). *Heat Exchanger Design Handbook (Mechanical Engineering Series)*. ISBN: 0824797876. Marcel Dekker, New York.

Linnhoff, B. and Hindmarsh, E. (1983). The pinch design method of heat exchanger networks, *Chem Eng Sci*, 38(5): 745–763.

Linnhoff, B. and Kotjabasakis, E. (1986). Process optimization: downstream paths for operable process design, *Chem Eng Prog*, 23–28, May.

O'Young, D. L., Jenkins, D. M. and Linnhoff, B. (1988). The constrained problem table for heat exchanger networks. *IChemE Symp Series* 109, 75–116.

O'Young, L. (1989). *Constrained Heat Exchanger Networks: Targeting and Design*, PhD. Thesis, University of Manchester (UMIST), UK.

Saboo, A. K., Morari, M. and Woodcock, D. C. (1985). Design of resilient processing plants: VIII. A resilience index for heat exchanger networks, *Chem Eng Sci*, 40(8): 1553–1565.

Saboo, A. K., Morari, M. and Colberg, R. D. (1986). RESHEX – An interactive software package for the synthesis and analysis of resilient heat exchanger networks, Part I: Program description and application, *Comp Chem Eng*, 10(6): 577–589. Part II: Discussion of area targeting and network synthesis algorithms, *Comp Chem Eng*, 10(6): 591–599.

Shenoy, U. V. (1995). *Heat Exchanger Network Synthesis; Process Optimisation by Energy and Resource Analysis*. Gulf Publishing Co, Houston, Texas, USA.

Sinnott, R. K. (2005). *Chemical Engineering Design*. Coulson and Richardson's Chemical Engineering, Vol. 6, 4th edition. Elsevier Butterworth-Heinemann, Oxford, UK.

Smith, R. (2005). *Chemical Process Design and Integration*. John Wiley & Sons Ltd, Chichester, UK.

Suaysompol, K. and Wood, R. M. (1991). The flexible pinch design method for heat exchanger networks, 1. Heuristic guidelines for free hand design, 2. FLEXNET – heuristic searching guided by the A* algorithm. *Trans IChemE Part A (Chem Eng Res Des)*, 69: 458–464 and 465–470.

Tjoe, T. N. and Linnhoff, B. (1986). Using pinch technology for process retrofit, *Chem Eng*, 47–60, April 28.

Trivedi, K. K., O'Neill, B. K., Roach, J. R. and Wood, R. M. (1989). A new dual-temperature design method for the synthesis of heat exchanger networks, *Comp Chem Eng*, 13: 667–685.

5 Utilities, heat and power systems

5.1 Concepts

5.1.1 Introduction

The utility systems supply the heating and cooling needs which cannot be met by heat exchange, and Section 3.4 introduced the concept of multiple utilities. In this chapter we take this one stage further by looking at the interaction of heating and cooling utilities with power requirements, covering the following:

- Combined heat and power (CHP) systems, which generate power while simultaneously providing hot utility (Section 5.2).
- Heat pumps, using power (or high-grade heat) to reduce both hot and cold utilities (Section 5.3.1).
- Refrigeration systems, providing below-ambient cooling requirements and consuming power (Section 5.3.2).
- Total site analysis, optimising the heating, cooling and power requirements of an entire multi-plant complex (Section 5.4).

Economic aspects are included, as are case studies (Sections 5.5 and 5.6).

5.1.2 Types of heat and power systems

Most chemical processes and their associated sites do not just require heat; they require power as well. This power may be used to drive electric motors, pumps, or compressors, power for instruments and visual displays, and lighting. Most sites pay to import this power in the form of electricity from an external supply company, but the power itself must ultimately be generated somehow. In some countries, a significant proportion is produced by hydroelectric or tidal power or using other renewable sources. However, in most cases, the vast majority of power is generated from heat engines.

A heat engine is a device for converting heat into power. The high-temperature heat is provided by burning coal, oil, natural gas or other fossil fuels or combustible materials; alternatively, it may be supplied from a nuclear reaction. In most power stations, the heat is used to evaporate water to make high-pressure steam. This steam is then passed into a turbine and exerts a force on the blade to rotate the turbine and produce shaft power. The exhaust steam emerges at low pressure. Often it is condensed and cold water is re-cycled to the boilers to be re-used. Latent heat of condensation of the steam is therefore wasted. As a result, the thermal efficiency of these processes (power produced divided by heat supplied from fuel) is at most 40%.

There are other kinds of heat engines. For example, the internal combustion engine burns diesel oil, petrol (gasoline) or natural gas, producing power and releasing heat in the exhaust gases and in the water required to cool the cylinders. Likewise, in the gas turbine, fuel is burnt in a stream of compressed air to produce hot gas at a high pressure; this is then passed through a turbine, produces power and emerges as hot gas at low pressure and about 500°C. Again, these processes are only about 40% efficient or less in producing power.

The low generating efficiency of heat engines means that a substantial amount of heat is produced and wasted. However, typical process plants have heat as well as power requirements. Why should we not use a heat engine on our site to produce power and simultaneously use the heat it rejects as hot utility on our processes, thus giving a much more efficient system?

This is the concept of CHP. However, such a system must be carefully designed to ensure that any heat produced is at a useful level. For example, the exhaust steam from a typical stand-alone power station is condensed at 40°C, which is far too low to provide any useful heating duties. By "backing off" the power production a little, steam can be condensed at 100°C instead, which is sufficient to supply hot water to a district heating scheme. Furthermore, steam can be drawn from the turbine at higher temperatures and used directly for process heating needs, albeit with a further loss in power production. We need to find ways of tailoring a CHP system to supply heat at the temperatures we require on the plant. This section tells us how.

It is also important to be aware of other systems which link heat and power needs. Heat pumps are the primary example. These generally work as a reversed heat engine, using a power input to upgrade heat from a low temperature to a higher one. Heat pumps also include vapour recompression systems.

Finally, we note that above-ambient hot utility requirements are provided, directly or indirectly, by some form of combustion process. Below-ambient heating and above-ambient cooling can use ambient conditions as a final heat sink or source (e.g. cooling water systems ultimately reject heat to ambient). This leaves below-ambient cold utility; there is no simple equivalent to combustion to provide this (although absorption can do so in some circumstances – Section 5.3.1). Instead, the heat must be upgraded from the process temperature (e.g. −20°C) to ambient temperature by a refrigeration system. Power will be required for this, and so it is a form of heat pump.

5.1.3 Basic principles of heat engines and heat pumps

The thermodynamic basis of the heat engine is quite simple, as shown in Figure 5.1(a). It operates between a heat source at temperature T_1 and a heat sink at some

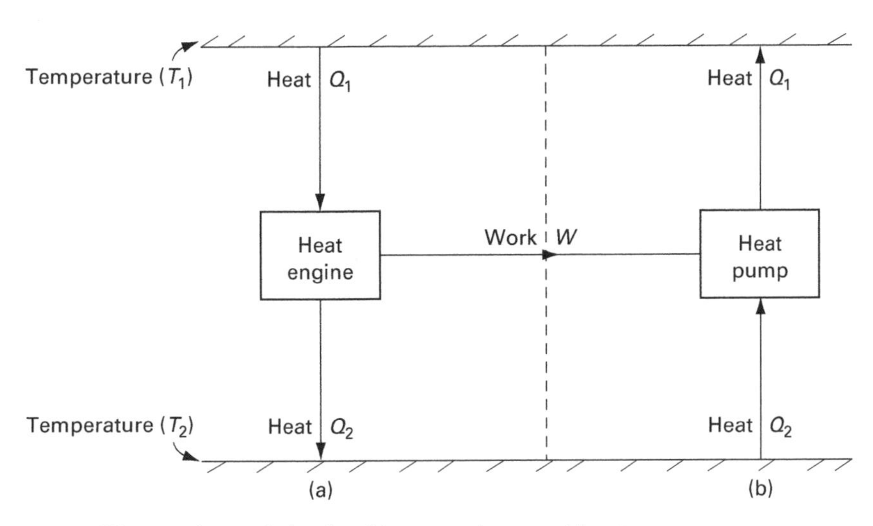

Figure 5.1 Thermodynamic basis of heat engines and heat pumps

lower temperature T_2. It takes heat Q_1 from the source, rejects heat Q_2 to the sink and produces work W. By the First Law of Thermodynamics:

$$W = Q_1 - Q_2 \tag{5.1}$$

The Second Law of Thermodynamics (Carnot's equation) states that all the heat cannot be converted into work, and there is a definite upper limit governed by the temperatures between which the heat engine operates:

$$W = Q_1 \eta_{mech} \frac{(T_1 - T_2)}{T_1} \tag{5.2}$$

η_{mech} in Equation (5.2) is the mechanical efficiency of the system, which is one for thermodynamically ideal or "reversible" engines. All real engines fall short of this ideal. The factor $(T_1 - T_2)/T_1$ is the Carnot efficiency η_c, which represents the maximum possible conversion of heat to work. T_1 and T_2 are absolute temperatures (in Kelvins or degrees Rankine).

It follows from the Carnot equation that for a given heat source temperature T_1, if the heat sink temperature T_2 is increased, the power produced falls. Since, in a heat and power system, T_2 becomes the level at which that hot utility is supplied, correct choice of T_2 is vital.

A heat pump is simply a heat engine running in reverse, as illustrated in Figure 5.1(b). It accepts heat Q_2 from the sink at T_2, rejects heat Q_1 into the source at T_1 and consumes work W. Again, by the First Law of Thermodynamics, Equation (5.1) applies and by the Second Law:

$$W = \frac{Q_1}{\eta_{mech}} \frac{(T_1 - T_2)}{T_1} \tag{5.3}$$

It is usual to define overall efficiencies for real heat engines and heat pumps, according to:

$$W = \eta Q_1 \tag{5.4}$$

with $\eta < \eta_c$ for heat engines, and $\eta > \eta_c$ for heat pumps. Equation (5.4) can be rewritten for heat pumps as:

$$W = \frac{Q_1}{COP_p} = \frac{Q_2}{COP_r} \tag{5.5}$$

where COP_p is the so-called "coefficient of performance" based on heat output Q_1 from a heat pump or vapour recompression system, and COP_r is the coefficient for a refrigeration system based on heat abstracted from the process Q_2. From Equations (5.1), (5.3), (5.4) and (5.5), we see that:

$$COP_p = \frac{Q_1}{W} = \frac{1}{\eta} = \frac{\eta_{mech} T_1}{(T_1 - T_2)} \tag{5.6}$$

$$COP_r = \frac{Q_2}{W} = \frac{Q_1 - W}{W} = COP_p - 1 \tag{5.7}$$

$$COP_r = \frac{Q_2}{W} = \frac{1 - \eta}{\eta} = \frac{\eta_{mech} T_2}{(T_1 - T_2)} \tag{5.8}$$

Hence, a low temperature lift $(T_1 - T_2)$ gives a high COP_p and a large amount of heat upgraded per unit power. However, lower values of T_2 reduce COP_r, so refrigeration systems need more power per unit of heat upgraded as the absolute temperature falls.

With the above understanding, we can begin analysing the combined power and heat recovery problem.

5.1.4 Appropriate placement for heat engines and heat pumps

The heat produced from a heat engine can become a process hot utility; the heat removed by a heat pump becomes a process cold utility (and the heat rejected can become a hot utility). So we can expect the principles outlined earlier on utility placement to apply here.

Figure 5.2 shows a schematic diagram of a process, divided in two at the pinch temperature. The region above the pinch requires less heating H and regions below the pinch requires net cooling C. We will try to supply some of the heating needs by a heat engine producing work W. If this is to produce heat at a useful temperature, its efficiency will be somewhat less than that of a power station; we will assume that it is 33%, so that 3 W units of fuel are required and 2W units of exhaust are produced. If the heat engines were run separately from the process, then, a total of $(H + 3W)$ hot utility and $(C + 2W)$ cold utility would be required. In Figure 5.2, the heat engine is releasing heat to the process instead of rejecting it

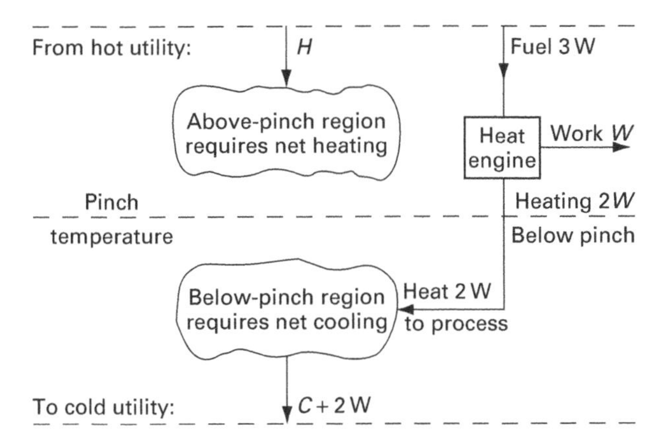

Figure 5.2 Inappropriate placement of a heat engine

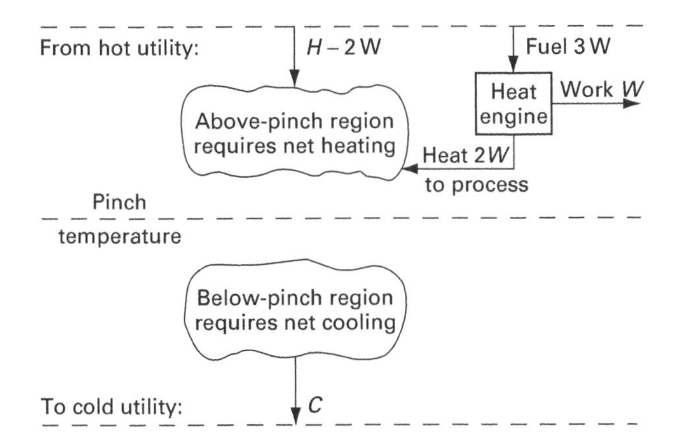

Figure 5.3 Appropriate Placement of a heat engine

to cooling water. However, if we add up the hot utility and cold utility requirements, we find that they are no different to the stand-alone case. The CHP system has made no net saving. What has gone wrong?

The answer is revealed when we see where the waste heat from the heat engine is going. It is being supplied to the process below the pinch. But we already know that the process does not require heating there – this breaks one of the three golden rules. So the heat engine is wrongly placed in thermodynamic terms.

Now let us change the system so that the heat engine rejects heat above the pinch, as shown in Figure 5.3. Now, the exhaust heat provided to the region above the pinch gives a direct saving in the heat which must be supplied to this region. The total hot utility requirement has now fallen to $(H + W)$ while the cold utility has fallen back to C. Since we are producing work W, we are in effect converting heat to power at 100% marginal efficiency! However, this does not violate the Second Law of Thermodynamics. The 100% comes from a *comparison* of two systems, one

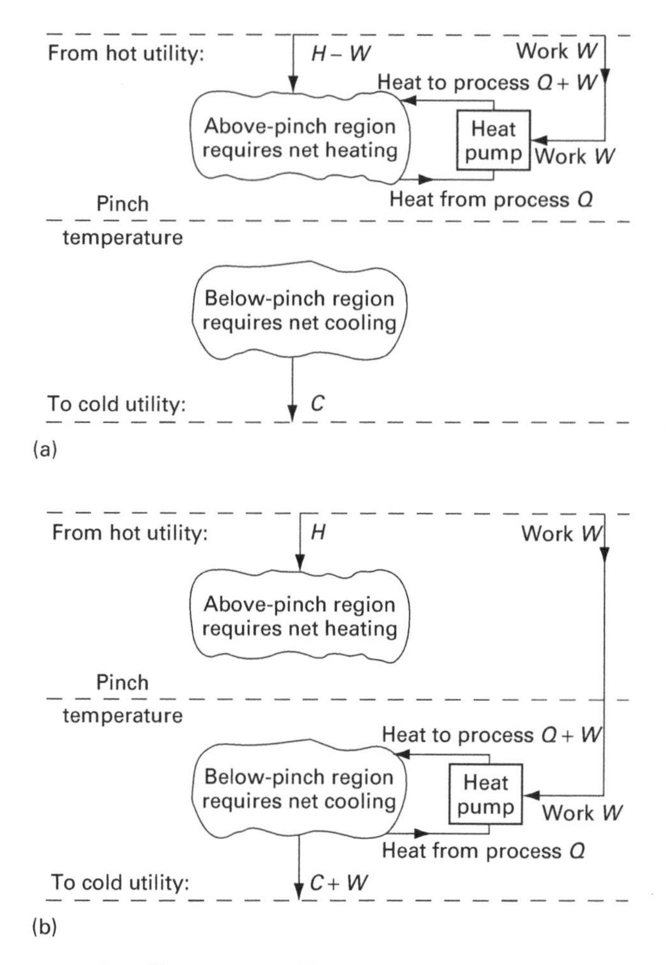

Figure 5.4 Inappropriate Placements of heat pumps

of which is more efficient in absolute terms than the other. (In practice, "secondary" energy losses, such as power conversion (generator) losses, mean that the conversion is not quite 100%; but the "100%" is useful for conceptual understanding.)

The only proviso is that the process must be able to make use of all the heat supplied by the heat engine at the exhaust temperature. In other words, the new low-temperature utility must obey the Appropriate Placement principle; it must be not only above the pinch, but also above the grand composite curve. So the choice of heat engine will depend on the required utility load and temperature level, and this is discussed in Section 5.2.

The Appropriate Placement concept is readily extended to heat pumps. In Figure 5.4(a), a heat pump is shown placed entirely above the pinch. It can be seen that all this heat pump succeeds in doing is replacing a quantity of hot utility W by work W, which will rarely (if ever) be a worthwhile swap. In Figure 5.4(b) a heat pump operates entirely below the pinch, converting work W into waste heat W. This has been

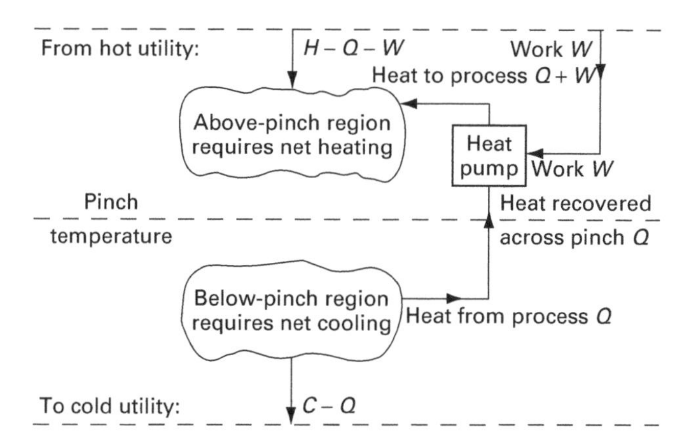

Figure 5.5 Appropriate Placement of a heat pump

likened to putting an immersion heater in a cooling tower! Energy is saved overall only when the heat pump operates *across the pinch* as shown in Figure 5.5, that is when it pumps *from the process source to the process sink.* Clearly, Appropriate Placement for heat pumps means placement across the pinch, when the "normal" efficiency is obtained. Placement on either side of the pinch is inappropriate, leading to energy wastage. The broader principle also again applies – heat must be removed below the grand composite curve and supplied above the grand composite curve.

5.2 CHP systems

5.2.1 Practical heat engines

The usual method of generating industrial CHP is to use a heat engine which burns fuel, generates shaft work and also produces heat which can be used to heat the process. Three types of machine may be used, all of which are illustrated in Figure 5.6:

(a) *Steam turbines (Rankine cycle)*: High-pressure steam is generated in a boiler and is let down to the lower temperatures and pressures required for the site processes through one or more passout turbines. Power is generated from the shaft of the turbine. A variety of fuels can be used to heat the boiler.

(b) *Gas turbines*: Fuel (usually natural gas) is burnt in compressed air in a furnace. The resulting gases, at high pressure and temperature (typically 1,000°C) are passed directly through a turbine and generate shaft work, although about two-thirds of this is needed to drive the compressor. The exhaust gases emerge at 450–550°C and can be used to provide process heating, sometimes directly but more commonly by raising steam in a waste heat boiler.

(c) *Reciprocating engines*: Fuel is burnt in an internal combustion engine, generating power by the usual piston-and-crankshaft arrangement. The hot exhaust gases, typically at 300–400°C, can provide process heat. The engine cylinders

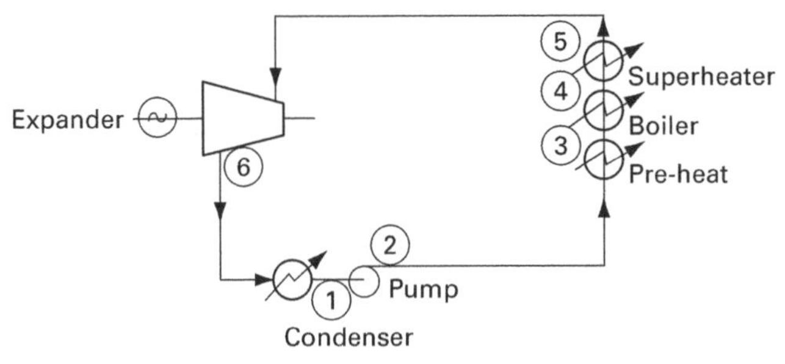

(a) Steam Rankine cycle

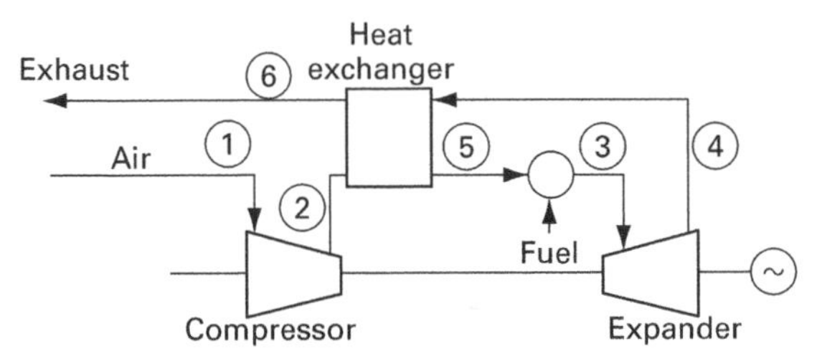

(b) Simple gas turbine with heat exchanger

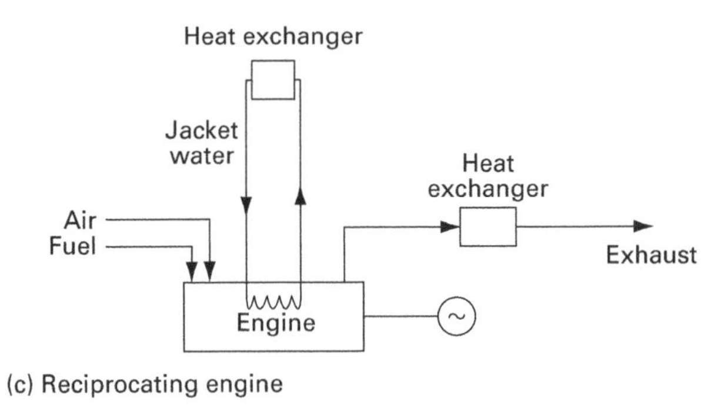

(c) Reciprocating engine

Figure 5.6 Schematic diagrams of heat engine systems

also require substantial amounts of cooling, almost invariably by water at 70–95°C. The fuel may be diesel oil, interruptible natural gas or a combination of these in a dual-fuel engine.

5.2.2 Selection of a CHP system

The three alternatives can be ranked in order of the ratio of power produced to exhaust heat supplied to the process. The differences are summarised in Table 5.1 below and can also be effectively illustrated by Sankey diagrams (Figure 5.7).

Examples of all three types are in industrial operation. The choice between them depends on the size and heat-to-power ratio of the site, the potential for exporting

Table 5.1 Combined heat and power systems

CHP system	*Power/ heat ratio*	*Heat/ power ratio*	*Typical size MW(th)*	*Typical size MW(e)*	*Main heating range (C°)*
Steam turbines	<0.2	>5	3.0–50.0	0.5–10.0	100–200
Gas turbines	0.67–0.2	1.5–5	2.0–30.0	1.0–20.0	100–500
Diesel/gas engines	1.25–0.5	0.8–2	0.2–5.0	0.2–5.0	100–300, <80

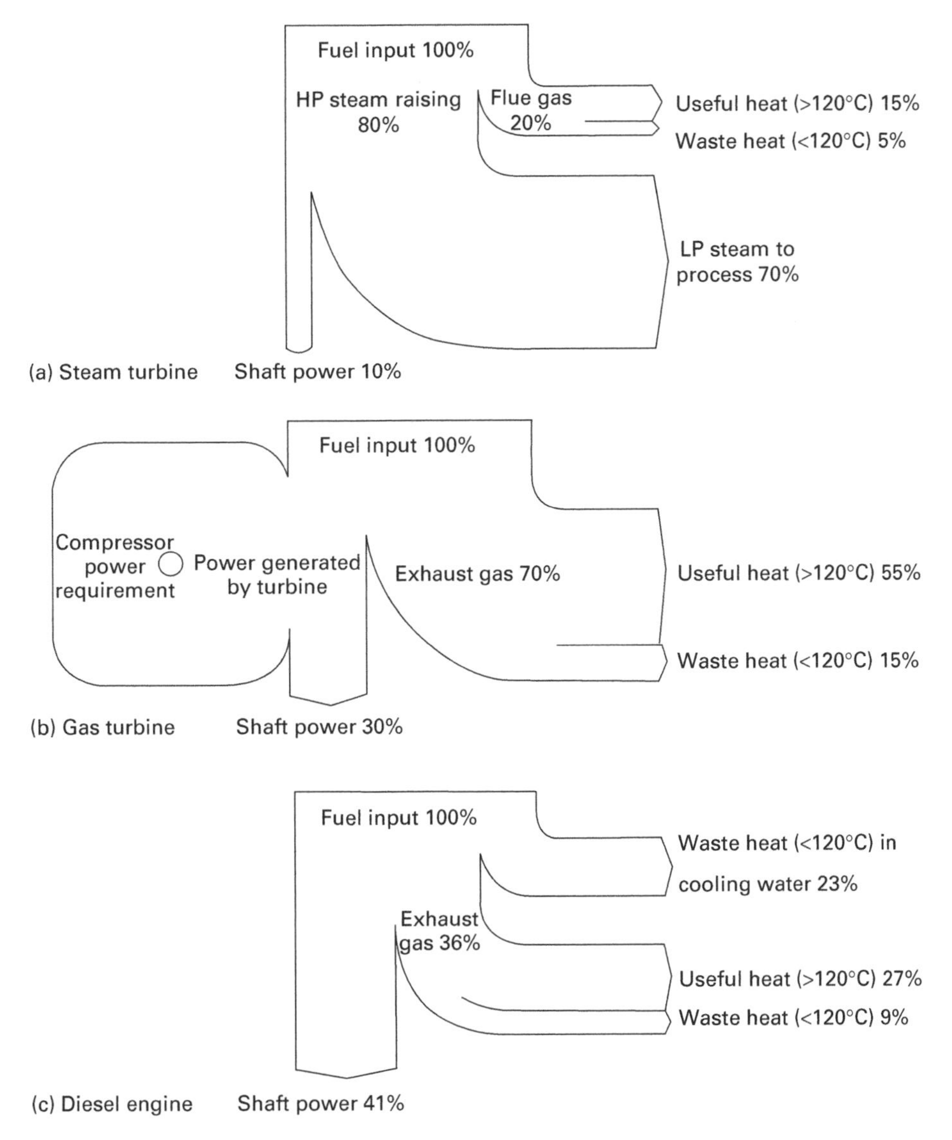

Figure 5.7 Sankey diagrams for CHP

power, and the temperatures at which the process requires heat. Again, the grand composite curve is a key tool.

Steam turbines produce large quantities of heat at moderate temperatures. Little power is generated by letting high-pressure steam down to pressures above about 15 bar, corresponding to saturation temperatures above 200°C. Hence, they are not a good choice where much of the hot utility is required above this level. Conversely, if most site heat requirements are for heat loads at fairly constant temperatures in the 100–200°C range, for example for evaporation loads or even heating by direct steam injection, a steam turbine will be very suitable. They are generally large units and are best suited to sites with heat loads above 10 MW. Steam systems plot on the grand composite curve as a set of multi-level constant-temperature utilities (Section 3.4.3).

Gas turbines release heat from their exhaust gas over a wide range of temperatures from 600°C downwards. They are thus especially suitable for sites with high-temperature heat requirements, for example for hot-air dryers or high-temperature reactions. For a site with a given heat requirement, they produce much more power than steam turbines. Until 1980, most commercially available units were in the 10–50 MW size range, but since then moderately sized gas turbines in the 1–10 MW range have appeared, derived from aero-engines, and these are very suitable for many industrial sites. However, performance on turndown is poor and gas turbines are not an option for some isolated sites where there is no adequate large gas main nearby. Gas turbines are variable-temperature utilities and, if the heat is used directly in the process, should be handled in the same way as flue gas (Section 3.4.5). For the more common situation where steam is raised in a waste heat boiler, this steam then acts as a constant-temperature utility; remember that there is a further ΔT penalty in raising the steam. The heat below the process pinch is wasted; it cannot even be used to preheat the air to the gas turbine as warm air has a lower density and the initial compressor then requires more power. Although Figure 5.6(b) shows a heat exchanger recovering heat from the turbine exhaust, this has to come after the compressor, when the air is already relatively hot and almost invariably above the pinch.

Diesel and gas engines also produce a large quantity of power, but relatively little process heat, although some of this is available at temperatures up to 400°C. For diesel engines in particular, heat recovery may be further limited as acid gas condensation and dewpoint corrosion can occur if the exhaust gas is cooled below about 200°C. Reciprocating engines are best suited to processes with a low pinch temperature and significant heat loads below 100°C (e.g. for space heating – central heating circulation systems typically operate in the 70–90°C range), as the heat from the water jacket can then be usefully used. Otherwise, only a small amount of heat available from the exhaust gas can be used and the overall thermal efficiency is little better than that of a stand-alone system. Originally, they were much smaller units than turbines, typically in the 50–500 kW range, and had to be used in multiple for large power outputs. Modern engines can be much larger; for example, the dual-fuel engine installed at Southampton (Section 5.6.2) is rated at 5.7 MWe. Even so, above about 5–10 MW, gas turbines would generally be preferred. The exhaust gas plots on the GCC as a variable-temperature utility like flue gas (Section 3.4.5); again, it is impossible to preheat the air input to the engine. The cooling water acts as a variable-temperature utility with recirculation, like a hot oil system.

Hence, a simple selection checklist could be:

1. What is the site power requirement? If it is below 1 MW, prefer reciprocating engines; above 5 MW, prefer gas and steam turbines.
2. Check the site heat-to-power ratio and compare with the values in Table 5.1. In recent years, there has been a tendency for electrical loads on sites to increase and for heat requirements to stay static or fall. As a result, the heat-to-power ratio of sites has risen, favouring reciprocating engines and gas turbines against steam turbines.
3. Compare the heat release profile of the CHP systems with the process GCC above the pinch to see which gives the best match. In particular:
   - If there are high-temperature heating loads above 200°C, this favours gas turbines and reciprocating engines.
   - Pinch temperatures above 70°C disfavour reciprocating engines.
   - Significant low-temperature heat loads, below 100°C, favour reciprocating engines. Note that these heat loads are often "non-process", for example space heating or domestic hot water, and are often omitted from the stream data set. It is worth checking to see whether these requirements exist.

Figure 5.8 illustrates examples of each type of system in a situation where they are well matched to the process heat requirements.

Steam turbines were once the dominant CHP system. However, virtually none have been installed recently, because of their low power production and the major advances in gas turbines. As a broad generalisation, many industrial processes have a pinch temperature at or slightly above 100°C, and gas turbines are now the normal choice for these plants. Where steam is preferred as the process heating medium, it may be raised in a waste heat boiler fed by the exhaust gases from the gas turbine. For large sites, this steam may be used to generate further power in a Rankine cycle, giving a so-called combined cycle system (Section 5.2.3.3); the high capital cost means that this is rarely economic below 10 MW. Conversely, some sites with an existing steam turbine CHP system have had a gas turbine retrofitted; the gain in power production justifies the capital cost.

If a gas turbine is the desired option but the required heat-to-power ratio is higher than the normal range, an easy and effective solution is to add supplementary firing. Burners are mounted in the exhaust gas duct to boost the temperature; either separate combustion air is supplied, or the burners can run on the remaining oxygen in the combustion gases. Temperatures of up to 850°C can be obtained, this being the practical upper limit due to materials considerations.

For buildings, or small industrial sites where much of the heat requirement is for space heating, the situation is completely different. The pinch is often at ambient temperature and the heat loads are rarely above 1 MW. Hence, reciprocating engines are the most popular form of CHP in this case.

Finally, we note that it is not necessary to generate electric power from the shaft work produced by our CHP system; it can also be used as direct mechanical drives for pumps, compressors, etc. Efficiency can be slightly higher than for generating electrical power, but the latter can be a more flexible option; the two are frequently mixed.

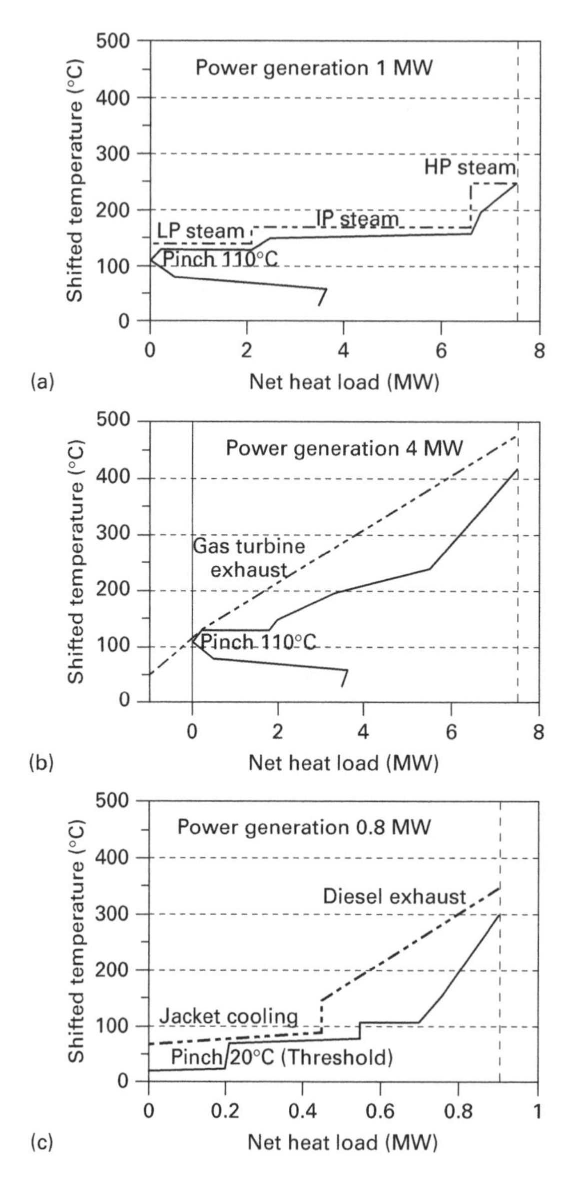

Figure 5.8 Typical well-matched utilities and GCC's

5.2.3 Refinements to site heat and power systems

5.2.3.1 *Optimising a steam Rankine cycle*

The classic steam Rankine cycle is shown in Figure 5.9. This has been used in large numbers of electricity generating stations, whether coal-fired, oil-fired, nuclear or based on other fuels, and has been refined to give the optimal level of stand-alone

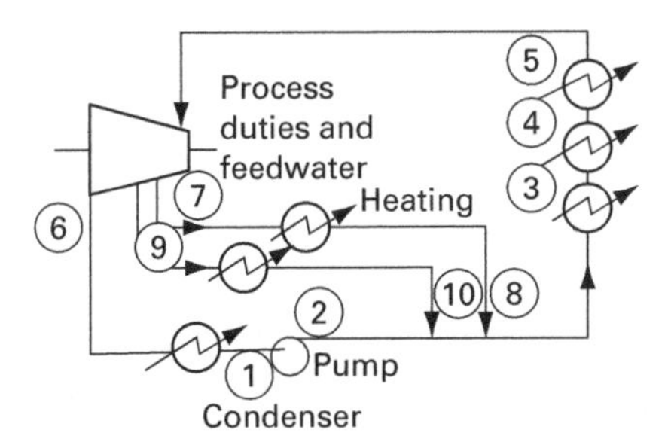

Figure 5.9 Flowsheet for steam Rankine cycle with intermediate steam levels and economiser

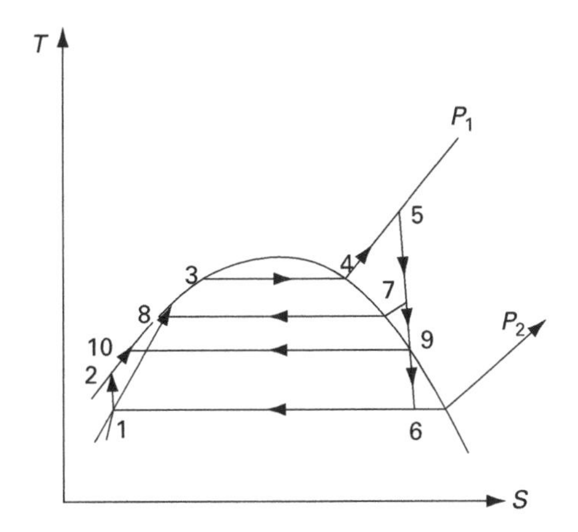

Figure 5.10 Temperature-entropy diagram for Rankine cycle with two intermediate steam levels

power-generation efficiency by increasing its complexity. Steam is produced superheated and at high pressure by the boiler, and then expanded through a cascade of back-pressure power turbine stages. For an on-site power station, IP and LP steam is drawn off as necessary to meet site needs. For a stand-alone generating station, most of the steam is expanded to sub-atmospheric pressure, and then condensed against cooling water; these condensing turbines are optional for a site CHP system.

Likewise, there are cold streams which need to be heated – the boiler feedwater and the combustion and secondary air to the furnace. The operating preference will be to heat these from the steam system itself, rather than by exchange with one of the processes supplied (another example of zoning and keeping sub-systems self-contained). Incoming fresh makeup water and any returned condensate is deaerated by the lowest back-pressure steam level, and may then be heated successively by hot condensate, heat from the boiler blowdown, LP and IP back-pressure steam, and finally the hot furnace flue gas in an economiser, as shown in the flowsheet in Figure 5.9 and the temperature-entropy diagram (Figure 5.10). The boiler feedwater will typically have been raised to the IP steam temperature level (say 200°C) before entering the economiser, so the flue gas leaving the economiser cannot be below this temperature. Since it still contains much useful heat, it is then matched against the incoming cold air streams. The eventual flue gas exhaust temperature is typically 140°C, giving an effective boiler efficiency of 90% or more.

Many "variations on the theme" introduced above are used in practice, and the reader can refer to standard texts on engineering thermodynamics, such as Haywood (1991), for more information. Additional back-pressure levels may be added solely for the purpose of feedwater preheating (often known as "bled steam"). This further increases the internal recirculation of steam in the system, thereby increasing power output and improving the cycle efficiency. Despite all this, the efficiency of stand-alone steam power generation rarely exceeds 40%, with the vast majority of the heat being thrown away in cooling towers after the steam has been condensed. The power generation of a site steam turbine CHP system will be lower, as less power is generated from IP and LP turbines than condensing sets, but this is outweighed by the opportunity to make good use of the heat in the IP/LP steam and reclaiming the "lost" 60%.

It is also possible to use waste heat from the process to heat boiler feedwater. This either reduces the high-temperature heat requirements or gives better power production for a given furnace use (because IP/LP steam previously used for BFW heating can now be let down to condensing pressure). However, the steam system is no longer self-contained. Another option is to heat incoming cold air with process waste heat; this has particular advantages if there are worries about lowering the flue gas temperature too far because of possible condensation or stack corrosion.

The power generated from a steam turbine working between given temperatures can be calculated by initially assuming an isentropic expansion and then allowing for reasonable entropy losses (15% is typical). Table 5.2 shows sample data (based on ESDU 1989) for a Rankine cycle supplied with 40 bar steam (saturation temperature 250.3°C) superheated to 500°C, with a turbine isentropic efficiency of 85%. Power production rises steadily as the exhaust steam level falls; in fact, the last column shows that power produced is roughly proportional to the fall in saturation temperature.

5.2.3.2 *Sizing a gas turbine system*

A given gas turbine has a fairly narrow range of heat-to-power ratio, which is unlikely to match exactly with the site requirements. Also, site heat and power needs are usually variable, whereas gas turbines tend to have poor turndown and run best at or near full load. Gas turbines can be sized either to meet the heat requirement or the power requirement. In effect there are four options:

1. Heat output of CHP system matched to site, CHP power output less than site power demand. Import additional power.
2. Heat output of CHP system matched to site, CHP power output greater than site power demand. Export excess power. Economics depend strongly on the price paid by external companies for power supplied in this way.

Table 5.2 Power produced from a steam turbine supplied with 40 bar superheated steam at 500°C

Steam pressure bar	*Saturation temperature (°C)*	*Steam exhaust temperature (°C)*	*Shaft work produced (kWh/tonne steam)*	*Power/ΔT_{sat} (kWh/°C)*
40	250	500, superheated	0	–
30	234	459, superheated	23	1.44
20	212	404, superheated	53	1.40
10	180	309, superheated	98	1.40
5	152	247, superheated	136	1.39
2.25	124	176, superheated	173	1.37
1	100	114, superheated	206	1.37
0.44	78	78, saturated	236	1.37
0.2	60	60, saturated	263	1.38
0.074	40	40, saturated	292	1.39

3. Power output of CHP system matched to site, CHP heat output less than site heat demand. Make up the heat deficit with package boilers, or by supplementary firing to add further heat to the turbine exhaust gases.
4. Power output of CHP system matched to site, CHP heat output greater than site heat demand. Normally undesirable as the excess heat must be thrown away and this part of the CHP system is then merely competing with stand-alone power generation. Sometimes worthwhile if power prices are very high, for example for "peak-lopping".

The choice between these alternatives will depend on the system economics (Section 5.2.4). The CHP system may operate in different modes at different times, depending on site demands.

The exact heat-to-power ratio for a gas turbine depends on the manufacturer and model. However, for initial sizing purposes, an average performance for typical gas turbines is as follows, assuming a power-generation efficiency of 30% and an exhaust gas temperature of 500°C:

- For fuel use of 1 MW, 3.6 GJ/h.
- Power produced (30%) = 0.3 MW.
- Heat released (66%) = 0.66 MW over temperature range 500–20°C; mean CP of exhaust gases = 1.375 kW/K = 0.001375 MW/K, corresponding to a gas flow of 1.25 kg/s at Cp = 1.1 kJ/kgK.
- Heat losses (4%) = 0.04 MW.

The temperature-entropy diagram for a gas turbine is shown in Figure 5.11.

5.2.3.3 *Combined cycle power generation*

The heat from a gas turbine exhaust often needs to be converted to steam as this is usually the most convenient way of supplying heat to plants (e.g. because of its excellent heat transfer coefficients when condensing, leading to compact and cheap

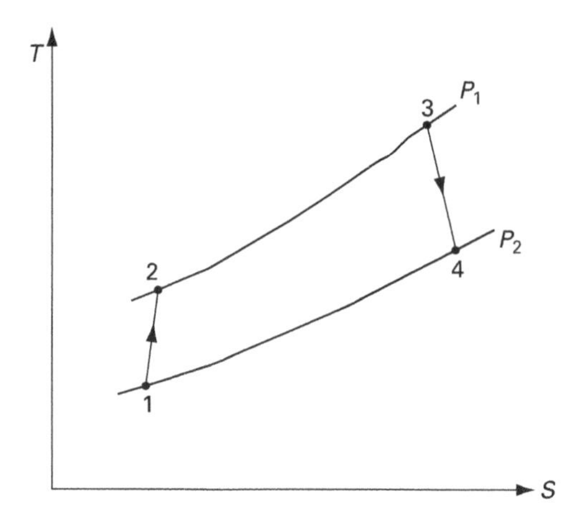

Figure 5.11 Temperature-entropy diagram for gas turbine system

heat exchangers). Since this heat is released at a range of temperatures from 500°C downwards, it can be used to raise steam at a variety of pressures including very high pressure (up to about 40 bar). If there is little or no site requirement for high-pressure steam, it can be let down through a steam turbine in just the same way as with a Rankine cycle. The power generated is usually a moderate proportion of that produced in the main gas turbine – typically 10–20% – but valuable nevertheless. The economic benefit of the extra power must be set against the additional capital cost of the steam turbine, and the payback is usually slightly longer. However, virtually all the modern gas-turbine stations installed in recent years by power-generating companies in the UK are combined cycle, due to their large scale and long projected life.

An alternative type of combined cycle is to duct the exhaust gases from the gas turbine directly into the steam-raising boiler for a Rankine cycle, where they act as a source of preheated air. These gases still contain around 14 vol% oxygen and can therefore support combustion, although some fresh air is normally mixed in. The strategy to be adopted in deciding on how to split the fuel and air flows between the gas turbine and steam boiler will depend on the required heat-to-power ratio.

5.2.3.4 *Diesel and gas engines*

As with gas turbines, the exact heat-to-power ratio depends on the machine used, but a typical performance can be given. Here, we assume a power-generation efficiency of 40% and an exhaust gas temperature of 400°C:

- For fuel use of 1 MW, 3.6 GJ/h.
- Power produced (40%) = 0.4 MW.
- Heat in exhaust gas (34%) = 0.34 MW over temperature range 400–20°C; mean CP of exhaust gases ~0.9 kW/K~0.0009 MW/K, gas flow ~0.8 kg/s at Cp = 1.1 kJ/kgK.
- Heat in cooling water (22%) = 0.22 MW.
- Heat losses (4%) = 0.04 MW.

5.2.3.5 *Distributed heating and power generation*

The traditional steam supply arrangement for a large site was to use a large centralised boiler house and distribute the steam produced via steam mains at a number of different pressures. A typical case would be to have HP steam at 600 psig (42 bar absolute, saturation temperature 253°C), IP steam at 100 psig (7.8 bara, saturation temperature 169°C) and LP steam at 10 psig (1.7 bara, saturation temperature 115°C). The perceived advantages of a central boiler house with multiple boilers (with or without an associated power-generation system) were:

(a) flexibility in operation – boilers could be shut down for maintenance or at periods of low demand while others carried the load;
(b) ability to burn a wide range of fuels, including cheap low-grade fuel oils or low-calorific-value waste-derived fuel;
(c) large absolute savings from economisers and other efficiency improvements on a large centralised system, giving acceptably short payback times for the extra capital cost;
(d) labour savings from centralised operation and maintenance.

What was often not realised was the extent of steam leaks and heat losses in the long steam mains. Kemp (1991) found that less than 50% of the steam generated in one site boiler house reached its destination a mile away, and this horrifying figure is believed to be not untypical of large sites. Obviously such losses completely outweigh any marginal gains in boiler efficiency from the centralised system. Moreover, there is the cost of maintaining the miles of steam mains and their associated steam traps (another frequent source of losses).

A distributed heating system, with smaller local package boilers supplying a single process or group of buildings, therefore has attractions. As well as the reduction or elimination of main losses, this may allow the heat to be supplied at the optimum level for that plant, rather than the levels imposed by the site for HP, IP and LP steam. In the same way, local distributed power generation can be used. Gas reciprocating engines, with their small size, have particular potential advantages for this.

There are also drawbacks to consider. A backup heating system has to be provided for use during planned or unplanned shutdowns; each plant requires separate maintenance; and the associated noise is no longer confined to a single location on site (a particular disadvantage with locating gas engines in occupied buildings, to set against their ability to provide space heating). Nevertheless, the use of decentralised heat and power systems has grown significantly in the last 20 years and this trend seems likely to continue. The study on the hospital site (Section 9.6) is a case in point.

5.2.4 Economic evaluation

The economic evaluation of processes which include CHP systems is difficult and there is some controversy on the best method. We do not just have to consider the economics of the CHP system itself; it can have a major impact on heat recovery projects because the benefits of saving fuel and process heat are altered significantly.

CHP economics depend strongly on the cost of heat, the cost of power, and the ratio between them. Unfortunately, all of these have fluctuated substantially since the 1970s. A period of high fuel costs in the 1970s and early 1980s was followed by a fall in energy prices, particularly for natural gas. In the early 2000s gas prices have abruptly risen again. CHP installations are major long-term projects, and uncertain rates of return can inhibit investment.

In describing CHP concepts initially, we referred to power being generated at close to 100% marginal efficiency. For economic evaluation, however, it is usually more helpful to calculate the marginal cost of hot utility. This is vital when deciding whether it is worthwhile to shift loads between utilities, for example HP and LP steam. Normally there is only a significant cost difference for (a) CHP systems (b) refrigeration levels. The basic definition is:

(Marginal cost of heat) = (Cost of fuel for CHP system) − (Cost of fuel for stand-alone heating) − (Value of power generated).

Power costs will also affect the result. Historically they have varied less widely than fuel costs. However, one important distinction is whether the power produced will replace imported power or be exported (if it exceeds site needs). The price obtained for exported power can be substantially lower than that paid for imported electricity. There can also be variations between day and night tariffs, and between winter and summer. This is discussed further in Section 5.2.4.2.

The ratio of heat-to-power costs also has a major effect. If power supplied from the grid is generated from fossil-fuel power stations, these will typically have generating efficiencies of 30–40% (up to 50% in the case of combined cycles). Primary energy requirements and cost per kW of power can then be expected to be 2–3 times the cost of fuel, and power savings will give a good rate of economic return. The CHP system will generally have a slightly lower efficiency than an optimised stand-alone power-generating system, but the marginal cost of the heat produced will still be much lower than if fuel is simply burnt in a furnace. See Section 5.2.4.6 for an example calculation. However, if electricity costs are low, for example where the grid can be supplied by cheap hydroelectric power, then both primary energy savings and power cost savings are far lower, and the CHP scheme will almost certainly be uneconomic.

Finally, capital costs will affect the payback time. Although these generally rise with inflation, this is not always the case, particularly for equipment where use is growing and economies of scale are beginning to be achieved in manufacturing. This applies to gas turbines and gas engines, for which capital cost increases over the last 30 years have been substantially below inflation.

5.2.4.1 *CHP and process heat recovery*

On many sites, the power station is sized to produce roughly the amount of power required by the site, especially where it is based on a gas turbine or combined cycle. If the site heat-to-power ratio is higher than that for the generating system (which is the usual situation), the excess steam demand is usually met by steam-raising "package" boilers. Any savings in steam demands by better heat exchanger network designs save fuel in the package boilers, leaving the operation of the power station unchanged.

However, if the CHP scheme meets the entire site heat load, the value of savings from better heat recovery can be considerably reduced on any utility produced from CHP (e.g. LP steam or turbine exhaust gas, but not HP steam direct from the boilers). This is because saving a tonne of steam not only saves the fuel required to raise it, but also eliminates the associated power output which is produced at 80–90% marginal efficiency. Hence, since power generated at 80% must then be replaced by power generated at, say 30% (either on the site station or by the external supplier), the saving accruing from a tonne of utility steam saved is much less than that accruing from a tonne saved in a simple boiler; if fuel cost is relatively low compared to power cost, it may even be zero or negative. It is useful to evaluate the "net cost" of low-pressure steam as described in Section 5.2.4.5.

Sometimes an alternative approach can be taken which means that heat recovery projects to save steam are still worthwhile. Considering LP steam, if we simply turn down the boilers and generate less steam, we produce less power and make no net gain. On the other hand, if we leave the boilers at their previous setting and run the steam down to a condensing turbine, we will generate more power than before for no extra fuel. This will be the case where the net cost of power generated in the condensing turbine is negative.

5.2.4.2 *Electricity tariff structures*

On sites with cogeneration, the cost of imported electricity has a profound effect on the operating policy of the site power station. Frequently, the cost of imported power varies according to the time of day and the season of the year. For example, in summer power may cost £15/MWh at night and £30/MWh during the day, and in winter these may rise to £20/MWh at night and £40/MWh in the day. In this situation it may only prove economic to run the power station for part of the day, and to switch it off at night when power is cheap and site heat demand may also be lower.

In recent years a much wider variety of tariff structures have been offered, particularly in the UK with the splitting up of the power generators into numerous competing companies. For example, in some cases there are major price hikes at peak periods (typically late afternoon), and on-site power generation should be maximised at that time. Or the user may pay extra for his peak usage in kilowatts, and he should therefore try to keep his net power use as steady as possible (a brief interruption in on-site power generation may have a serious effect in this case). Clearly, the user should check whether the operating strategy for his power station is compatible with the current tariff system. It may be difficult to vary the power output effectively (especially for units with poor turndown such as gas turbines, or where the associated heat load required by the site is constant and inflexible). In this situation in particular, it may be worthwhile to renegotiate the power supplier's tariff.

Steam turbines can offer the most flexibility of operation, as high-pressure steam can not only be passed through passout turbines to generate power and low-pressure steam, but can also be let down through condensing turbines if further power is required. However, will this be economic?

Suppose the fuel available is fuel oil, costing £150/ton, and having a calorific value (net) of 39,900 kJ/kg, then the cost of heat from this source (ignoring boiler and distribution losses) is:

$$£150 \times 3{,}600/39{,}900 = £13.53/\text{MWh steam}$$

If the "condensing" part of the power cycle has a cycle efficiency of 30%, then the marginal cost of in-house power is:

$$£13.53/0.3 = £45.1/\text{MWh power}$$

This is likely to be more expensive than imported power at any time, even on a winter day, hence it never pays to run the condensing sets. The power deficit should always be made up by importing.

In contrast, the marginal cost of generating power where useful low-pressure steam is produced, assuming a boiler efficiency of 85%, (and ignoring the cost of the steam which is needed anyway), will be:

$$£13.53/0.85 = £15.92/\text{MWh}$$

This is almost always worthwhile, the only question mark arising during summer nights when it might be just as cheap to import electricity. The power station could then simply be used as a source of process heat, with the turbines switched off or just "ticking over". But this assumes that all the extra heat which would have been used to generate power is recovered, either by giving additional LP steam or as a reduction in fuel use in the boilers. This is not always true if HP steam is generated and then simply let down through a valve; superheating and heat losses can occur.

Finally, the power contract may include a "load-shedding" agreement. At certain times of day when the load on the external utility supplier is high, the site may have to reduce its electricity demand at short notice to an agreed minimum. Any electricity imported over and above the minimum during a load-shedding period incurs a severe cost penalty (£20,000/MWh is a typical figure!). It might well not be feasible in this situation for the site to shed enough power demand at the short notice available, and in this case it pays for the power station to generate not only in the condensing sets, but also by blowing off back-pressure steam to atmosphere!

5.2.4.3 *Exporting power*

If a CHP scheme is capable of producing more power than the site needs at any time, savings can only be obtained through sale of surplus power to the external utility supplier. In the past, rates obtainable for exporting power were low, and sometimes the supplier simply would not accept it. Recent changes in legislation have altered this situation considerably, at least in the UK. However, if the extra power cannot be used or exported, substituting low-grade steam demand for high grade cannot produce energy savings, neither can improving the internal power cycle efficiency. All that happens is that fuel burnt in the power station is replaced by fuel burnt in the package boiler!

5.2.4.4 *Fuel value*

Next, we must never lose sight of the fact that a unit of heat (i.e. 1 MJ or 1 kWh) can have quite widely differing costs depending on the fuel source used.

Normally, a site will have more than one fuel available, with (on a common energy basis) different prices. Over recent years in the UK, fuel gas has had a considerable cost advantage over fuel oil. Hence the trend has been to use interruptible natural gas as a "base load" fuel (burnt to the maximum quantity allowed by the gas contract) and fuel oil for any excess heating duty. In some cases coal is cheaply available as well. If the whole of a site demand can be met by a cheap fuel like gas, it may become economically attractive to generate power in-house using condensing sets during the high daytime power tariff period. Previous government policy in the UK was to discourage this by raising the gas price, but this policy was abandoned in the 1990s, causing a "dash for gas". Now that cheap reliable sources are becoming scarcer, the pendulum has swung back at the time of writing (2006). Past experience shows that attempting to forecast future energy prices is an unproductive exercise!

Availability of cheap fuels will greatly improve the economics of a CHP scheme. In some cases the limitation becomes not economic, but practical – the maximum supply rate of gas through the existing mains (if a gas main exists at all – by no means guaranteed in rural locations).

5.2.4.5 *Marginal cost of process heating*

If a CHP system exists or is being considered, it is very useful to evaluate the net cost or marginal cost of process heating. This will often be much lower than the cost evaluated simply in terms of fuel burnt, and can have a profound effect on the economics of heat recovery schemes.

Consider the cost of 1 MWh of process steam generated by different methods.

(a) From a gas-fired package boiler at 80% efficiency.
Take the cost of gas as £3.33/GJ (approximately 33p/therm), which is £12/MWh. At 80% efficiency the cost will be £15/MWh.

(b) With cogeneration via a steam turbine, typically 10% power is produced for a corresponding increase of 11.5% in fuel (marginal efficiency 87%). So fuel cost rises to £16.70, but we produce 0.1 MWh power which is worth £4 at £40/MWh (daytime prices).

The net cost of 1 MWh process heating has therefore fallen to £12.70/MWh.

For a gas turbine, a typical balance is that fuel produces 50% useful heat and 30% power. Hence, to produce 1 MWh process heat requires 2 MWh fuel (costing £24) which generates 0.67 MWh power worth £26.70. So the net cost of 1 MWh process heating is now actually negative, at – £2.70/MWh. In this situation there is no incentive to install heat recovery projects! But remember that this only applies if the CHP system is satisfying all the site heat needs; if package boilers are also being used, steam saved from these is worth the full £15/MWh.

Precise figures for steam costs can be obtained from methods by Varbanov *et al.* (2004) and Smith (2005).

5.2.4.6 *Example: economics for a gas turbine project*

Consider a gas turbine installation generating 6 MW of power and 10 MW of useful heat and consuming 20 MW of fuel. An aero-engine derivative gas turbine or alternator set of this size would cost about £1 million at typical historic prices, typically requiring an installation factor of about two. Hence the total installed capital cost would be about £2 million.

The savings made by the project depend strongly on the electricity tariff and the number of hours worked per year. For single-shift operation, 8 h/day, 5 days a week, the working year is 2,080 hours. Producing 10 MW fuel in package boilers would cost £312,000 p.a., and for a gas turbine CHP scheme the net cost is – £56,000 p.a., giving a net annual benefit of £368,000 and a payback of the order of five and a half years, which is unattractive. However, for continuous three-shift operation, 24 h/day, 7 days a week, 8,000 h/yr (allowing for shutdowns) the net annual benefit rises to £1.47 million with a payback of less than 2 years. Realistically, much of the power saved in this situation will be at the "night rate" which will be nearer £20/MWh than £40/MWh. Assuming that half of the power saved is at this rate gives a final balance sheet as follows:

(a) Fuel burnt in package boilers, 10 MW, £15/MWh, 8,000 h/yr:	£1,200,000
(b) Fuel consumed by gas turbine, 20 MW, £12/MWh, 8,000 h/yr:	£1,920,000
(c) Power generated at day rate, 6 MW, £40/MWh, 4,000 h/yr:	£960,000
(d) Power generated at night rate, 6 MW, £20/MWh, 4,000 h/yr:	£480,000
(e) Net annual benefit from gas turbine, (a − b + c + d):	£720,000

This yields a simple payback of about 3 years. This result is typical of historic gas turbine schemes.

Further sizing and economic calculations can be found in Section 5.5 for the organics distillation unit.

5.2.5 Organic Rankine cycles

There is another possible way of using a heat engine. If the process pinch temperature is high, the waste heat below the pinch may be used as the high-temperature heat source to drive a heat engine which produces power. An example of this is the organic Rankine cycle, which is effectively a steam turbine system but using an organic compound as working fluid.

Figure 5.12 shows one possibility for such a cycle. They usually work below 200°C and hence the Carnot efficiency is lower than for the heat engines previously described; in the example shown it is assumed to be 20%, so that 5 W units of waste heat are required to produce *W* units of work. In Figure 5.12, the heat engine makes no saving compared to the stand-alone case. This is because the heat has been taken from the region above the pinch and transferred across the pinch. The net effect is simply to produce power very much less efficiently than if that heat had been used in a normal stand-alone heat engine.

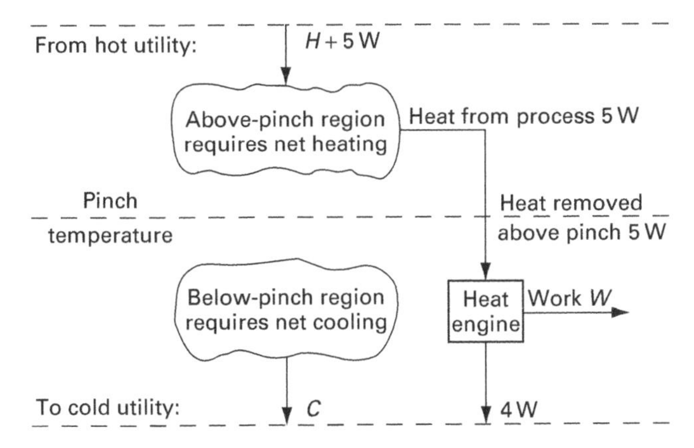

Figure 5.12 Inappropriate placement of an organic Rankine cycle

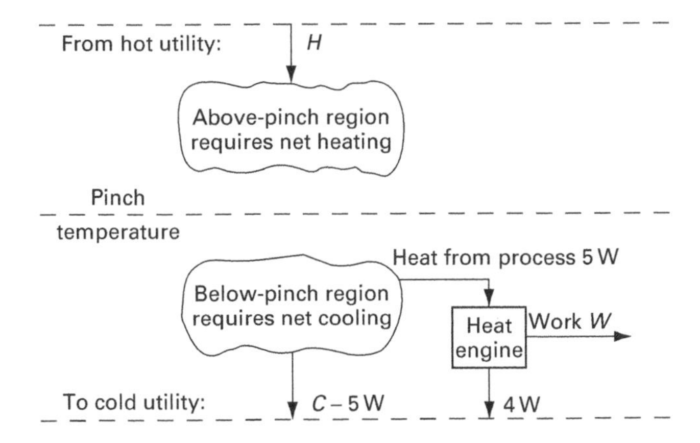

Figure 5.13 Appropriate Placement of an organic Rankine cycle

Figure 5.13 shows the correct placement of an organic Rankine cycle. Here, it is taking heat from below the pinch. Work W is being produced and the cold utility requirement has fallen by W. This sounds very attractive, but the condition is that the 5 W units of heat must be extracted below the pinch and below the grand composite curve. It is rare to find a process with enough waste heat available at a high enough temperature to supply such a cycle. Even where it is possible, the power generated is virtually always insufficient to give an adequate payback on the capital cost of the heat engine. At present, the only hope of economic power generation below the pinch is where a highly exothermic reaction takes place at high temperature – say 500°C – when the heat can drive a standard steam turbine (as with a combined cycle system). One interesting idea has been to use the temperature difference between the upper and lower levels of the ocean – the OTEC (Ocean Thermal Energy Conversion) concept; the extremely "dilute" nature of the energy driving force is compensated for by its magnitude, but would require correspondingly huge capital equipment.

5.3 Heat pumps and refrigeration systems

5.3.1 Heat pump cycles

Heat pumps may be used in two contexts; either as a refrigeration system to perform cooling below-ambient temperature, or as a heat recovery system to pump heat backwards across the pinch, as outlined in Section 5.1.4. However, the equipment used in both cases is similar.

Heat pumps cover a wider range of equipment than is often realised; they can be divided into five principal types.

1. *Closed-cycle heat pumps (including most refrigeration cycles)*: A working fluid (typically ammonia, a hydrocarbon-based refrigerant or – in the past – a fluorocarbon) takes in heat and evaporates, is compressed and then condensed to give out heat at a higher temperature, and returned to the evaporator via a let-down valve.
2. *Mechanical vapour recompression (MVR)*: A compressor is driven by electricity or the output from a plant turbine, and compresses some process vapour to a higher pressure and temperature.
3. *Thermal vapour recompression (TVR)*: High-pressure steam is passed into a venturi-type thermocompressor, and mixes with lower-pressure steam to give a larger flow at an intermediate temperature and pressure. This also includes ejectors, mainly used for drawing a vacuum.
4. *Absorption refrigeration cycles*: These take in some high-grade heat (or above-ambient waste heat) and extract some below-ambient heat from the process, rejecting all the heat at a median temperature close to ambient.
5. *Heat transformers*: These take in waste heat, upgrade some of it to a useful temperature and cool the rest, thus acting as "heat splitters". They are in effect a reversed absorption refrigeration cycle working entirely above-ambient temperature.

Types 1, 4 and 5 are closed cycles, in which the heat pump working fluid is in a separate loop. Types 2 and 3 are open cycles, in which the heat pump working fluid is one of the process streams.

A simple closed-cycle heat pump is illustrated in Figure 5.14(a). A moderate amount of electrical or mechanical power is used to upgrade a larger amount of heat. An absorption system is shown in Figure 5.14(b). Here, it is the "work potential" of above-ambient heat (usually steam) which effects the heat pumping without actually converting heat into shaft power. These systems tend not to be favoured nowadays due to their high capital costs (two columns required, one of them a high-pressure column) and heavy heat demand.

The choice between heat pump systems also depends on the working temperatures and on the relative heat loads below and above the pinch.

5.3.1.1 *Operating temperature*

Closed-cycle heat pumps, using a refrigerant as working fluid, normally operate at temperatures below 80°C, although systems using steam have been operated at up to

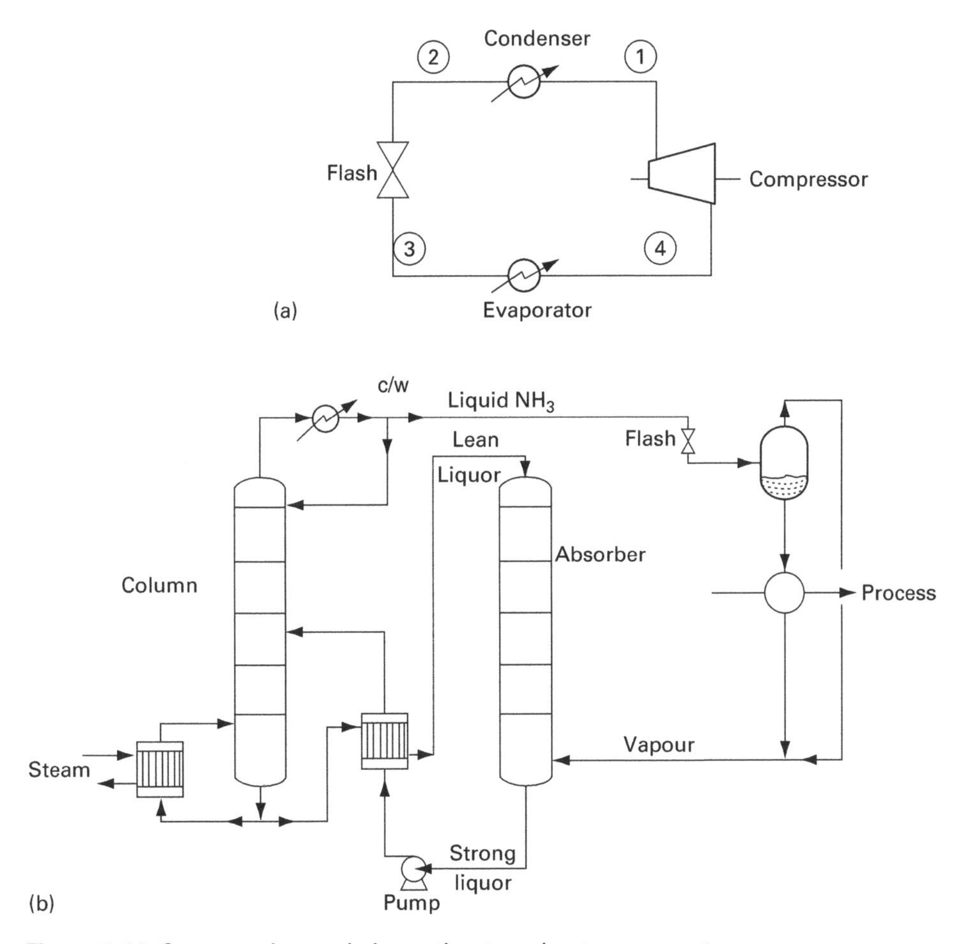

Figure 5.14 Compression and absorption-type heat pump systems

120°C. For a process with below-ambient cooling, requiring refrigeration, and a low-pinch temperature near-ambient conditions, the heat released from the refrigeration cycle can be used for process heating. Heat transformers and absorption heat pumps are only proven at near-ambient temperatures for a cycle based on ammonia and water. Cycles based on lithium chloride and water have been used at temperatures up to about 120°C, and Jeday *et al.* (1993) gives detailed calculations for a unit producing steam at 180°C from working fluids at 65–106°C, but successful applications in industry are very rare.

In contrast, mechanical and TVR are common where the pinch is at 100°C or above, when steam at atmospheric pressure or higher can be used as the working fluid. Below this temperature range, steam has a low density and the compressor required is usually large and inefficient. Steam compressors have been used as low as 60°C, notably on evaporators in dairies, but the temperature lift is only a few degrees. An alternative is to use as a working fluid a process fluid with a lower boiling point than steam. For example, in a distillation column processing an organic compound, the top vapour can be recompressed and used to heat the reboiler. Special safety precautions may be necessary to avoid fire hazards in the compressor.

5.3.1.2 *Ratio of absorbed and released heat loads*

When matching against the process, it is important to remember that the load limit on *either* the process source *or* the process sink can limit the total energy saving, since Appropriate Placement for a heat pump means placement *across* the pinch.

For closed-cycle heat pumps and MVR, the heat loads above and below the pinch are similar but the heat released above the pinch is slightly greater, due to the energy put in by the power drive. For TVR, the driver steam flow is much greater than the flow of vapour sucked in, and so the waste heat recovered is less than the heat released above the pinch. The opposite applies for heat transformers, where less than 50% of the waste heat is usually upgraded. The shape of the GCC therefore suggests which system will be most suitable; Figure 5.15 shows the different types fitted to their ideal GCCs.

Table 5.3 summarises the differences between the different types of heat pump.

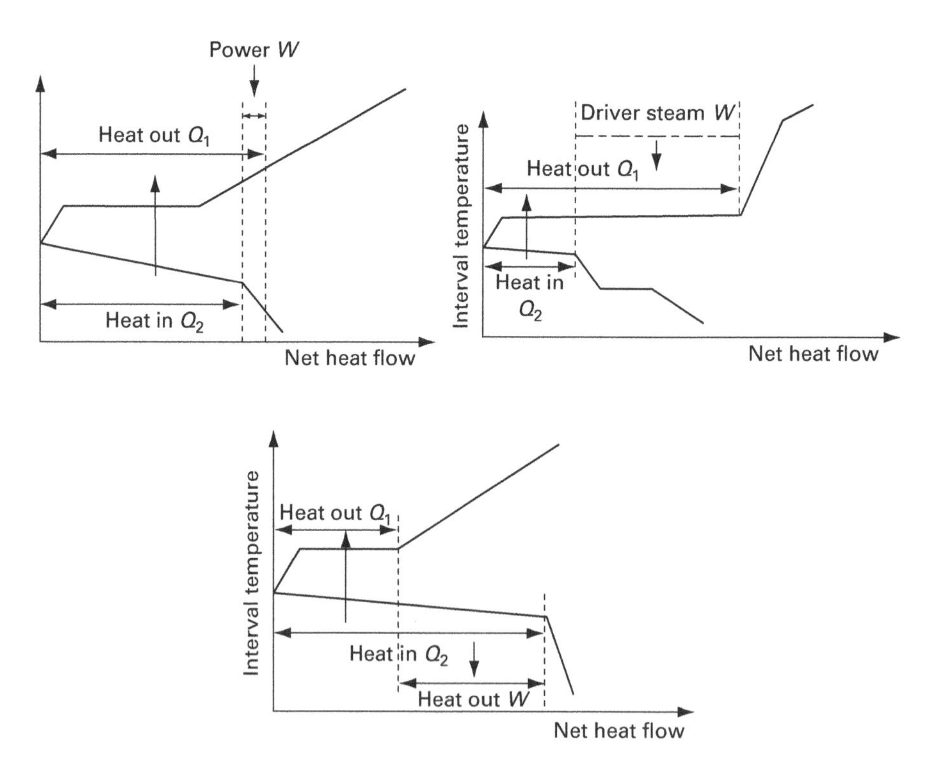

Figure 5.15 Heat pumps ideally fitted to different shapes of GCC

Table 5.3 Summary of different heat pump types

Type description	*Cycle type*	*Whether usable for:* Heat recovery	Refrigeration	Vacuum drawing	*Ratio Q out Q in*	*Ratio W in Q in*	*Usual temperature range (°C)*
1. Closed cycle	Closed	✓	✓		~=1	<0.2	Below 80
2. MVR	Open	✓			~=1	<0.2	Above 60
3. TVR	Open	✓		✓	>1	1–5	Above 80
4. Absorption	Closed		✓		>1	1–1.5	Below 50
5. Transformer	Closed	✓			<1	–	Below 80

5.3.1.3 *Economics*

Whether a heat pump is worthwhile in cost terms, as well as thermodynamic ones, depends on the temperature lift involved and in the relative cost of heat and power. The power requirement is governed by the Carnot efficiency:

$$W = \frac{Q_1}{COP_p} = \frac{Q_1(T_1 - T_2)}{\eta_{mech} T_1} \tag{5.9}$$

Typically, η_{mech} is 50% and so, for a temperature lift of 30°C at ambient temperature (300 K), W is about one-fifth of Q_1. Since power typically costs 3–4 times as much as heat energy, the cost savings from upgrading 10 kW of heat are only equivalent to the supply cost of about 2–3 kW, but the equipment still has to be sized to handle 10 kW. At higher temperature lifts, the economic advantage of heat pumping can disappear altogether, especially as the temperature drop across both the condenser and the evaporator must be allowed for in a closed-cycle system. In Western Europe and the USA, only MVR systems have regularly yielded cost-effective projects; the temperature lift is usually low, there is no separate evaporator with its associated pressure drop, and the equipment is simpler and cheaper. TVR has sometimes been economic due to its low capital cost (see the case study in Section 9.4) if the low ratio of recovered heat to steam supplied has been acceptable.

Heat pumps were strongly advocated during the 1970s and 1980s, when the cost of fossil fuel and heat energy was relatively high in the UK and Europe. Even then, projects with a worthwhile payback period were hard to find because the high-capital cost of the equipment gave unacceptably long payback periods. During the 1990s, fuel prices were low (especially for gas) while power prices did not fall so much, so heat pumps were generally a very unattractive investment proposition. Project economics should always be checked using both current prices and the historic range, to explore sensitivity. Heat pumps are most promising where cheap power is available (e.g. hydroelectric) or there are no convenient local sources of fossil fuel. This is of course the direct opposite of the situation for CHP schemes.

5.3.2 Refrigeration systems

A refrigerator is simply a heat pump, but one where the ultimate destination of the rejected heat is the ambient sink. However, of the five alternative cycles described in Section 5.3.1, only closed cycle (a) and absorption systems (d) are feasible for refrigeration. A simple "mechanical" refrigeration system is the same as the closed-cycle system shown in Figure 5.14(a), except that the working fluid is condensed against cooling water. Figure 5.14(b) shows the corresponding absorption system. Since the latter is more complex and expensive, it would not normally be preferred.

The grand composite curve, once again, can give the clue as to when the absorption system might be favoured over the ubiquitous compression system. The absorption system requires a large above-ambient heat input. If there is sufficient waste heat from the process below the pinch at a sufficiently high temperature, then the absorption system could be run completely on "free" energy, whereas the competing compression refrigeration system has to run on expensive imported power (the alternative, using the below-pinch waste heat in an organic Rankine cycle as in Section 5.2.5, almost never produces enough power to be economic.) Conversely, compression refrigeration will be favoured if there is little waste heat available, or the pinch temperature is close to ambient, or the refrigeration load is required well below ambient, or a CHP system can be installed above the pinch.

Refrigeration systems tend to be the most expensive of all site utilities per unit of heat load. The reason for this can easily be understood from the Carnot efficiency. The work required to absorb heat from the below-ambient heat source is given by a rearrangement of Equation (5.8):

$$W = \frac{Q_2}{COP_r} = \frac{Q_2(T_1 - T_2)}{\eta_{mech} T_2} \tag{5.10}$$

The upper temperature, T_1 is usually fixed at ambient, but as the lower (refrigeration) temperature T_2 falls, the final term rises exponentially and becomes infinite as T_2 approaches absolute zero. Hence, the power consumed in refrigeration rises sharply as the required refrigeration temperature falls, and is very sensitive to irreversibilities in the system design. Table 5.4 illustrates this for a system rejecting heat in the condenser at 27°C (300 K) – this has to be slightly above-ambient temperature to provide a temperature driving force. Likewise, the refrigerant in the evaporator must be cooler than the process from which it is abstracting heat; here ΔT is taken as 5°C, though even lower values are possible. For a typical refrigeration system with a mechanical efficiency of 50% (most losses occur in the compressor), to remove 1 kW of heat from a process at 0°C already requires 0.24 kW power. If power is three times as expensive as heat, cold utility costs more than hot utility below −10°C, which is a far cry from above-ambient cooling which we normally assume is virtually free! The energy and cost penalty increase sharply as temperature falls.

Because of this, any possible ways to increase the energy efficiency of a refrigeration system are worth considering. Multi-level refrigeration systems are common, and the spacing between levels and the temperature difference across the coolers is much lower than for hot utility and cooling water systems, in order to minimise

Table 5.4 Power consumption of a refrigeration system rejecting heat at 300 K condenser temperature

Process temperature (°C)	*Refrigerant temperature in evaporator*		*Ideal* COP_r	*Practical* COP_r (η_{mech} *= 50%)*	*Power used per unit cooling kW/kW*
	(°C)	*(K)*			
32	27	300			0
20	15	288	24.00	12.00	0.08
0	−5	268	8.38	4.19	0.24
−20	−25	248	4.77	2.38	0.42
−40	−45	228	3.17	1.58	0.63
−60	−65	208	2.26	1.13	0.88
−80	−85	188	1.68	0.84	1.19
−100	−105	168	1.27	0.64	1.57
−150	−155	118	0.65	0.32	3.08
−200	−205	68	0.29	0.15	6.82
−250	−255	18	0.06	0.03	31.33

power consumption. Since the process cooling is always done by evaporating the refrigerant, refrigeration cycles plot on the GCC as a series of constant-temperature utility levels, and loads and levels are matched in the usual way.

One way of reducing the power required by the simple cycle in Figure 5.14(a) is to incorporate an "economiser", as shown in Figure 5.16(a). The compression and flash expansion are split into two stages, with flash vapour from the first expansion stage being returned to the suction of the highest pressure compressor stage. In this way, the quantity of vapour flowing through the lowest pressure part of the system is reduced, saving power.

Matching of refrigeration cycles against the grand composite curve is illustrated in Figure 5.16(a)–(f). Because one never cools above-ambient duties using refrigeration (for obvious reasons!) in any process, a utility pinch always exists at cooling water temperature. This is point A in Figure 5.16(b), (d) and (f). The cooling duty, below the cooling water pinch temperature must be handled by refrigeration. With the process source profile ABCDEF, the load QR could all be handled by the system in Figure 5.16(a), with all process duties supplied from a single level 1–2 (Figure 5.16(b)). However, with this design, large loss of driving force exists between AB and the refrigeration utility. Considerable power saving is achieved by the design in Figure 5.16(c) where the process duty due to AB is moved to the higher level 3–4, shown in the T/H diagram in Figure 5.16(d). Shifting load upwards in temperature like this reduces vapour flow in the low pressure part of the refrigeration cycle, although increasing it in the higher pressure part. The net result, however, is reduced power consumption.

It is also possible to exploit pockets in the grand composite curve, as shown in Figure 5.16(e) and (f). A level 5–6 is added *to recover* refrigeration from DD′ and a level 7–8 is added to replace the cooling previously supplied by DD′ to BB′ in

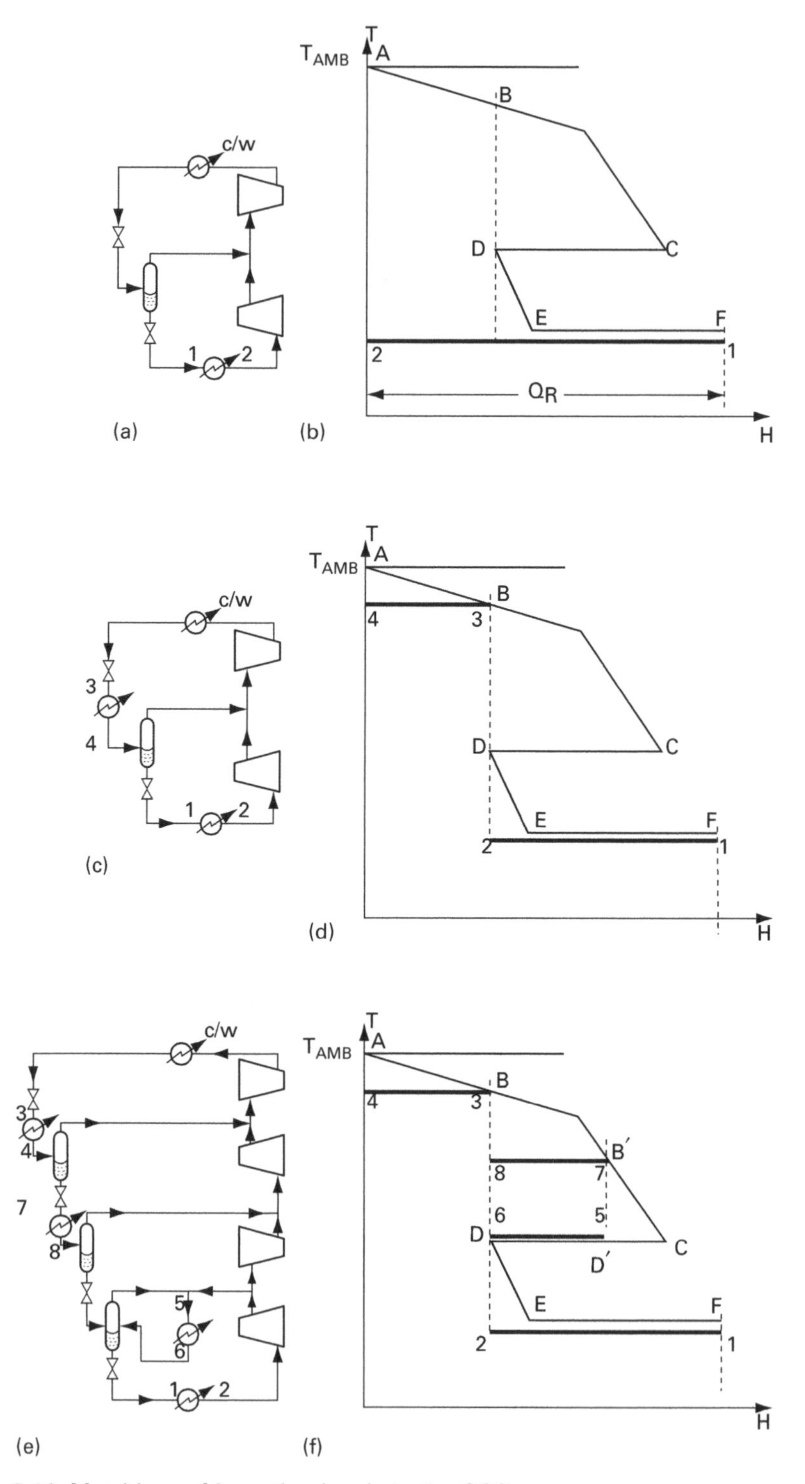

Figure 5.16 Matching refrigeration levels to the GCC

process interchange. A further power saving is obtained, because now, part of the heat rejected by the process into the lowest level 1–2 can be disposed of at below-ambient temperature, that is over 5–6. This more than compensates for the increased load 7–8 at higher level. Note the sharp increase in design complexity in going from the design in Figure 5.16(c) to that in Figure 5.16(e). The economics will depend entirely on the loads and levels involved; this evolution would clearly be uneconomic at near-ambient temperatures, but designs of the type shown in Figure 5.16(e) are commonly seen in low-temperature gas separation plants.

Refrigeration system design is a large and complex subject, beyond the scope of this book. Different refrigerants can be used, with various boiling and freezing points; it is even possible to use a mixture of refrigerants, which generally evaporates and condenses over a range of temperatures. Multi-stage systems can be used to reduce compressor power requirements. The subject is covered thoroughly by, for example, Haywood (1991) and Smith (2005). The key point in terms of pinch analysis is that the below-ambient grand composite curve can be used in the systematic exploration of the design options.

5.3.3 Shaft work analysis

Ethylene production is a typical low-temperature process, which extends significantly below-ambient temperature with several major distillations carried out either across or below ambient. There are many heat sources and heat sinks below ambient in the process, with numerous integration opportunities. A complex refrigeration system supplies and removes heat to and from the process below ambient. The refrigeration system will usually consist of two cycles (ethylene and propylene), and will operate at several levels. There are only two utilities: cooling water and shaftwork. We have a simultaneous design problem of low-temperature distillation, of other process operations, of process heat recovery, and of the refrigeration system, and complex tradeoffs exist. The effect of process changes or new refrigeration levels on the power consumption could only be assessed indirectly, a somewhat cumbersome procedure.

Linnhoff and Dhole (1992) showed how to obtain shaftwork targets for overall low-temperature systems. Consider Figure 5.17. A process grand composite curve is shown below ambient with the Carnot Factor as vertical axis instead of temperature. Due to this substitution, the area in the construction represents exergy (Linnhoff 1990). Specifically, the cross-hatched area between the process grand composite and the refrigeration levels represents exergy loss. Consequently, a change in refrigeration system design as shown in Figure 5.17 (the example relates to the introduction of an additional level) is easily assessed in terms of the consequent reduction α of the exergy loss and therefore of the exergy supplied by the refrigeration system. The reduction in overall shaftwork is equal to (α/η_{ex}) where η_{ex} (the exergetic efficiency of the refrigeration system) is approximately constant. Hence it is possible to by-pass the design of the refrigeration system in targeting.

Use of this approach is demonstrated in Figure 5.18 (Linnhoff and Dhole 1992). A base case design (case A) is compared with an alternative design (case B). Cases

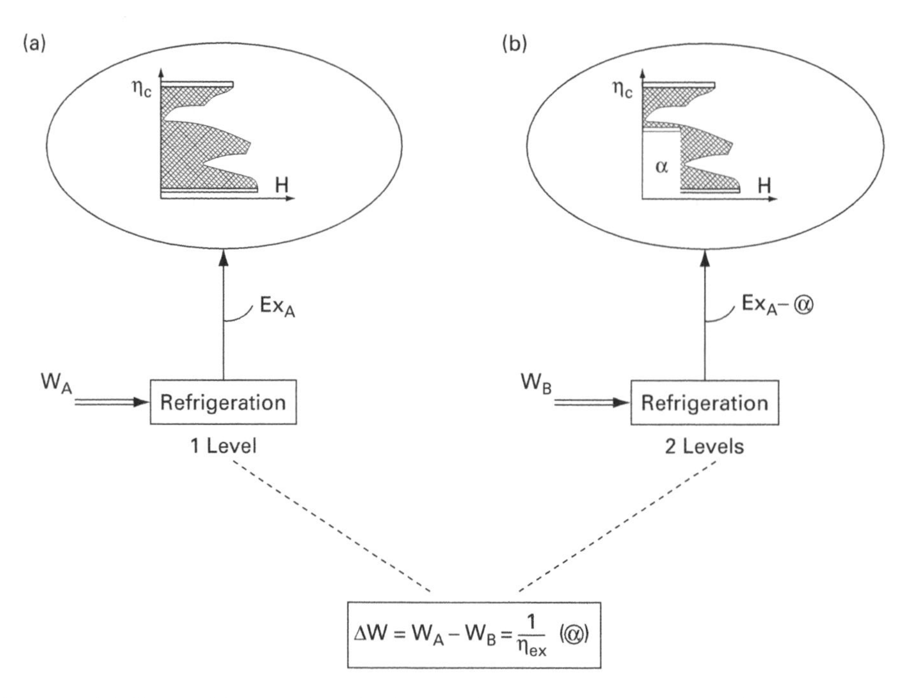

Figure 5.17 Effect of 1 and 2 level refrigeration systems on shaftwork

A and B differ in terms of heat exchanger network and refrigeration system design. The shaftwork target predicts an improvement in overall power consumption of $\Delta W = 3.83$ MW. Detailed design and simulation of cases A and B identify an improvement of $\Delta W = 3.76$ MW. This implies a discrepancy of just 1.9% between simulation and the targeting approach. Dhole and Linnhoff (1993a) extended these techniques to process changes, using the concept of an "exergy grand composite curve" (EGCC).

The combined benefit of distillation column profiles and of low-temperature shaftwork targets in the design of low-temperature distillation-based processes (such as ethylene) has been significant. Results achieved with these techniques offer significant improvements over results achieved using the previously established principles of pinch analysis (Morgan 1992).

5.3.4 Cooling water systems

Cooling water has been something of a "poor relation" among utilities as it generally costs substantially less than hot utilities or refrigeration. However, optimisation can still be important, particularly if it is desired to cope with an increased cooling load without additional investment, or there is a limit on water use or discharge temperatures. Cooling water systems have been studied in detail by Kim and Smith (2004).

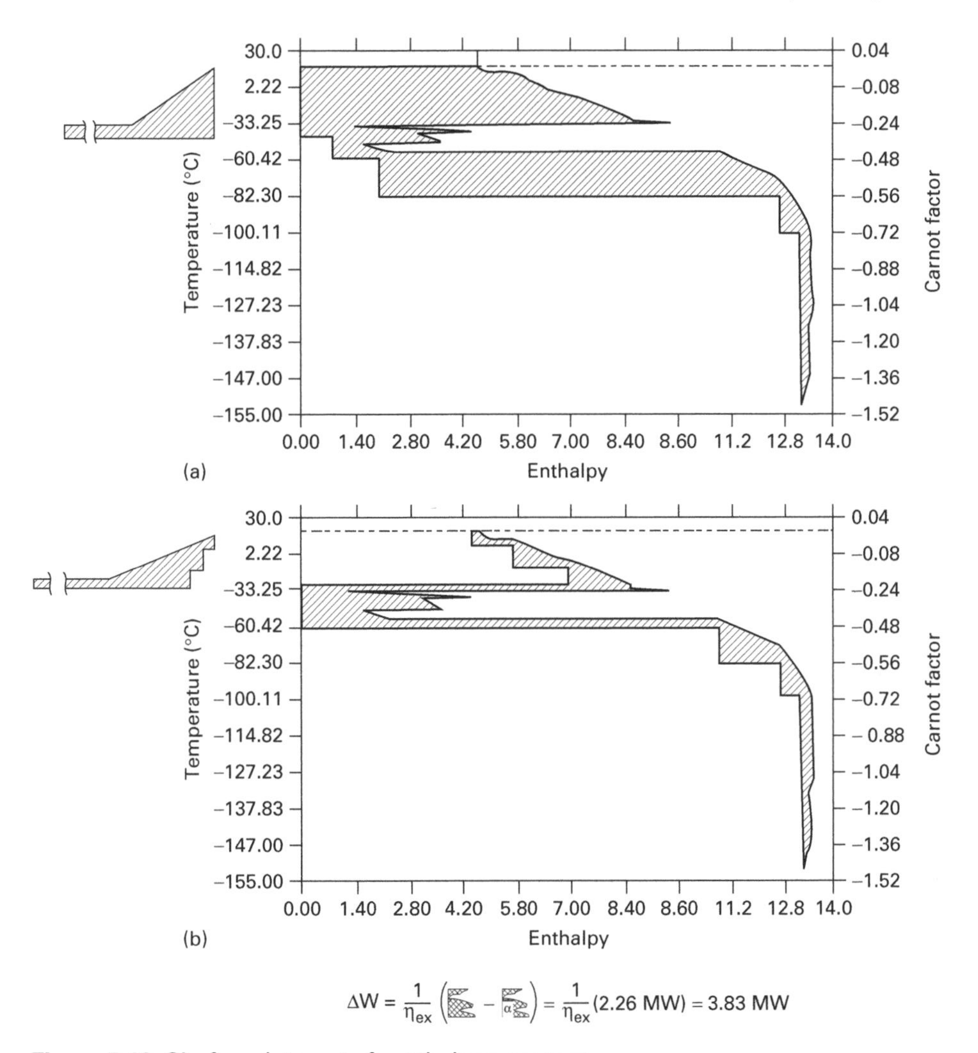

$$\Delta W = \frac{1}{\eta_{ex}}\left(\ldots - \ldots \right) = \frac{1}{\eta_{ex}}(2.26\ \text{MW}) = 3.83\ \text{MW}$$

Figure 5.18 Shaftwork targets for ethylene process

5.3.5 Summary

This section of the guide has shown how the Appropriate Placement concept can be applied to heat engines and heat pumps, and has shown how it can be used to maximise power output of practical CHP systems at 100% marginal efficiency. Some final points may be made:

(i) Although only steam Rankine cycles, gas turbine cycles, reciprocating engines and compression heat pumps were discussed explicitly above, any practical power cycle profile can be matched by inspection against the grand composite curve.

(ii) Different power and heat pump cycles produce very different results when matched against the same process. Hence the choice between the main options for cycle type can easily be made, with approximate utility levels, prior to detailed design. Optimisation, to study the effect of varying ΔT_{min} and temperature levels on power output, can be carried out at a later stage.

(iii) Gas turbines and gas/diesel engines have been gaining ground over steam turbine CHP systems because of their higher power output.

(iv) Conventional closed-cycle heat pumps are only economic for low-temperature lifts and where power is cheap relative to heat. Mechanical and TVR have lower capital cost and are applicable in some circumstances.

(v) Low temperature refrigeration is extremely expensive and complex multi-level systems are justified.

(vi) Once the power cycle profile has been determined, the heat exchanger network is designed by adding the utility streams due to the cycle working fluid to the process stream data, as described in Section 4.6.

5.4 Total site analysis

So far we have considered utility systems for individual plants (Chapter 3), incorporation of utilities into networks for individual plants (Chapter 4) and the principles of heat and power systems (Section 5.2). Now we need to consider the complex interactions between them, which is a non-trivial task on even a moderately complex site, as can be seen from the schematic diagram in Figure 5.19. The questions we might want to ask include:

- How much steam will we be using at each of the steam levels (HP, MP and LP)?
- How can we maximise the power generation from letdown to the various levels?
- Is it worthwhile to install additional turbines instead of producing MP steam by throttling through a valve?
- Is it worth installing heat exchangers between streams in separate processes or zones?
- Can we raise steam from any of the site processes below the pinch, and can it be transferred practicably to other processes?
- What is the real cost of steam at each level? Is it worthwhile to install a heat recovery project which saves LP steam when we consider the loss in power generation?
- Are our chosen temperature levels optimal considering the whole site rather than individual plants?

Total site analysis gives the most effective way of highlighting the key points in these interactions. The techniques were mainly developed in UMIST's Department of Process Integration in the early 1990s.

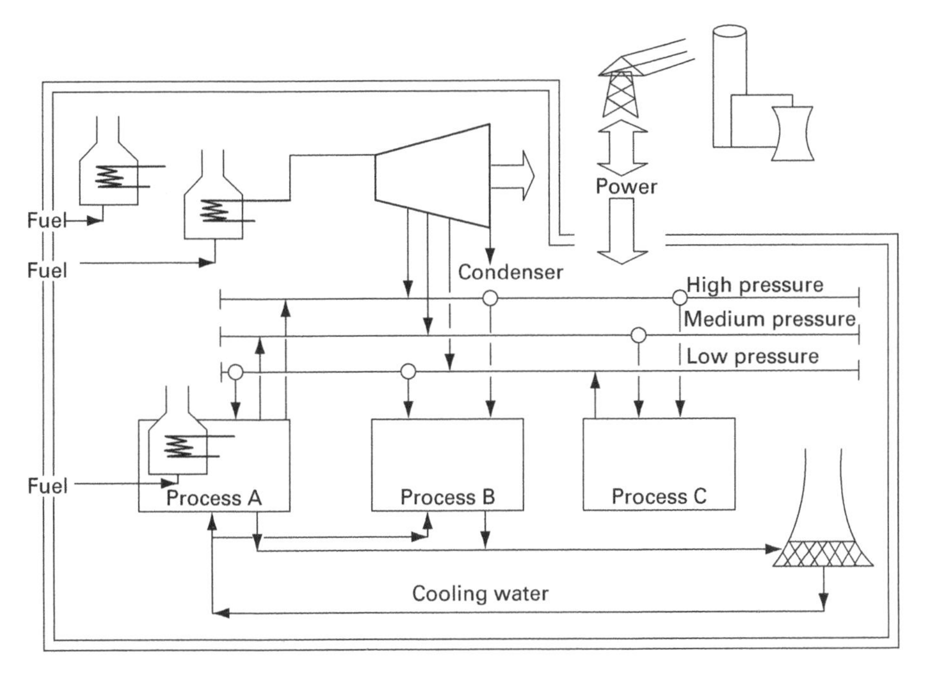

Figure 5.19 Layout of a typical total site

5.4.1 Energy targeting for the overall site

Energy targets and pinch temperatures for the individual processes can be obtained by the Problem Table method as already described, using the stream data set of each process. So an easy way to obtain a heat recovery target for the overall site is to combine all the stream data sets together and obtain targets, the pinch temperature and the grand composite curve for the combined processes. The potential for heat recovery between the processes is then given by the difference between this "overall target" and the sum of the individual process targets. The concept is analogous to that of "zoning" for subsections of a process, as described in Section 3.5.1.

However, there are some problems with this simple approach.

(i) It makes no allowance of the feasibility of heat recovery between different processes, or of practical ways in which this can be achieved. Often there will be an additional temperature penalty for such matches, which this method does not allow for.

(ii) It assumes that heat can be recovered from within a "pocket" on a plant; implementing this in practice may require heat to be transferred away to a stream on another plant at one temperature and heat transfer back from the same plant at another temperature. Such matches will often need long pipe runs and have low driving forces, and will usually be totally uneconomic compared with a simpler system ignoring energy recovery within the pocket.

(iii) There is no real information on the interaction with the site heat and power system.

Hence, an alternative method is needed.

5.4.2 Total site profiles

The net heat required by or available from each plant at any temperature is given by the grand composite curve or the problem table. The external heating which is required in practice is given by the heat demand above the pinch, *ignoring the pockets*. The process is acting as a heat sink for this amount of heat at the given temperatures, and this can be described as a process sink profile. Likewise, below the pinch, the heat which is released in a convenient way for transfer to a separate plant is the net heat flow below the pinch ignoring the pockets, or the process source profile. For example, in our standard four-stream example, the process sink profile would simply include the 20 kW of net hot utility and the process source profile the 60 kW of cold utility. The construction is illustrated and explained by Klemes *et al.* (1997) and Smith (2005).

The heat required by all the plants across the site, added together at the various temperatures, gives the site sink profile; this is analogous to the cold composite curve. Likewise, adding together all the process source profiles for heat release gives the site source profile, analogous to the hot composite curve (see Figure 5.20). Like the composite curves, these can if desired be plotted together and the overlap shows the possibilities for heat recovery. Furthermore, an additional ΔT_{min} can be imposed for heat transfer between separate processes; this is the temperature separation between the site source and sink profiles, and can be called ΔT_{site}.

In practice, it is often more convenient to bring the profiles together by subtracting $\Delta T_{site}/2$ from the source profile temperatures and adding $\Delta T_{site}/2$ to the sink profile temperatures. This is analogous to the construction of the shifted composite curves

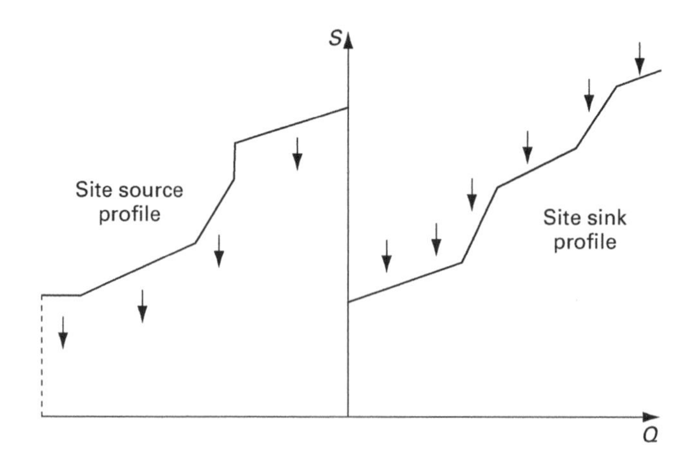

Figure 5.20 Site source and sink profiles

(or the grand composite curve) and the advantages will become clear in the following section.

It is important to note that there are two approaches to developing such profiles. The first, as explained above, uses the grand composite curves of the individual processes. This is valid, but assumes that all potential heat recovery projects will be implemented, which is rarely the case (especially if the economics are adversely affected by the low utility costs resulting from CHP). Therefore, a more common approach, especially in retrofit situations, is to use the current utility requirements for the individual processes. This data can be extracted more readily and accepts the heat integration arrangements that are already in place in the individual processes. Predicted utility consumptions by this second approach will be higher, but more realistic; it is known as top level analysis (Varbanov *et al.*, 2004).

Overall, we can find three targets for heat recovery; the combined grand composite curve, the site profiles based on grand composite curves, and the site profiles based on current utility needs, increasing in that order. This gives us a feasible range, depending on the amount of additional heat recovery we choose to implement within processes.

5.4.3 Practical heat recovery through the site steam system

As previously mentioned, there are strong arguments against linking separate processes together by means of heat exchangers. Apart from the problem of physical separation (sometimes by long distances) and piping costs, it causes problems if one process is shut down while the other is running; an alternative heating or cooling source must be found to replace the heat previously exchanged.

However, heat may already be transferred around the site using the steam mains. So a more convenient alternative presents itself – raise steam on one plant (replacing steam previously generated in the boiler house) and use the extra steam on another plant. No new piping is required and if one plant is shut down, the change in demand is simply compensated for by altering the load on the site boilers. In essence, we are achieving heat recovery via the site utility system. An example can be seen in the case study in Section 9.5.

The site steam levels can be drawn on the total site profiles and show how the heat available can be equated to the steam generated and used by processes at the different steam levels. The total site profiles become steam system composites, as shown in Figures 5.21 and 5.22. The advantage of applying the further temperature shift is now clear; if $\Delta T_{site}/2$ is chosen as the temperature difference contribution required to raise steam, the steam levels plotted on the total site profile will be at their actual temperatures whether steam is being generated or used, and the possibilities for heat transfer can be seen at a glance.

Let us take an example. Suppose we have a hot stream at 130°C and a cold stream at 100°C and the ΔT_{min} is 20°C (contribution 10°C for each stream). It is easily feasible thermodynamically for these streams to exchange heat; on the shifted or grand composite curves, the hot stream appears at a shifted (interval) temperature of 120°C and the cold stream at 110°C. However, they are on separate plants and direct

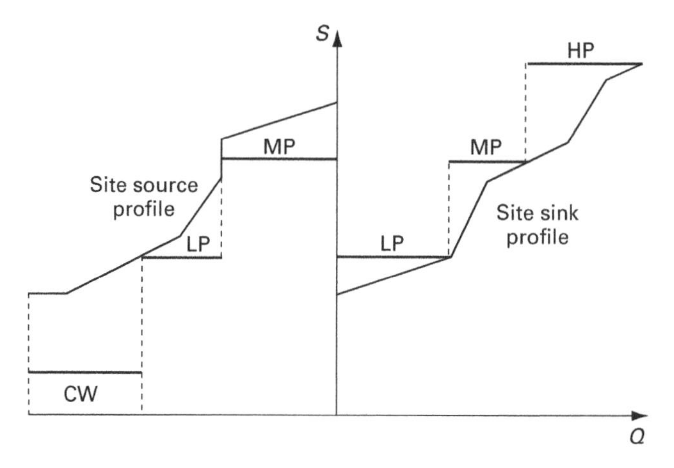

Figure 5.21 Site steam composites based on source and sink profiles

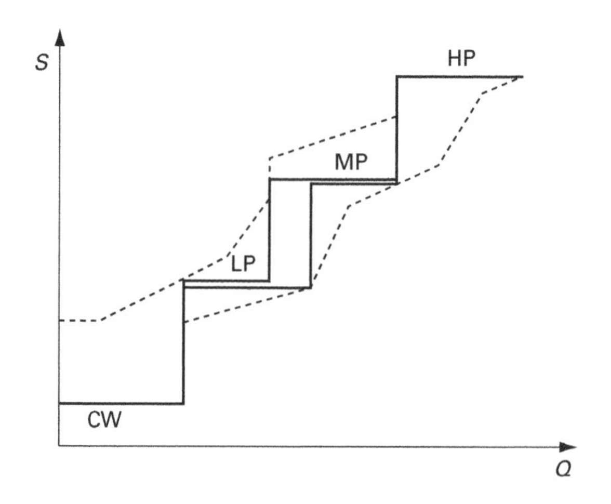

Figure 5.22 Heat recovery using site steam composites

exchange is undesirable. Now if ΔT_{site} is set at 10°C, so that the stream contribution ΔT_{cont} for steam raising or condensation is 5°C, it would be possible to raise LP steam at 115°C from the hot stream, transfer it along the steam main and use it to heat the cold stream, with the total ΔT on each of the steam/process exchangers being 15°C.

This simple analysis is sufficient for latent heat streams at constant temperatures. For sensible heat streams, things are more tricky.

5.4.4 Indirect heat transfer

Consider two streams on different plants which have potential to exchange heat. The hot stream runs from 200°C to 160°C with a CP of 2 kW/K and the cold stream

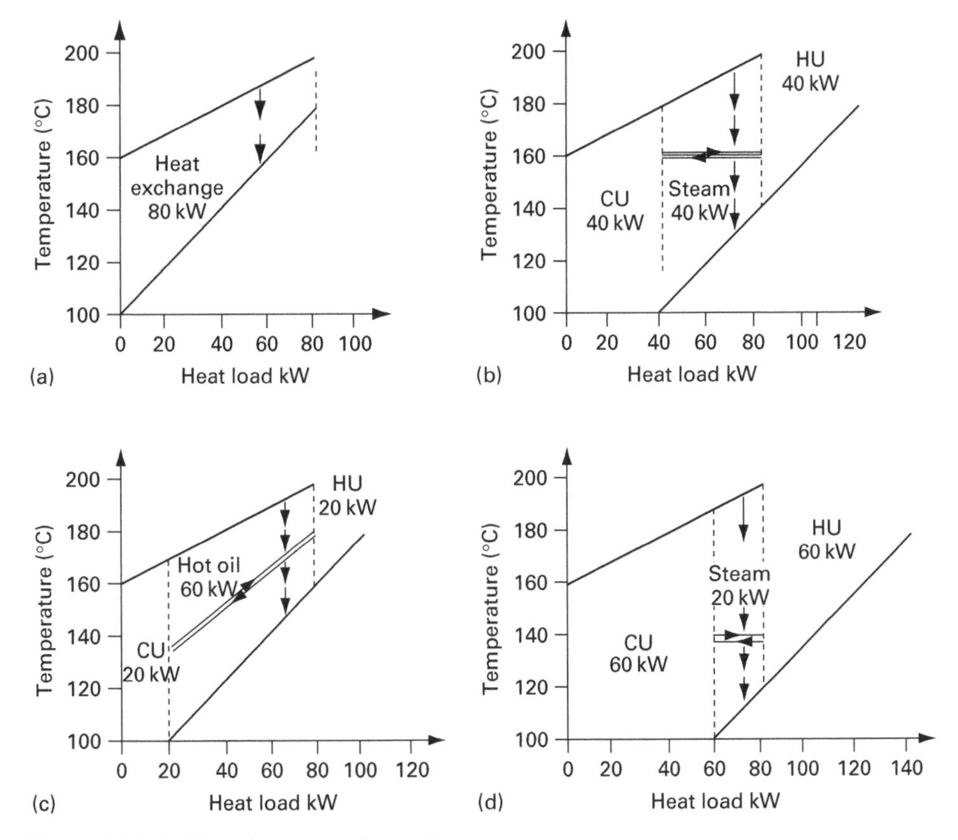

Figure 5.23 Indirect heat transfer options

from 100°C to 180°C with a CP of 1 kW/K. Both streams have a heat load of 80 kW. ΔT_{min} on all streams will be taken as 20°C. Several different options for heat recovery are possible.

(a) One stream can be piped over to the other plant and exchange heat directly with the other stream. The full 80 kW can then be recovered, as shown in Figure 5.23(a). This of course requires that the two plants operate simultaneously and incurs additional pressure drop through the pipework.
(b) The hot stream can be used to raise steam at an intermediate temperature level (here 160°C) and this steam can then be transferred to the second plant and used to heat the cold stream. This gives more process flexibility and no additional pressure drop in process lines. However, temperature degradation occurs and therefore only 40 kW can be recovered – see the composite curves in Figure 5.23(b).
(c) The heat can be transferred to a recirculating stream, for example a hot oil circuit, which then transfers the heat to the second plant. Here, since the heat transfer fluid undergoes sensible heating, less temperature degradation occurs

and 60 kW can be recovered, as shown in Figure 5.23(c). Note that the flow rates and CP in the two halves of the hot oil circuit must be the same.

Method (c) has been used, for example, in a food processing factory where an incinerator was used to remove volatile organic compounds from the exhaust gases. The hot flue gas was then used to heat a hot oil circuit which recovered the heat for process heating duties elsewhere in the factory; this was far more convenient than direct heat exchange. For temperature-sensitive products where hot water is used as the hot utility in preference to steam, heat recovery via the site hot water system has been successfully achieved in brewing and edible oil processing; see Section 8.7.

Method (b) is increasingly used in large and complex sites which already have a network of steam mains. However, the mains pressures will usually already be set and this will constrain heat recovery further. For instance, if the intermediate steam temperature in our example had to be 140°C, only 20 kW could then be recovered, as shown in Figure 5.23(d), because only one end of the steam heat exchanger achieves ΔT_{min}.

It can be seen that applying the additional penalty ΔT_{site} works well for case (c) but does not give the full story for (b).

5.4.5 Estimation of cogeneration targets

Now that we have the steam system composites, we can see how much steam is generated and used at each temperature level. Can we use this to estimate, quickly and easily, how much power will be generated by letting down VHP (very high-pressure steam) generated in boilers to HP, MP or LP steam, or to see the power benefits of using LP steam on the plant instead of MP steam?

Fortunately, a simple approximate relationship between heat load, temperature and power generation exists, as shown by Raissi (1994). He showed that:

$$W = m \times w = (Q/h_{sat})\, e\, (T_{in,sat} - T_{out,sat})$$

where W is the work produced, m is the mass flow through the turbine, w is the work produced per unit mass flow, Q is the heat load, h_{sat} is the specific enthalpy at saturation, e is a constant and T_{in} and T_{out} are the temperatures at the inlet and outlet of the turbine. This is consistent with Table 5.2. In other words, the work produced is proportional to the heat load and the temperature difference between the levels – that is the area of a rectangle on the total site profiles, as shown in Figure 5.24. (A more accurate estimate can be generated using Willans' line, which is a similar relationship but involves a constant so that work is not directly proportional to mass flow and heat flow through the turbine.)

So we can now see how much power is likely to be obtained by cogeneration – add together the areas of the rectangles for levels between which a turbine exists (including mechanical pump drives, etc.) If the steam is simply throttled, the area of that rectangle must be ignored for power generation purposes; on the other hand, it shows the incentive to retrofit a new turbine.

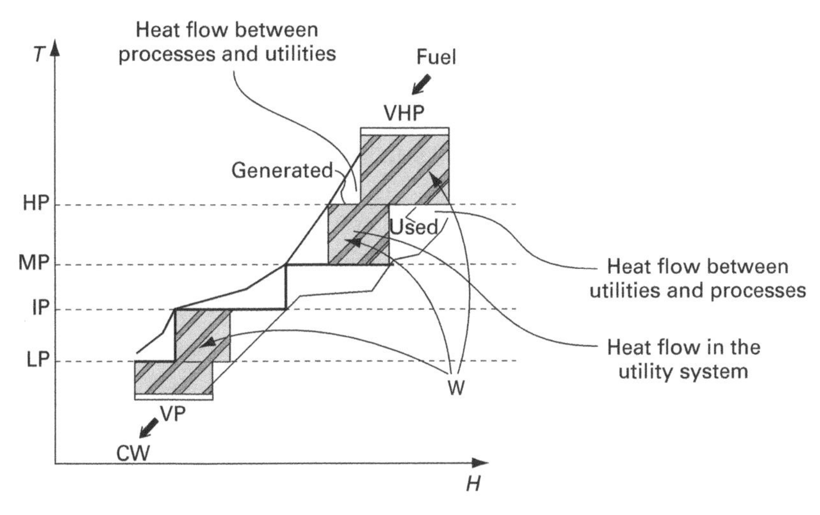

Figure 5.24 Estimating power generation from the total site profiles

It can be seen easily from this plot that if steam is not recovered to the maximum possible extent via the steam mains, the steam system composite curves will move apart. Hence more steam will have to be generated – but the cogeneration work will also go up.

The plot is very similar in concept to the exergy grand composite curve noted in Section 5.3.3 and indeed it could be plotted with Carnot factor as the vertical axis rather than temperature if desired. Dhole and Linnhoff (1993b) described this approach, including a case study where use of the total site profiles allowed a substantial saving by lowering the MP steam level. For most purposes, however, temperature is the simplest and most convenient plotting parameter.

5.4.6 Emissions targeting

Emissions reduction and emissions targeting have both a process and a total site dimension. For example, consider two alternative process designs, scheme A with a simple purge and scheme B with a more sophisticated separation/recycle concept reducing process waste. However, as is often the case, better separations involve additional energy, and the reduction in process emissions needs to be assessed relative to the increase in fuel related emissions. This dilemma is increasingly recognised both by designers and legislators (Linnhoff March 1991). There have been instances where regulations requiring excessively low ppm-limits on process emissions led to additional fuel consumption such that *overall* emissions deteriorated. Pinch analysis can assess the overall picture and help designers, planners and legislators to come to a rational assessment of trade-offs between process related and fuel related emissions and to agree on achievable targets.

Total site analysis has a significant part to play by giving targets for (1) central site combustion, (2) total site electric power import or export and (3) de-centralised combustion on site. Global CO_2 emissions follow and can be targeted, for example as a function of investment. Emissions targeting is discussed in more detail by Linnhoff and Dhole (1993) and Rossiter *et al.* (1993).

5.5 Worked example: organics distillation unit

Let us see how we might go about selecting and sizing a CHP system for the organics distillation unit example. The atmospheric and vacuum units will be considered together (there seems little reason to replace one furnace but not the other). From the data collection and targeting described in Sections 3.2 and 3.8, we know the current hot utility use is 8.5 MW but the target for the combined plants is only 6.085 MW. The average site power requirement is approximately 3 MW.

Applying the various criteria given in Section 5.2.2:

1. As the site power requirement is above 1 MW, gas and steam turbines are preferable to gas and diesel engines, for which multiple units would be required.
2. The site heat-to-power ratio is approximately 2. Comparing with Table 5.1, reciprocating engines and gas turbines are preferred to steam turbines.
3. The process GCC above the pinch is sloping and heat needs to be supplied up to 329°C shifted temperature:
   (i) The high-temperature heating loads above 200°C favour gas turbines and reciprocating engines over steam turbines, as otherwise a separate furnace would be required.
   (ii) and (iii) The pinch temperature is above 70°C, so all the hot jacket water from reciprocating engines would be wasted (there are no local offices or buildings with a significant space heating load).

The decision seems relatively clear-cut in this case; if a CHP system is to be adopted, it should be a gas turbine, with a diesel/gas engine as a second choice. The question now is whether CHP will be economic.

The heat-to-power ratio for the plant is between 2 and 3, which is a little high for a gas turbine. A high exhaust gas temperature (over 500°C) may be used to maximise heat output and reduce power generation. Alternatively, the unit may either be sized to meet the typical site power needs, with additional heat being supplied by supplementary firing (or the existing furnaces); or it may be designed to produce 6 MW(th), in which case surplus electricity can be exported.

The grand composite curve for the combined process is shown in Figure 5.25(a). We can also construct a site sink profile based on the current utility heat loads and process temperatures, and this is shown in Figure 5.25(b). Table 5.5 shows the data for this (from Sections 3.2 and 3.8).

In both halves of Figure 5.25, we now add the exhaust profile for a gas turbine, based on (i) matching the heat requirement (ii) calculated heat load for matching the power requirement. We can see from this that the heat produced by the CHP system is substantially less than the site heat demand when the site power demand is matched. In all cases, there is a utility pinch, at 162°C shifted temperature in (a) (corresponding to an exhaust gas temperature of 202°C), but at 102°C in (b) (exhaust gas actual temperature 142°C).

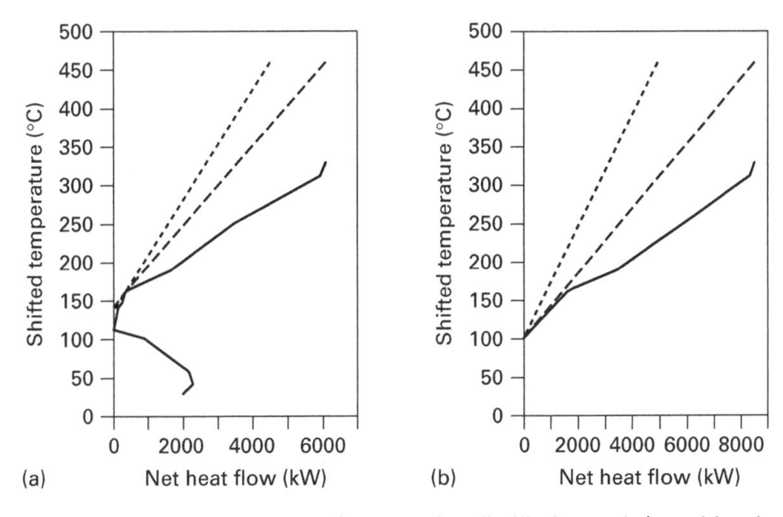

Figure 5.25 Grand composite curve for organics distillation unit (combined units)

Table 5.5 Data for site sink profiles based on cold streams heated in furnaces

Stream name	*Initial temperature, °C*	*Target temperature, °C*	*Heat capacity flow rate CP (kW/K)*	*Heat flow rate, kW*
Crude feed	92	180	26.8	−2360
Dehydrate	152	302	30	−4500
Vacuum crude	155	319	10	−1640

Table 5.6 shows the calculated heat rates and savings for gas turbines meeting either site power needs or site heat needs, with a value for gas engines as a comparison. Power generation and useful heat per kW fuel are taken from Section 5.2.3.2 for gas turbines and Section 5.2.3.4 for gas engines. The cost of gas is £3.33/GJ (£12/MWh), and the efficiency of package boilers is 80%, giving an effective cost of £15/MWh of heat actually delivered to the process. Assuming that imported power is charged at £40/MWh but only £20/MWh is paid for exported power, the economics are evaluated in the lower part of the table. Annual savings are obtained by multiplying the net saving by the number of hours worked in the year (5,000 for this site). Installed capital cost is taken as £500/kW for both gas turbines and gas engines (possibly an overestimate for the latter).

In this case, the economics are rather disappointing. If we base the CHP system on the target energy consumption and process GCC, the best payback time is barely 5 years, for the gas turbine matched to the site power needs. Contributory factors are the relatively low number of hours worked per year (5,000, as against 8,860 for full 24/7 operation) and the high temperature of the utility pinch, which gives large

Table 5.6 Performance and economics calculations for gas turbine and gas engine

	Based on energy targets and grand composite					*Based on current utility use and site sink profile*				
	Gas turbine			*Gas engine*		*Gas turbine*			*Gas engine*	
	Without CHP	*Power match*	*Heat match*	*Power match*	*Heat match*	*Without CHP*	*Power match*	*Heat match*	*Power match*	*Heat match*
Exhaust temperature (°C)		500	500	400	400		500	500	400	400
Utility pinch (°C)		202	202	202	202		142	142	142	142
Heat input (kW)		10,000	14,851	7,500	34,348		10,000	17,268	7,500	36,822
Power (kW)		3,000	4,455	3,000	13,739		3,000	5,180	3,000	14,729
Losses (kW)		400	594	1,950	8,930		400	691	1,950	9,574
Useful heat (kW)		4,098	6,085	1,329	6,085		4,923	8,500	1,731	8,500
Waste heat (kW)		2,503	3,716	1,221	5,593		1,678	2,491	819	4,019
Site heat demand (kW)	6,085	6,085	6,085	6,085	6,085	8,500	8,500	8,500	8,500	8,500
Package boilers (kW)	7,606	2,484	0	5,945	0	10,625	4,472	0	8,461	0
Gas cost (£K/yr)	0	600	891	450	2,061	0	600	1,036	450	2,209
Power cost (£K/yr)	600	0	−146	0	−1,074	600	0	−218	0	−1,173
Coal cost (£K/yr)	456	149	0	357	0	638	268	0	508	0
Total site bill (£K/yr)	1,056	749	746	807	987	1,238	868	818	958	1,036
Net saving (£K/yr)		307	311	250	69		369	419	280	201
Capital cost (K)		1,500	2,228	1,500	6,870		1,500	2,590	1,500	7,364
Payback time (yr)		4.9	7.2	6.0	99.0		4.1	6.2	5.4	36.6

waste heat losses in the flue gas. CHP systems based on the current energy consumption and site sink profiles have better economics, as the utility pinch has fallen to 142°C, but the payback is still 4 years at least. Matching the heat needs and exporting power gives worse economics because of the relatively low price paid for exported power. Indeed, for the gas engine there is a gross mismatch in this case and the vast majority of the power is exported, so that the net saving is barely positive and payback time can be up to 100 years! The gas engine matched to the site power requirement does better, even though all the heat rejected to cooling water must be wasted (hence the high figure for heat losses). If the capital cost were 20% less, payback times would be comparable to the gas turbine. Hence, if gas turbines and gas/diesel engines both seem feasible on a plant, it is worth getting actual manufacturers' quotes for both types.

If we do the calculation the other way round, in terms of marginal cost of heating, we find that the cost of 1 MWh gas burnt in a gas turbine @ £12/MWh is £12, and generating 0.3 MWh power @ £40/MWh saves £12, so the net cost of process heat from the gas turbine exhaust is 0! Hence there is no incentive to install heat recovery projects if these would reduce the heat load on the CHP system. However, the preferred CHP system here is based on site power needs and the heat output is always less than the hot utility target. Therefore any heat recovery on the plant will save steam at the full price (£15/MWh) or reduce fuel consumption in supplementary firing.

5.6 Case studies and examples

5.6.1 Whisky distillery

The majority of whisky production is grain whisky produced in the large-scale continuous processes. These distilleries have an excellent record on waste minimisation in general. Although a large amount of waste solid is produced from the spent grain from which the alcohol has been extracted, it is dried and sold as high-grade animal feed. Thanks to the application of pinch analysis, several of the largest distilleries have also optimised their energy systems and achieved considerable overall savings. The description that follows is representative of more than one distillery.

The process flow diagram is shown in Figure 5.26 and the grand composite curve is Figure 5.27. The basic hot utility requirement is no less than 48 MW and the pinch is at 95°C. The main heat load is for steam to heat the distillation columns but there is also a substantial requirement for high-temperature heat for the hot air dryers; the latter was typically provided by a gas-fired burner, while the steam was raised in package boilers.

The sharp pinch suggests that heat pumping would be possible. In fact, thermocompressors are already in use, taking sub-atmospheric pressure steam flashed from the slurry emerging from the base of the columns and upgrading it to heat the columns. With best available equipment, 1.6 kg of high-pressure driver steam is needed to upgrade 1 kg of flash steam at 88°C. However, there is plenty of heat demand just above the pinch, so this imbalance is not a problem.

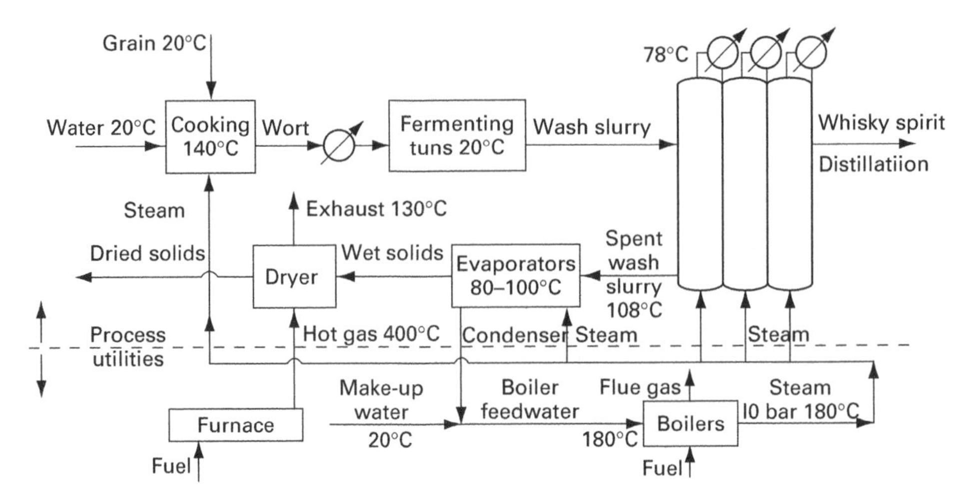

Figure 5.26 Process flow diagram for typical grain whisky distillery

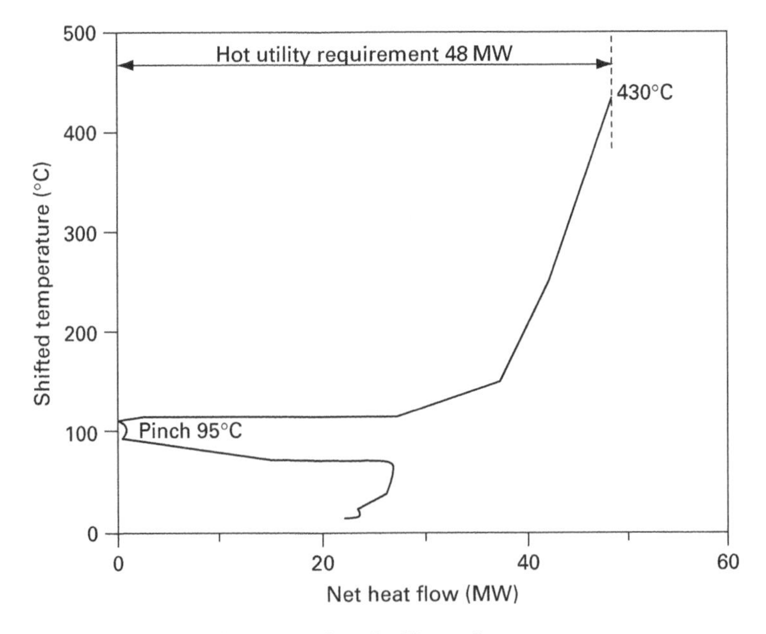

Figure 5.27 Grand composite curve for distillery site

Can we consider CHP? The size of the heat and power loads ruled out reciprocating engines at the time; about 20 parallel units would have been needed! A steam turbine system would effectively supply the large heat loads in the 100–120°C range, though not the heat needs of the dryer. However, the power demand of the site was approximately 12 MW. In the 1980s, when the majority of these schemes were installed, the cost of buying power from the national grid was far higher than that which the grid would pay to independent power generators for exported power

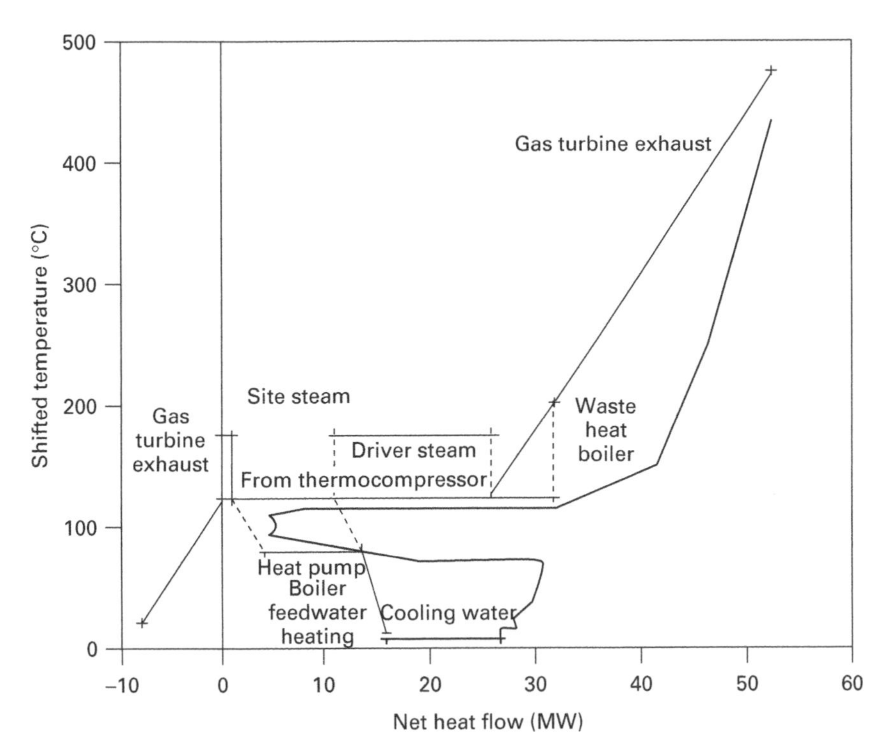

Figure 5.28 Final allocation of utilities for grain distillery

(this has since changed), so a CHP scheme which would exactly satisfy the site's mean power requirement was preferred. The heat available from a gas turbine generating 12 MW was approximately 30 MW, while that from a steam turbine was over 100 MW – far too great for the site's needs. Moreover, the capital cost of such a large steam turbine would have been much higher. The gas turbine was therefore selected, and this had the further advantage that the exhaust was clean enough to be used directly as the carrier gas for the dryer.

The heat from the gas turbine and the heat from the existing thermocompressors roughly balanced the total site heat requirements. The package boilers were still required to provide the driver steam for the thermocompressors. However, their efficiency could be increased by using below-pinch waste heat to preheat the boiler feedwater. Where other heating needs had to be met by steam, this was raised in waste heat boilers heated by the exhaust gases from the gas turbine.

Finally, we should remember that the GCC is for an ideal on-target process. In fact, unavoidable constraints on heat exchange and the relaxing of heat exchanger networks meant that the actual hot utility requirement was 52 MW instead of 48 MW. The GCC is therefore shifted to the right by 4 MW. The additional heat was supplied by steam from the package boilers. The final result is shown in Figure 5.28.

Malt whisky distilleries, which produce distinctive high-quality products on a much smaller scale, have also been active in energy saving. In several cases, the heat from the evaporated vapour from the batch stills has been recovered. This heat is below the pinch, and therefore requires upgrading using a heat pump. An open-cycle system is used, with MVR. The vapour temperature is 78°C, and if steam were raised at this temperature it would be at a maximum of 0.4 bar and occupy a large volume, which would give a bulky, expensive and inefficient compressor. Hence, the vapour from the still itself is used as the working fluid and recompressed to a temperature which is high enough to heat the still via a bottom reboiler.

5.6.2 CHP with geothermal district heating

An elegant example of a CHP system, albeit not on a process plant, has been developed at Southampton, UK. The system began in 1986 with the boring of a well to extract geothermal energy from hot brine present in rocks a couple of miles below the earth's surface. The brine is pumped to the surface, passed through a heat exchanger and discharged to the sea. The resulting hot water was used to heat the Civic Offices.

The successful operation of this scheme showed that there were opportunities for extending the hot water mains to provide a more extensive local district heating scheme. The geothermal well provided the base load, but could not cope with times of peak heat demand. Therefore, two 400 kW(e) gas engines were installed, and heat from the exhaust and jacket water was used to supplement the heat from the brine. This also produced a CHP system, with the gas engine supplying the power needs of the local buildings.

During summer, there is no significant demand for process heating. However, it is still worthwhile to run the diesel engine to generate power. Therefore, an absorption refrigeration system has been installed, driven by the exhaust heat from the diesel. This provides the cooling load for the air conditioning systems of the offices. Hence, a blend of different heat and power systems is able to satisfy the highly varying demands of the locality throughout the year.

The scheme has been highly successful and has been successively expanded. In 1998, a dual-fuel gas/oil engine of no less than 5.7 MW(e) was added. Ice storage is being planned to cope with peak cooling demands on hot summer days; the ice would be produced overnight using surplus CHP power, when demand and electricity export prices are both low. Heating capacity is now seven times that of the original geothermal well, which is still in use. Package boilers are required for peak heating top-up, but a new biomass-fired boiler (using wood chips) is planned. Another proposed initiative is an anaerobic digestion plant to produce biogas, suitable for the CHP plant, from household waste. By 2005, 70 GWh of energy per year was being produced, and associated carbon emissions reduction was 11,000 tonnes per year. A detailed case study has been produced by IEA (2005).

There is also a distributed heating aspect; a nearby estate was given its own local CHP plant (a 110 kWe gas engine with additional boilers) rather than supplying it from hot water mains. The existing mains are thoroughly insulated to minimise heat losses. A further scheme was planned for the Millbrook housing and industrial area; this would have generated no less than 48 MW and a gas turbine was the preferred option. However, this plan has currently been shelved because the major gas price increases in 2005–2006 have seriously affected the economics.

5.6.3 Tropical power generation and desalination

In the Middle East, the demand for water now exceeds available supplies in many areas (e.g. the United Arab Emirates). Desalination plants are used to make up the shortfall, and as this is an evaporation process, large amounts of heat are needed.

Integrated power and water projects have become common in recent years, using combined cycle gas turbine systems. The steam turbine is set up to release low-pressure steam at a suitable temperature and pressure to heat the first evaporator effect. The steam from each effect is condensed, heating the next effect, and the condensate becomes the desired fresh water product. The latent heat released in each effect is roughly equal to the heat released from the gas turbine exhaust, so the amount of desalinated water produced will be roughly proportional to the number of evaporator effects. There is a multi-way trade-off between the power produced from the steam turbine, the exhaust temperature of the steam, the number of evaporator effects, the temperature differences in each effect and the amount of pure water produced for a given level of fuel use and power generation. Analysis must allow for both the utility system profiles and the integration of the evaporator as described in Section 6.5.1.

Figure 5.29 shows a number of alternatives in schematic form. Alternative 1 has a high exhaust steam temperature so that power generation from the steam turbine is

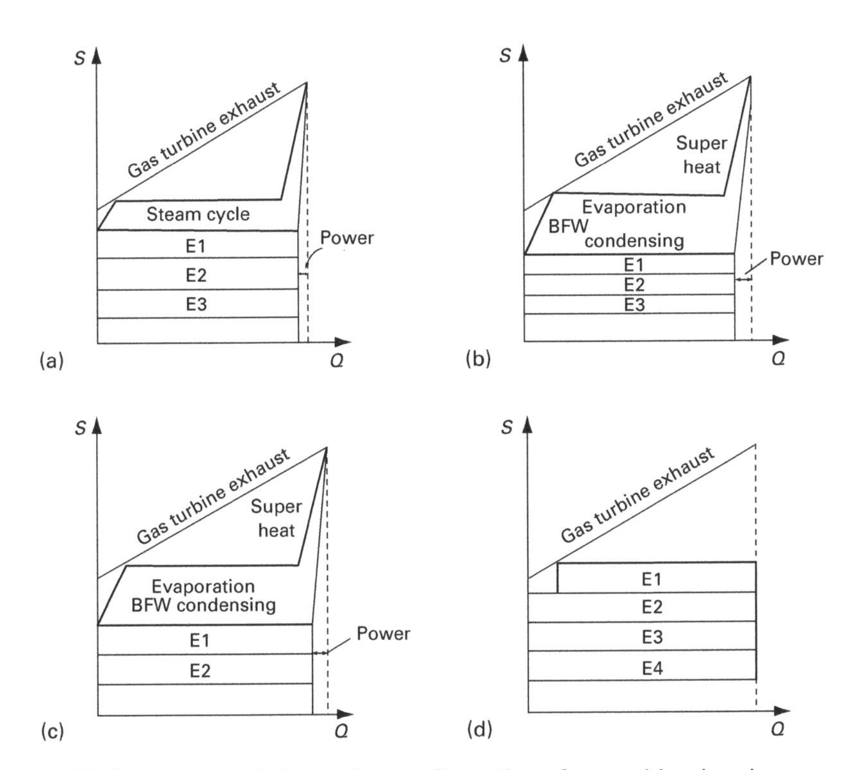

Figure 5.29 Schematics of alternative configurations for combined cycle desalination plant

relatively low, but a three-effect evaporator can be fitted in. Alternative 2 uses a lower exhaust temperature, giving higher turbine power; to obtain three evaporator effects, ΔT in each effect must be squeezed so capital cost is high. In three, a low exhaust temperature is again used, and the ΔT is restored so that capital cost is lower, but only two effects can be fitted in and water output falls. Option four sacrifices the steam turbine altogether but allows four effects to be fitted in, with increased water production. Any of these could be the preferred option, depending on local water and power needs and investment limits.

Theoretically, these situations could be studied by exergy analysis. In practice, the values assigned to heat and power in different situations do not reflect their exergetic efficiency, and this method will not give an economic optimum.

Increasing population and water demand in many countries means that these techniques could become more widespread. In the UK, a controversial desalination plant has been proposed for London.

5.6.4 Hospital site

The case study on a hospital site described in Section 9.6 is a typical example of how CHP can be applied to a building complex. The main heat demands were at temperatures below 100°C, for central heating, domestic hot water and air heating. Instead of the existing site steam system, which incurred huge distribution losses (less than 50% of the steam generated reached its destination), using gas engines to provide local CHP in each building gave far higher efficiencies. In the years since this study was performed, such "distributed systems" have been successfully installed in a significant number of factories, offices and similar complexes.

Exercises

Consider the desalination plant design described in Section 5.6.3. Using typical figures for heat, power and exchanger costs from other case studies in this book, and assuming the steam is raised at 40 bar, make an estimate of the relative economics of the four alternative configurations. What factors affect the comparison?

References

Dhole, V. R. and Linnhoff, B. (1993a). Overall design of subambient plants, paper presented at *ESCAPE-III Conference*, Graz, Austria, July. Also in *Comp Chem Eng*, 1994, 18(Suppl): S105–S111.

Dhole, V. R. and Linnhoff, B. (1993b). Total site targets for fuel, co-generation emissions, and cooling, paper presented at *ESCAPE-II Conference, Toulouse, France, October 1992*. Also in *Comp Chem Eng*, 17(Suppl): S101–S109.

ESDU International (1989). Application of process integration to utilities, combined heat and power and heat pumps. ESDU Data Item 89001.

Haywood, R. W. (1991). *Analysis of Engineering Cycles*, 4th edition. Pergamon, Oxford, UK.

IEA (2005). Urban community heating and cooling; the Southampton district CHP scheme. International Energy Agency, District Heating and Cooling programme. Downloadable from www.iea-dhc.org or www.southampton.gov.uk.

Jeday, M. R., Labidi, J. and Le Goff, P. (1993). A heat transformer for upgrading the waste heat of an industrial sulphuric acid plant. ChERD (Trans IChemE Part A), 71:A5, 496–502.

Kemp, I. C. (1991). Some aspects of the practical application of pinch technology methods, *ChERD (TransIChemE)*, 69(A6): 471–479, November.

Kim, J. -K. and Smith, R. (2004). Cooling water system design. *Chem Eng Sci*, 56: 3641.

Klemes, J., Dhole, V. R., Raissi, K., Perry, S. J. and Puigjaner, L. (1997). Targeting and design methodology for reduction of fuel, power and CO_2 on total sites, *Appl Thermal Eng*, 17: 993.

Linnhoff, B. (1990). Pinch technology for the synthesis of optimal heat and power systems, Trans ASME, *J Energ Resour Technol*, 111(3): 137–147.

Linnhoff, B. and Dhole, V. R. (1992). Shaftwork targets for low temperature process design, *Chem Eng Sci*, 47(8): 2081–2091.

Linnhoff, B. and Dhole, V. R. (1993). Targeting for CO_2 emissions for total sites, *Chem Eng Technol*, 16: 252–259.

Linnhoff March Inc., Leesburg, Va. (1991). Expanded pinch analysis procedure for pollution prevention at Amoco's Yorktown, Va. refinery. (Report prepared for Amoco Corp., Chicago, Ill. and United States Environmental Protection Agency, Washington, DC, 6 June publicly available.)

Morgan, S. (1992). Use process integration to improve process designs and the design process, *Chem Eng Prog*, 62–68, September.

Raissi, K. (1994). *Total Site Integration*, PhD. Thesis, UMIST, Manchester, UK.

Rossiter, A. P., Spriggs, H. D. and Klee, H. (1993). Apply process integration to waste minimization, *Chem Eng Prog*, 30–36.

Smith, R. (2005). *Chemical Process Design and Integration*. John Wiley & Sons Ltd, Chichester, UK.

Varbanov, P., Perry, S., Makwana, Y., Zhu, X. X. and Smith, R. (2004). Top level analysis of utility systems, *TransIChemE*, Part A, 82: 784.

6 Process change and evolution

6.1 Concepts

So far we have looked at ways of finding energy and cost targets, designing heat exchanger networks and choosing appropriate utilities for a given set of stream data, either from an existing process or from a proposed flowsheet for a new plant. But what would happen if we change the operating conditions on the plant? For example, we may wish to run a distillation column at a slightly different pressure and temperature. The streams associated with the column will also change in temperature, so the set of stream data will change. As a result, the energy and cost targets and the heat exchanger network will also be different. Indeed, can we find a way to systematically change the process conditions in order to reduce the overall energy targets?

Whether we are designing a new plant or trying to improve an existing one, there will often be a range of operating conditions that could be used. Of course, it would be possible to obtain targets for all the different possible conditions by simply repeating the problem table analysis each time. However, this trial and error method is time consuming and does not really improve our understanding of the process.

We want to know how a change in the conditions of one stream will affect the energy use of the overall plant, and to find the optimum new conditions. Pinch technology allows us to achieve this, using what is known as **process change** analysis.

Process change is defined as altering operating conditions or otherwise changing the process flowsheet to change the stream data, thus giving more opportunities for heat recovery; this can reduce energy targets or give simpler cheaper networks. Examples would include:

- changing the temperature of a distillation column;
- adding a pumparound, intermediate reboiler or intermediate condenser;
- changing the number of effects (stages) in an evaporator or flash system;
- modifying a dryer to use low-temperature heat or a different drying gas;
- changing a reaction temperature slightly.

Chapters 2–5 of this User Guide have covered heat exchanger networks and utility systems. However, these are only part of the overall process. A process plant generally consists of all or some of the following sections:

- A reactor, where the main chemical processes take place and impure product is formed from the raw materials.

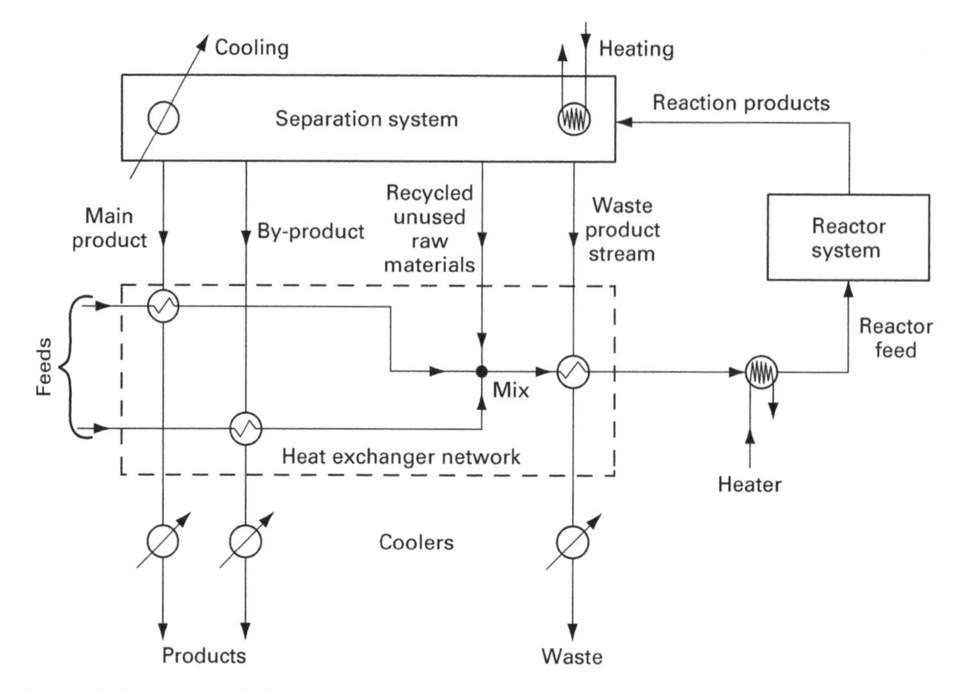

Figure 6.1 Schematic flowsheet of a typical process plant

- A separation system, which divides up the mixture of products, waste and unreacted raw materials emerging from the reactor using some separating agent, such as heat or a solvent; usually includes recycle streams.
- A heat exchanger network, which supplies the heating and cooling needs of the streams and may include heat recovery from hot streams to cold streams.
- Site utility systems, supplying the external heating and cooling requirements of the plant by fuel, steam, cooling water, refrigeration, etc.

A schematic flowsheet of such a process is shown in Figure 6.1.

Conventional design methods start by designing the reactor, then the separation system, then the heat exchanger network, and using utilities to supply the residual needs. This approach was illustrated in the "onion diagram" (Figure 1.3). Design starts at the centre and works outwards. The version shown in Figure 6.2 (in effect, a "bag of onions") reminds us that there may be heat integration between processes and that they often share site utility systems.

Pinch analysis goes a stage further by allowing the designer to work from the outside to the inner layers of the onion as well. This allows the reaction and separation systems to be looked at in the context of the heat flows of the overall process and site. Often, changes to the process can be found (usually in the separation system) which increase overall heat recovery and give a better integrated system.

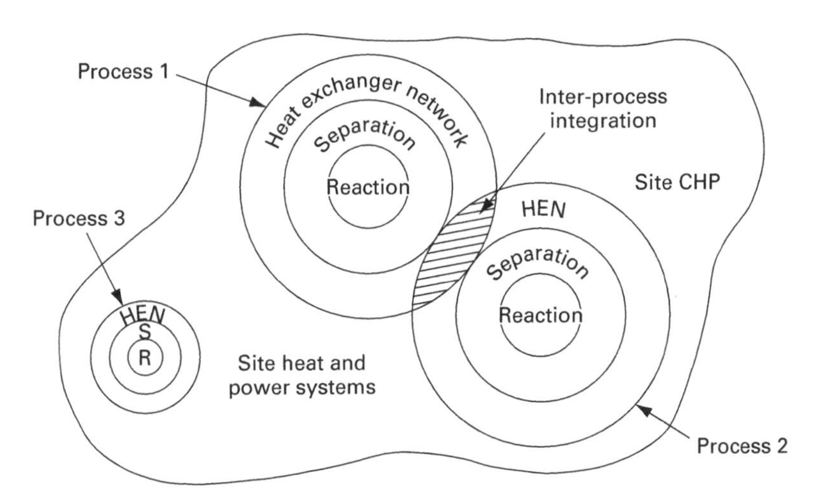

Figure 6.2 The onion diagram

6.2 General principles

6.2.1 The basic objective

So far we have always assumed that the temperatures of the streams are fixed and we cannot change them. Suppose that we were allowed to change their temperature; what would we want to do?

Consider a hot stream releasing 100 kW heat between 100°C and 50°C. What would happen if we were able to alter it to release the 100 kW between 110°C and 60°C, leaving all other streams unaltered? Firstly, new matches could become possible with cold streams whose temperature was too high to be heated by this hot stream at its original temperature, and additional heat recovery and lower-energy targets may be possible. Secondly, even if we kept to the original matches, the temperature driving forces have been increased, so heat exchanger area and capital cost go down. We are in a win-win situation; there is no trade-off. So ideally, we should **keep hot streams hot** – maximise their supply and/or target temperatures.

By applying the same logic to cold streams, it is clear that the ideal there is to **keep cold streams cold.** In both cases, the potential gains are threefold:

(i) actual reductions in energy targets, where heat exchange becomes possible with streams which were previously unsatisfied, especially in the pinch region;
(ii) more choices in stream matching, potentially reducing network complexity by allowing fewer separate matches, fewer loops and avoidance of "undesirable" matches;
(iii) increased temperature driving forces, giving smaller cheaper exchangers.

Thus, keeping hot streams as hot as possible and keeping cold streams cold can lead to either energy or capital cost savings, or even both!

6.2.2 The plus–minus principle

Consider the simple process whose composite curves are illustrated in Figure 6.3. The hot composite curve consists of only two streams; A between 20°C and 40°C, and B between 40°C and 120°C. If we wish to increase the heat exchange and reduce the utility requirements, how should we modify the composite curves? The answer is to increase the heat load of hot streams above the pinch or cold streams below the pinch, and conversely to decrease the cold streams above the pinch and hot streams below the pinch. This is the so-called "plus–minus principle" stated by Linnhoff and Vredeveld (1984). In Figure 6.3, the sections of the composite curves where heat should be added are marked with a plus, and sections where the heat load should be reduced are marked with a minus.

The formal statement of the plus–minus principle is that a process change will reduce the utility targets if it does one of the following:

(a) increases the total hot stream heat load above the pinch;
(b) decreases the total cold stream load above the pinch;
(c) decreases the total hot stream load below the pinch;
(d) increases the total cold stream load below the pinch.

Changes (a) and (b) will reduce the hot utility requirement. Changes (c) and (d) will reduce the cold utility requirement.

A more concise statement of the principle is that energy will be saved by a process change which:

- increases the proportion of the hot composite curve above the pinch, or
- decreases the proportion of the cold composite curve above the pinch.

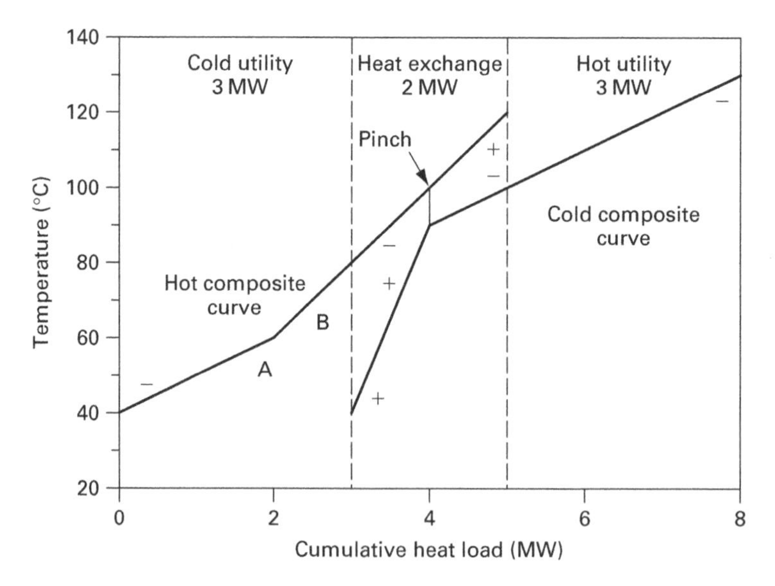

Figure 6.3 Illustration of the plus–minus principle

Suppose we have an opportunity to change the temperature of hot stream A. This stream is at present below the pinch, in a "minus" region. If we can move it to a "plus" region above the pinch, we should increase the overlap between the composite curves, increase the heat recovery and reduce the utility requirements at the same ΔT as before.

Let us change stream A to operate between 140°C and 120°C. Figure 6.4 shows that, as expected, moving 2 MW of heat load from below the pinch to above it has saved 2 MW of both cold and hot utility and has increased the heat exchange correspondingly.

The plus–minus principle is useful for first-stage screening. However, it tells us nothing about whether the proposed change to stream A is feasible and, more importantly, whether it will affect other streams in the plant. Often, the temperatures of a group of streams in a sub-section of the plant (e.g. around a distillation column) are linked together, and changing the temperature of one will affect all the others. Hence, the simple plus–minus principle on its own cannot identify process changes with certainty. However, it can point the way to potential savings and act as a simple check on whether a proposed change will be beneficial overall.

Note that process changes can include many simple energy-saving methods as well as complex plant modifications. For example, consider a stream which is piped between two parts of a plant and falls from 150°C to 140°C due to heat losses from poorly insulated pipework. If these heat losses are reduced by better insulation, the heat load and supply temperature of the following stream S will change. This is a process change because it alters the stream data. If the stream S is a cold stream, its heat load will decrease; if it is a hot stream, more heat will be released.

In either case, if the pinch is below 140°C, S is above the pinch. Hence, by statements (a) and (b) of the plus–minus principle, heat recovery will increase and the

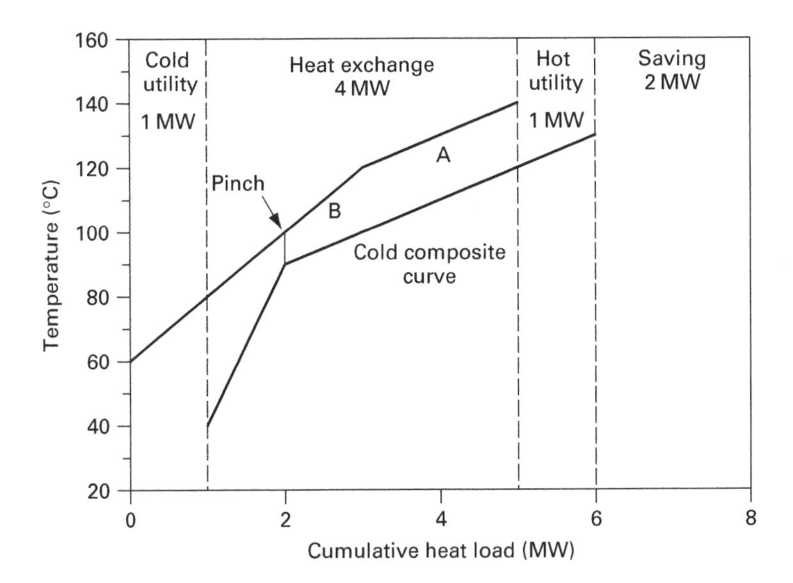

Figure 6.4 Composite curves of altered process

hot utility target will fall. Strangely, if the pinch temperature is higher and S is below the pinch, more insulation will merely increase the cold utility target! Admittedly, insulating a hot stream which is just below the pinch and being used for heat exchange will increase temperature driving forces and can again decrease the area requirement. However, there is no point at all in insulating a stream at relatively low temperature which is simply going to a cooler or to drain. This discovery has been exploited in some factories by removing the insulation from drain lines or even installing finned tubes, so that the heat which would otherwise be wasted is used for space heating of the factory.

6.2.3 Appropriate Placement applied to unit operations

The Appropriate Placement principle was described in Section 5.1.4. It was stated in a general form, in terms of heat sources and sinks. However, any hot process stream can be considered to be a heat source and any cold process stream is a heat sink. In particular, if a set of linked streams release heat below the pinch and require heat above the pinch, the system which they comprise is, in effect, transferring heat across the pinch. If the process conditions could be altered so that the system released heat above the grand composite curve (GCC) or received heat below it, the overall energy target would be reduced.

Here, too, we want to see how the heat sources and sinks from one part of the process relate to the GCC of the rest.

Figure 6.5 shows the GCC of a process which includes a distillation column. The reboiler at the bottom of the column is working at a shifted temperature of 130°C; the condenser at the top operates at a shifted temperature of 70°C. Both reboiler and condenser have heat loads of 3 MW; they are shown as thick horizontal lines in Figure 6.5.

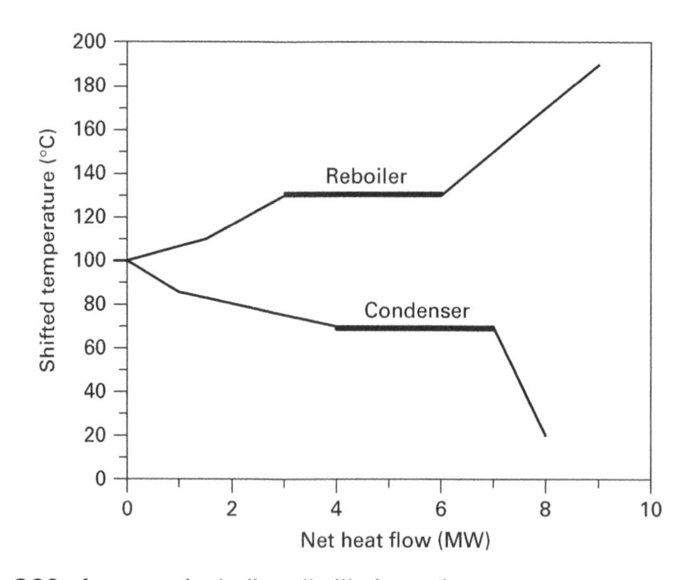

Figure 6.5 GCC of process including distillation column

It would be interesting to see how the distillation column relates to the rest of the process. To do this, we take out the reboiler and condenser streams and plot them separately. This is known as "splitting the GCC", a concept introduced by Hindmarsh and Townsend (1984). The streams which had been removed are known as the extracted process and streams which remain are known as the background process. Figure 6.6 shows the distillation column plotted as a box; it lies on the left-hand side of the GCC of the background process.

Looking at Figure 6.5 and Figure 6.6, we see that the reboiler requires heat above the pinch of the overall process and the background process, and the condenser rejects heat below the pinch. Therefore, this distillation column is working across the pinch. Figure 6.6 can be compared with the misplaced heat engine in Figure 5.3. We want to change the operating conditions of the distillation column so that it lies either above or below the GCC.

Figure 6.7 shows how this may be done. If the pressure of the distillation column is raised so that the condenser and reboiler temperatures increase by 60°C, the column fits entirely above the pinch and above the GCC. Alternatively, lowering the column pressure so that its temperature drops by 55°C will make the column fit below the pinch and below the GCC. In both cases, the distillation column can now exchange heat with the background process. Placing the column above or below the pinch is another application of the Appropriate Placement principle.

Of course, a change of this magnitude in the operating pressure and temperature of the distillation column will rarely be possible on an existing plant, and often will not be allowable even in new design, because of product decomposition, loss of separation efficiency, safety or other considerations. However, there are other ways of reducing heat transferred across the pinch.

One option is to install an additional "intermediate condenser" in the central part of the column, so that it works at a higher temperature than the main condenser at the

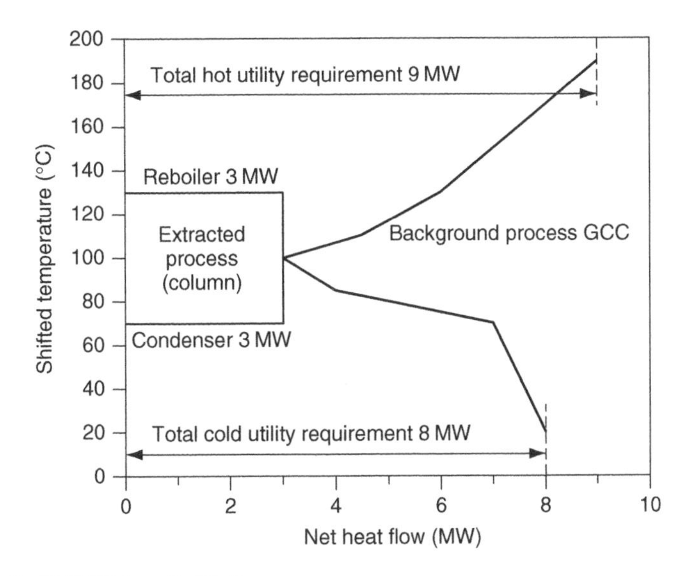

Figure 6.6 Split GCC for distillation process

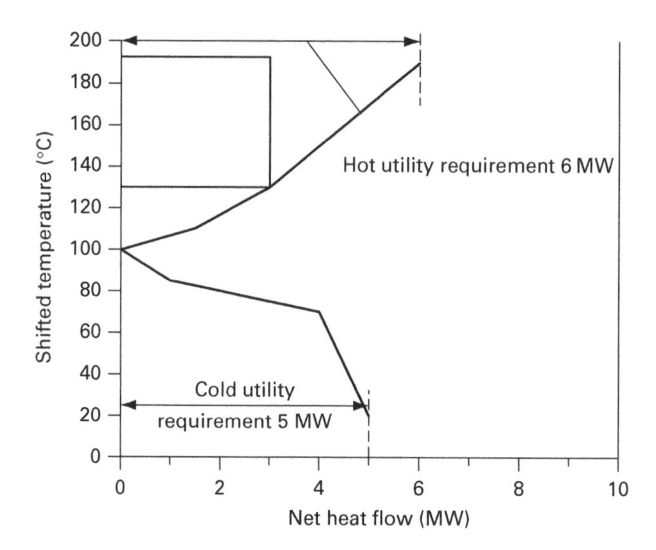

Figure 6.7 Distillation column fitted above background process GCC

top of the column. The split GCC then shows us instantly what temperature the intermediate condenser must be at to be above the pinch, and the maximum heat that can be recovered at any given temperature. For example, in Figure 6.6 an intermediate condenser at 110°C can recover 1.5 MW. Alternative methods are to use intermediate reboilers or pumparounds, and these will be discussed in more detail in Section 6.4.1.

The ideas outlined in this short section are the basis of all process change. It is clear that it is a very powerful tool which allows us to optimise the operating conditions and energy use of the entire process. In terms of the onion diagram, it means that we are no longer confined to working in one direction in design, from reaction system to separation system, heat exchanger network and site heat and power systems. Instead, we can have a two-way interaction. In many studies, the savings from process change analysis far outweigh those from heat recovery projects. In design of new plants in particular, it obviously makes sense to get the plant "right first time" and produce the most elegant and efficient solution. The detailed methods used and the results are different for the different kinds of separation systems and for reactors, and will be discussed in depth in the following sections.

6.3 Reactor systems

The reactor lies at the heart of the process, and often it is the reactor conditions which are chosen first when developing a process flowsheet, so as to maximise yield, selectivity and product quality. Understandably, designers are often reluctant to make major changes in the reaction conditions, even for a new plant. However, pinch analysis may suggest minor refinements which allow the process to integrate better. For an existing process, it is very unlikely that a change in reaction conditions will be allowed, except in the rare cases where it will actually give processing benefits (e.g. producing a greater proportion of a desirable component of a mixture).

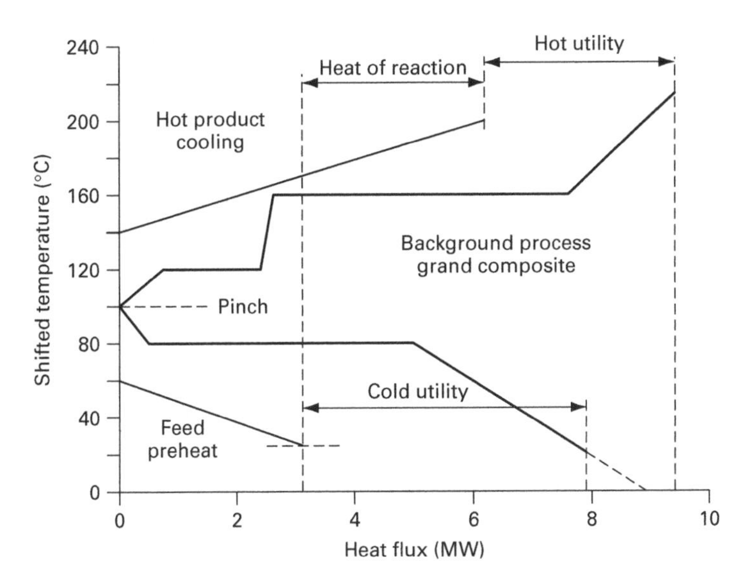

Figure 6.8 Exothermic reactor ideally placed above the pinch

The possible extent of heat integration of the reactor with the rest of the process depends mainly on the reaction temperature. Changing this temperature usually has three effects on the reaction system:

1. Altering the speed of the reaction – a 10°C temperature increase typically doubles the rate.
2. Altering the proportions of components produced in the output mixture; these depend on the competing reactions occurring, are highly case-specific and can again change greatly for a 10°C difference.
3. Altering the heat load of the reactor – usually a less significant effect than the other two.

If we wish to perform a pinch analysis of the reactor system, the first question is how to incorporate the heat of reaction into the stream data. Often, the heat absorbed or released in a reactor is absorbed directly by the liquid contents. The effect of the heat of reaction is that the temperature of the emerging product is different from the temperature of the feed streams at the reactor inlet. In effect, the reactor creates a "pseudo-stream" which is balanced exactly by the reaction heat.

However, some reactors must be maintained at a constant temperature or within strict limits, and therefore require external heating or cooling. This can be done by utilities or by heat exchange with the rest of the process. The Appropriate Placement principle applies; if the reaction is exothermic, the heat should ideally be released above the pinch and above the GCC; if it is endothermic, it should come below the pinch so that it can be driven by waste heat from the process. The placement can be studied by taking the reactor stream out of the process flowsheet, giving a split GCC. Figure 6.8 shows an exothermic reactor appropriately located so that it releases heat above the GCC. A feed preheat stream is included; this comes below the pinch so can be heated by waste heat.

Since only small changes in reactor conditions can usually be tolerated, there is little opportunity to change the reactor temperature to integrate better with the rest of the process unless it is very close to the pinch temperature or the temperature of a large latent heat load. Any changes would alter the product mixture significantly and would probably require a complete redesign of the separation system. The losses could outweigh the benefits from moving the reactor, for example if reflux ratios had to be increased. If the reactor has a reflux condenser, it may be possible to choose the flow and temperature of the recycle to integrate better with the rest of the process, without affecting the actual reaction conditions.

However, instead of moving the reactor, why not move the pinch? In many cases there is the opportunity for changing other parts of the process to fit better with the reactor. For example, in Figure 6.8 a further refinement has been added; instead of feeding the reactor with cold feed, it is preheated below the pinch. This will increase the heat that needs to be removed from the reactor, but since this is done above the pinch, it can be usefully used to heat the process.

Reactor systems have been analysed in depth by Glavic *et al.* (1988) and Smith and Omidkhah Nasrin (1993).

6.4 Distillation columns

Distillation columns are of great importance in process analysis, as they are both the most common and the most energy-intensive separation system, and hence they were also the first separation system to be analysed specifically from a pinch viewpoint, by Linnhoff *et al.* (1983). Since then, many additional insights have been found.

6.4.1 Overview of basic analysis method

In distillation columns, a two- or multi-component mixture of volatile liquids is separated by application of heat. The difference in relative volatility makes the composition of the vapour phase different to that of the liquid phase. However, the difference is almost never enough to allow effective separation in a single stage. A tower with multiple trays is generally used and a large proportion of the top vapour is condensed and recycled to the column. Generally, distillation is a very energy-intensive operation. It is also the most common liquid-phase separation system and is therefore treated in some depth here.

The method of splitting the GCC, described in Section 6.2.3, is generally very effective for analysing distillation columns. However, important assumptions were made in the simple example given there which simplified the analysis considerably. These were:

1. Both the reboiler and the condenser were at a constant temperature.
2. The reboiler and condenser loads were equal.

3. There was only a single distillation column in the system.
4. All column heating and cooling was provided via the reboiler and condenser.

In general, these assumptions do not apply. Let us see how to handle the general case.

6.4.2 Refinements to the analysis

6.4.2.1 *Sensible and latent heat loads*

The reboiler and condenser will only be at constant temperature if the whole of the heat loads are due to latent heat. In practice, this is often not the case. Two-phase mixtures frequently exist, and these can condense over a range of temperatures, according to the vapour pressure curve. Sensible heat may also be involved. A good example of this is the superheater which follows the boiler of a power plant.

The net result of these situations will be that the reboiler and condenser work over a range of temperatures. However, this does not cause major problems. The extracted process simply plots as a trapezium instead of a rectangle, with the top and bottom lines sloping. Possible changes are then explored in just the same way as before.

6.4.2.2 *Unequal reboiler and condenser loads*

This is a common situation. Heat losses from column, sensible heat brought in by feed streams and removed by hot product streams may all lead to differences between the reboiler and condenser heat loads. This may be handled in two ways.

If the difference between the reboiler and condenser loads is small, it may be ignored. The lower of the two heat loads is taken, and the trapezium is plotted as before. The small surplus heat load on the other unit which has not been balanced is left in the background process.

The advantage of this method is its simplicity. The effect of changes in column temperatures can still be found very rapidly by moving the box. However, the small unbalanced heat load has been left in the background process in the wrong position. When the eventual column temperature is selected, the stream data must be modified and the process re-run to give a corrected GCC.

The second method is to plot the actual column heat loads back-to-back with the GCC of the background process. Some care then needs to be taken in reading off hot and cold utility targets, but the representation is realistic. Both alternatives are shown in Figure 6.9, for a column which also incorporates a condensing stream working over a range of temperatures.

An alternative proposed by Kemp (1986) is to include the feed and product sensible heat streams into the column heat loads and turn the rectangle into a polygon. However, in most cases this additional complication is unnecessary. Remember that we are performing a preliminary search, not a highly precise optimisation. If we do identify a possible process change, it will be necessary to recalculate the stream data and energy targets anyway.

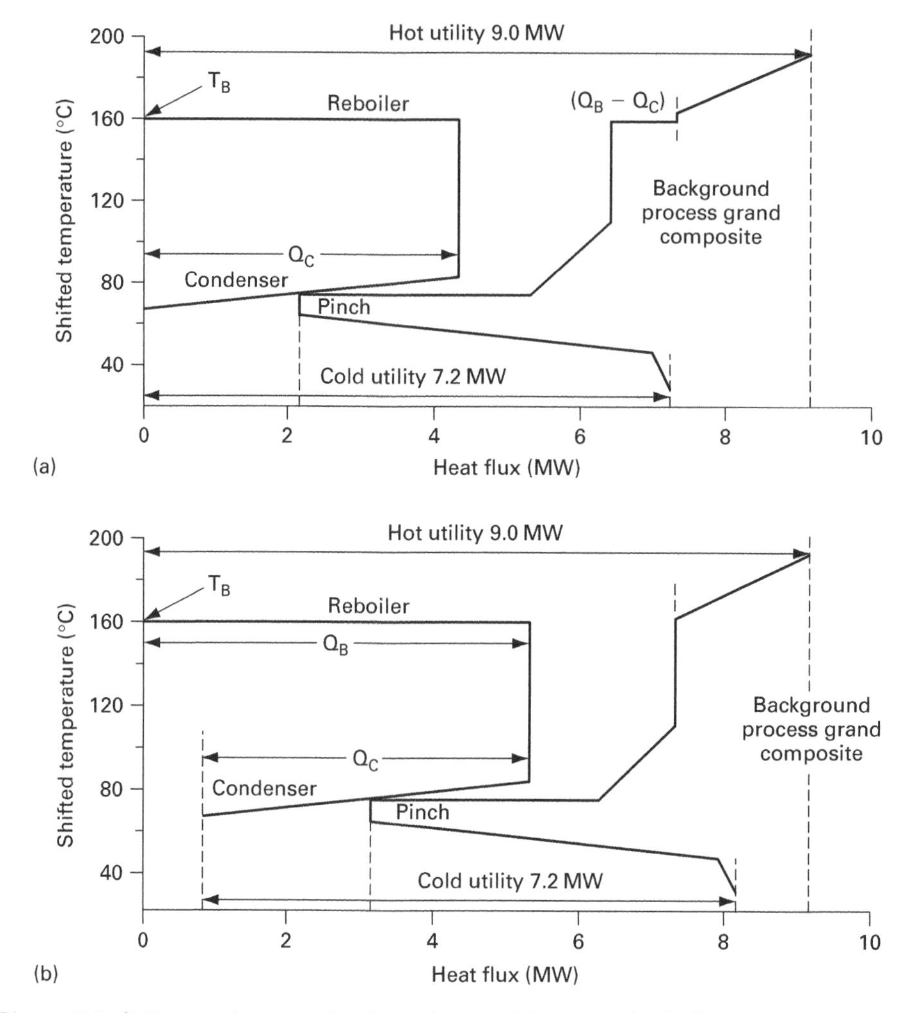

Figure 6.9 Split grand composite for column with unequal reboiler and condenser loads

6.4.3 Multiple columns

If the box representation can be used for one column, it can be used for several. So in a plant with multiple distillation columns, all can be plotted as separate boxes around the background process GCC. This can suggest quite a number of alternatives for altering column conditions. For example, in Figure 6.10 three columns are plotted and A and C lie to the side of B. They could be made to lie above each other by several methods:

- raise the pressure of B slightly and A considerably;
- lower the pressure of B slightly and C considerably;

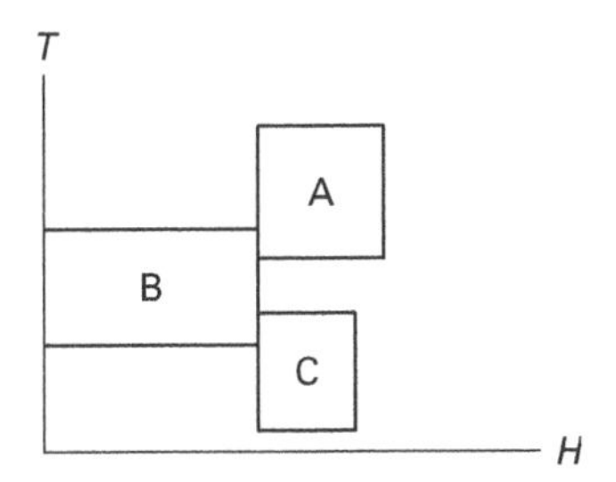

Figure 6.10 Multiple distillation columns illustrated by box representation

- raise the pressure of A slightly and lower that of C slightly;
- raise the pressure of A slightly (gives less energy saving as C still lies to the side of B).

Two warnings must be included. Firstly, the reboiler and condenser loads should be assumed equal; the representation becomes unworkable if multiple columns with unequal loads are plotted. Secondly, the box is only an approximate representation of the column and there may be alternatives to shifting the temperature and pressure of the whole column, as discussed below.

6.4.4 Distillation column profiles

It is not essential to do all the heating of a distillation column at the bottom, where the temperature is highest, or all the cooling in the top condenser. Several alternatives exist.

A **pumparound** is a large flow of liquid drawn off from a distillation column which releases sensible heat above the condenser temperature and is returned to a higher tray in the column. They are most commonly used in oil refineries.

An **intermediate condenser** recovers heat at a higher temperature than the condenser by condensing some vapour on an intermediate tray in the column.

An **intermediate reboiler** supplies heat at a lower temperature than the reboiler to evaporate liquid on an intermediate tray in the column.

Feed preheating will increase the temperature at which the liquid enters the column; some of it may flash to vapour on the feed tray.

The Appropriate Placement principle applies in all these cases, so, to save energy, a pumparound or intermediate condenser should lie above the pinch and the background GCC while an intermediate reboiler or feed preheating should lie below the pinch and GCC. Figure 6.11 shows the shapes of the modified boxes which would result in each case. The temperature of the column feed relative to the pinch of the background process is an important factor. If it is above the pinch, some of the stripping section of the column will be above the pinch and an intermediate condenser or pumparound is possible. Conversely, if it is below the pinch, an intermediate reboiler below the feed tray can also be below the pinch, as can feed preheating. If the feed tray is close to pinch temperature, little can be done unless the feed tray temperature itself is altered.

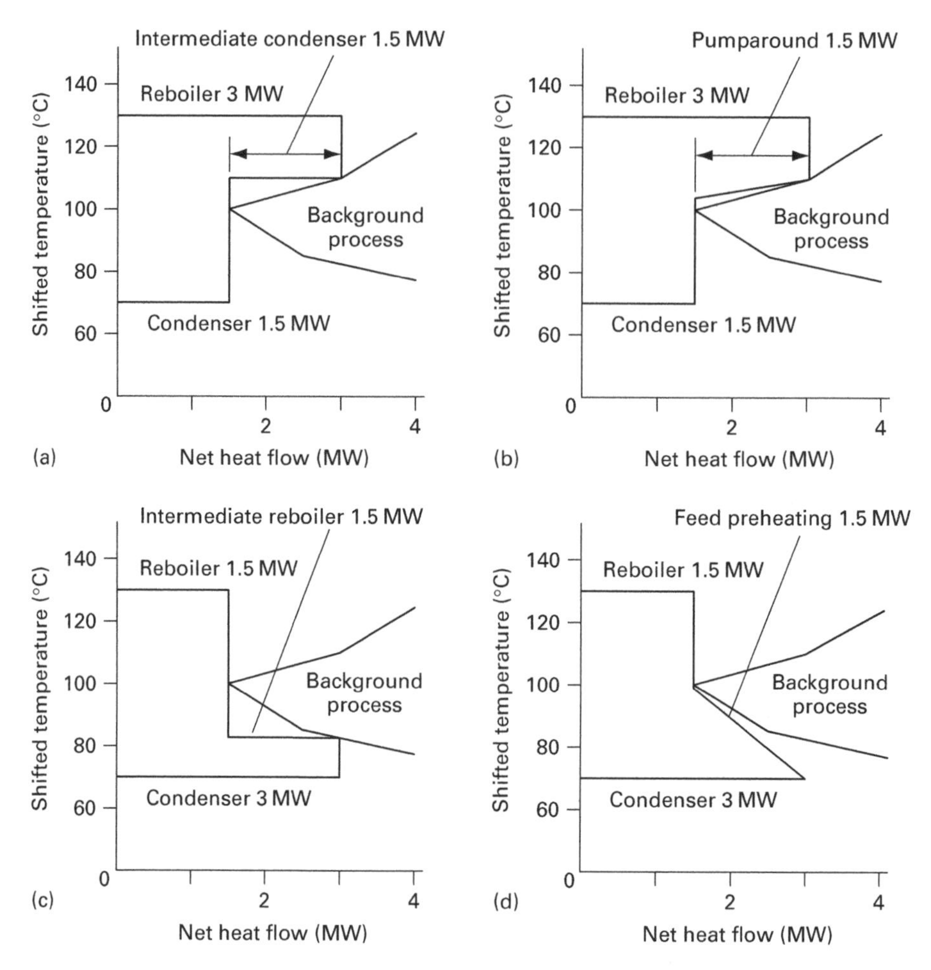

Figure 6.11 Distillation columns with additional heating or cooling

However, although Figure 6.11 shows us how much heat could be exchanged with the process at a given temperature, how do we know whether the distillation column itself will still operate effectively under the new conditions? All the suggested changes tend to reduce the liquid-to-vapour ratio L/V in the top part of the column (the stripping section) and increase L/V in the bottom part (the extractive section). This can adversely affect the separation efficiency. To restore the original product compositions, the reflux ratio may have to be increased, reducing the benefits from the energy integration. Originally there was no clear way of determining how great this penalty would be. However, work by Dhole and Linnhoff (1993) indicated how accurate **column profiles** could be generated which would allow the trade-offs to be assessed rapidly. Column profiles are not unlike GCCs in appearance, and behave in a similar way. They indicate at what temperature heat needs to be supplied and rejected up and down the column. The pinch point of

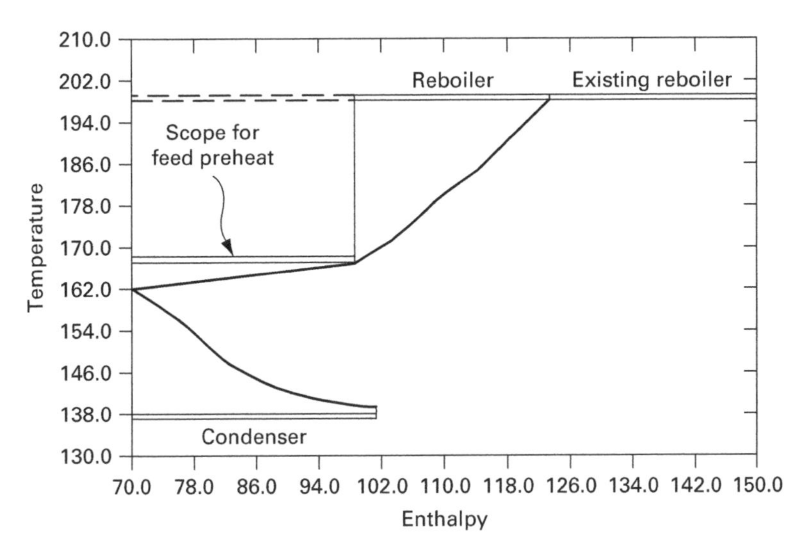

Figure 6.12 Column profile for a petrochemical distillation

the column profile is located at the column feed. Not all heat needs to be supplied at reboil temperature. Some can be supplied at lower temperature. Likewise, not all heat needs to be removed at condensing temperature. Partial heat removal at higher temperatures may be appropriate. The similarity to multiple utilities is clear.

Figure 6.12 shows a typical set of column profiles based on a real case; note that this also successfully handles unequal reboiler and condenser loads. There is scope for significant heat supply directly above the feed point. An intermediate reboiler would be possible, but as the heat is mainly required just above the column pinch, feed preheating is an excellent alternative, probably with much lower capital cost. Preheating, in this case, should result in a reduction in reboil duty more or less on a one-to-one basis.

Column profiles or similar graphs were described by Kaibel (1987) and Fonyo (1974), but only as theoretical concepts which could not realistically be computed for anything but ideal binary mixtures. Their computation for real columns was near impossible. However, Dhole and Linnhoff's techniques can generate graphs of this type with reasonable accuracy from a single converged tray-by-tray simulation, even for multi-component non-ideal mixtures.

A column grand composite curve (CGCC) and column composite curves (CCCs) can be produced in a tray-by-tray fashion, as shown in Figure 6.13. The CCCs describe vapour and liquid travelling up and down the column and depict available driving forces. The CGCC helps to assess the use of external heat sources and sinks, and indicates what heat loads might be placed on intermediate condensers and reboilers; the CCCs help interpret column internal processes, driving forces and capital costs. Both allow the designer to include the consideration of economics (reflux ratio) hand in hand with technical feasibility.

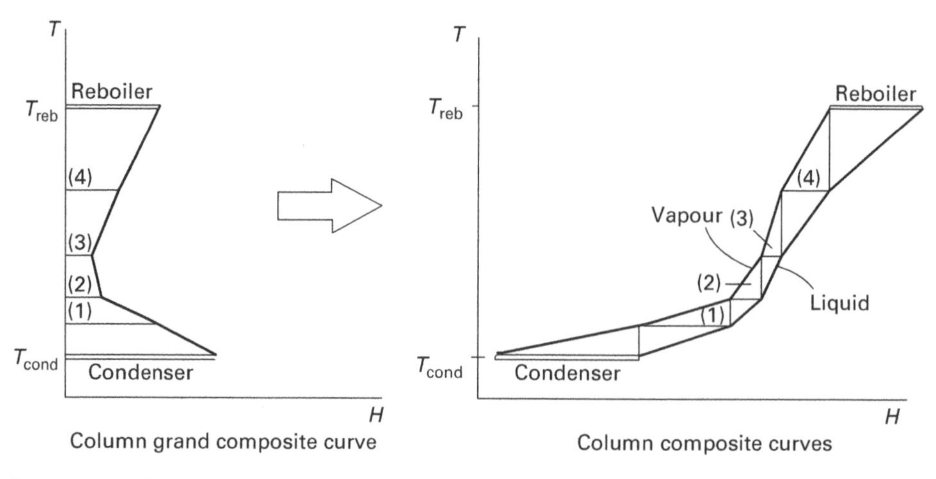

Figure 6.13 Column grand composite and composite curves

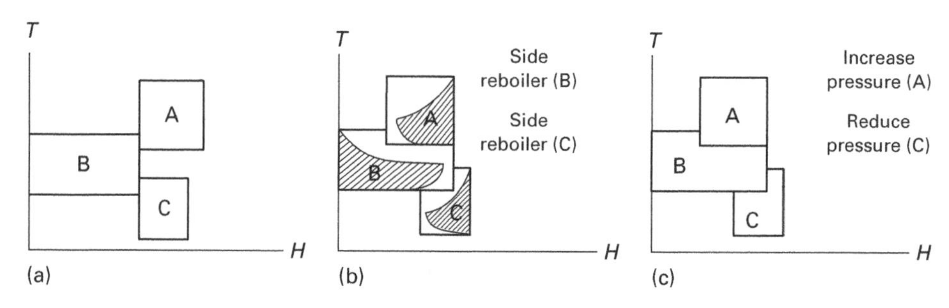

Figure 6.14 Multiple distillation columns: (a) box representation, (b) column profiles representation, (c) modified box representation

How should this information be included into the analysis? One possibility is to substitute the CGCC for the box representation of the distillation column directly. Returning to the multiple columns case from Section 6.4.3, the box representation (Figure 6.14(a)) indicates that for integration, the pressure in column A should be increased and the pressure in column C should be reduced. Consideration of the same problem in terms of column profiles (Figure 6.14(b)) gives a completely different assessment of the situation; a side reboiler in column B would enable integration to take place between columns A and B and a side reboiler in column C allows integration between columns B and C, and none of the column pressures need to be changed at all. Clearly this is a more attractive proposition. However, the column profiles do represent an ideal thermodynamic case with all heat being used precisely at the temperature at which it is generated, which would require very complex equipment to achieve in practice. So it may well be easier to deduce a suitable temperature for an intermediate reboiler from the CGCC and then simply modify the box representation, as shown in Figure 6.14(c).

One important warning must be included. When the possibilities for column modification have been considered and an alteration has been chosen, it is **essential** to re-simulate the column in detail. Changing the column configuration will frequently alter one or more of the reflux ratio, top and bottom composition and level of impurities in the products. Sometimes the extent of the changes is unacceptable and an iterative process is required to obtain the final optimum column design.

Heat pumping between the condenser and reboiler is another way of saving energy on a distillation column. Unfortunately, the temperature lift is usually too high to allow an economic system to be obtained. Intermediate reboilers or condensers may help to reduce the temperature lift. Mechanical vapour recompression (MVR) has been successfully applied to whisky stills.

6.4.5 Distillation column sequencing

Where a number of columns are to be used to split a multi-component mixture (say ABCD), the splits can be done in a number of different orders:

(i) AB/CD followed by A/B and C/D;
(ii) A/BCD followed by B/CD, then C/D;
(iii) A/BCD followed by BC/D, then B/C;
(iv) ABC/D followed by AB/C, then A/B;
(v) ABC/D followed by A/BC, then B/C.

Even for a simple three-component mixture, there are two alternative sequences, as shown in Figure 6.15. Choosing the best order can give even greater energy savings for multiple columns than optimising the relative column temperatures. A number of heuristics have been developed over the years for deciding sequences. Good rules of thumb are:

(i) Perform the easiest separations first, that is where there is a high relative volatility between adjacent components.
(ii) Remove the lightest components one-by-one (i.e. prefer the "direct sequence").
(iii) Remove a component which is a large fraction of the feed first.
(iv) Favour near-equimolar splits between top and bottom fractions in each column.

The last three heuristics quite often come into conflict and an alternative which has been proposed to replace them is:

(v) Prefer separations between components where the top product is 20–50% of the total flow.

These rules were developed for stand-alone sequences of columns without heat integration. However, Smith and Linnhoff (1988) found that the best stand-alone sequences also tended to heat-integrate better with the rest of the process, so there is a double benefit. Nevertheless Smith (1995) re-emphasised the limitations of simple heuristics, which often tend to conflict with each other and merely leave the designer confused.

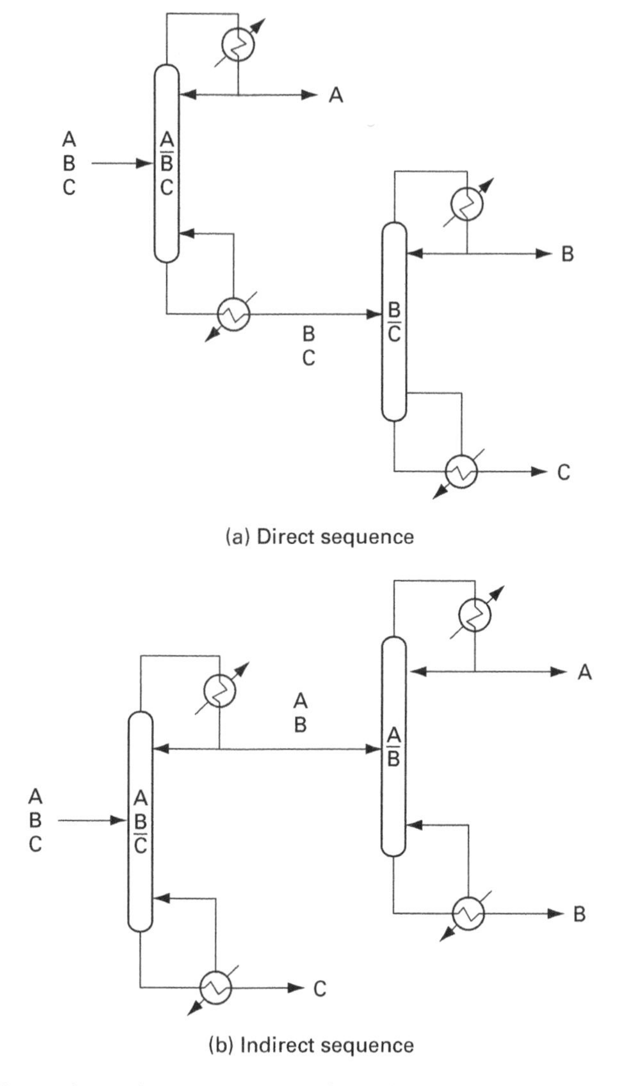

Figure 6.15 Alternative column sequences for a three-component separation

In recent years, increases in computer power have allowed the development of software which evaluates all possible sequences, even for problems with a large number of components. A good example is the CHiPS package developed at Edinburgh University (Fraga and McKinnon 1994). The most difficult and time-consuming step is to evaluate which of the sequences is actually best in practice. Not only energy, but also features such as fouling, operability and safety, must be allowed for. It is important that engineers still know and understand their plant and do not simply abdicate the responsibility to the computer.

Rajah and Polley (1995) suggested that heat integration and preferred separation schemes could be easily deduced by making a Heat Load Table – a list of the expected heat loads and temperatures of the reboiler and condenser for each split at the expected pressures. Comparing these values would show rapidly whether any possibilities for integration existed, and whether a change in column pressure would be useful or necessary.

Smith (1995) suggested an alternative approach. The cost of the energy and capital for a single column tends to be linked to the vapour load (flow rate of vapour from the top of the column). So by summing the vapour loads for all the columns, an indicator of total cost for a sequence can be found, and this can be compared with other sequences. What is now needed is a simple expression for vapour load, and this has been provided by Porter and Momoh (1991). By simplifying the Underwood equation, they obtain the following expression for a single column:

$$V = F_L + FR_F/(\alpha - 1)$$

where F is the total molar flow rate, F_L is the molar flow rate of all components lighter than and including the light key, α is the relative volatility of light key and heavy key, and R_F is the ratio between the actual reflux ratio and the minimum theoretical reflux ratio; R_F can typically be taken as 1.1. It is easy to evaluate this expression for all the possible sequences using a spreadsheet.

Smith also pointed out that there may be 20–50 possible sequencing options in a typical multi-component process, and there is usually little to choose in energy terms between the best five or so. So the choice between them should be made using other factors, for example plant layout, safety or heat integration.

6.4.5.1 *Complex columns and side strippers*

So far we have considered only simple columns, with a reboiler and condenser and with the key components adjacent in volatility. However, more complex arrangements can be considered. For example, the vapour from one column can be fed directly to the following column, without condensing it first. This will give capital savings (smaller condensers and reboilers required) and frequently also energy savings (the loss of temperature driving forces in the condenser and reboiler is avoided). There are several other possibilities:

(i) Take three products from a single column – often attractive if the middle component has the largest mole fraction and one of the others is small.
(ii) Use a prefractionator before a single column where three products are being produced, as in Figure 6.16(a) – this can allow a pure middle product to be obtained and typically saves 30% on energy compared to separate simple columns.
(iii) Thermal coupling of columns – with the top or bottom product from one column being passed directly to another column with a corresponding return flow, as in Figure 6.16(b).
(iv) Take a side-product from the middle of a column and purify it in a separate column known as a side-rectifier (for a top product, as in Figure 6.16(c)) or a side-stripper (for a bottom product).

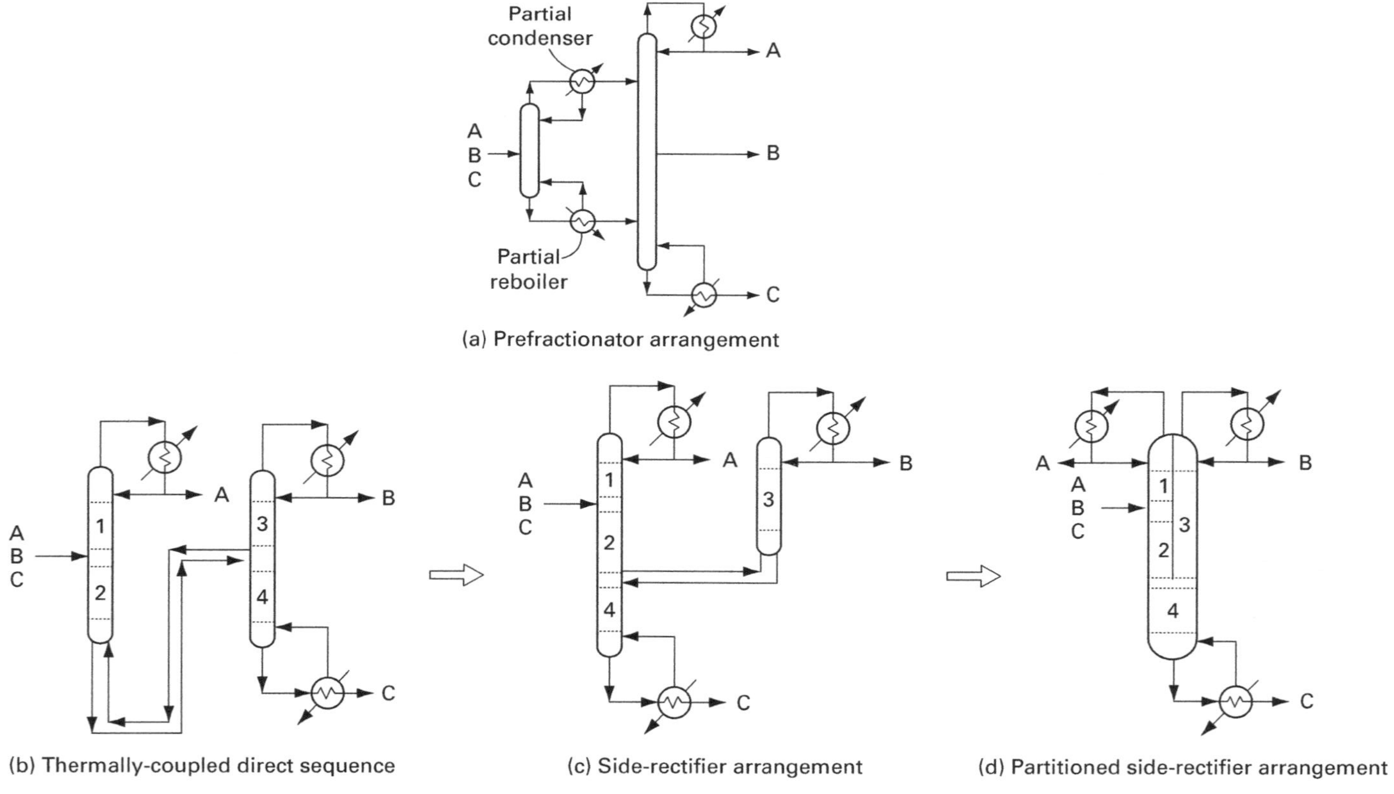

Figure 6.16 Thermally coupled columns and side-rectifier

(v) Combine a side-column such as a prefractionator directly into the same shell as the main column, producing a "dividing wall" or "partitioned" column, Figure 6.16(d). Since only one shell is required, this has a lower capital cost than two separate columns.

Thus, for a separation between three components, there are several alternatives; three simple columns, a prefractionator/column system (with or without thermal coupling), a column/side-stripper or column/side-rectifier pairing or a single dividing wall column may be used.

Smith (1995) reviews the various alternative arrangements in some depth. In general, complex columns give reduced latent heat loads and an energy saving compared with simple columns. However, the overall temperature difference between reboiler and condenser tends to be greater than that for the individual simple columns (on the split GCC, two or more small boxes are replaced by one tall one) and heat integration of the remaining loads will be more difficult. Hence the overall energy consumption of the fully heat integrated system may not be lower for complex columns. Smith therefore recommends that thermal coupling should not be considered until an initial overall design has been established, and the heat integration targets with simple columns have been found.

6.5 Other separation systems

6.5.1 Evaporator systems

Evaporators are one of the simplest and most clear-cut separation systems. A volatile solvent, often water, is vaporised to remove it from a solution containing an involatile solute. The separation is thus much sharper than in a distillation column and an evaporator is very close to an ideal stage.

The heat required by an evaporator is dominated by the latent heat of evaporation of the liquid. There may also be smaller contributions from sensible heating/cooling of feed streams and condensate.

Typically, the heat required for evaporation is provided by condensing steam or hot vapours in a heat exchanger immersed in the evaporating liquid, known as a calandria. As with all heat exchangers, the fluid inside the calandria must be hotter than the liquid in the evaporator, so that heat transfer can occur. However, quite low temperature differences down to 1–2°C may be employed, especially in plate units.

The amount of water to be evaporated, and hence the heat required for vaporisation, is usually fixed by the process flowsheet. There are three standard methods for reducing the energy requirement of the system while maintaining the required evaporation rate:

1. Use multi-stage evaporation. The vapour from one evaporator is condensed to provide heat to another stage at a lower temperature. These stages are generally called effects. The heat is being re-used in each effect and it is easily seen that the energy consumption of the system is inversely proportional to the number

of effects. However, if a single effect is replaced by multiple effects, either the total temperature difference over the system must be greater or the temperature driving forces in each effect will be lower. Multiple effect systems therefore have higher capital cost to offset the lower energy cost.

2. Use MVR. The vapour from the effect passes to a compressor and its pressure is increased. It therefore condenses at a higher temperature than that at which it was evaporated, and can be used to reheat the same effect. This is a form of mechanical heat pump, and can give very large savings – a 1 MW compressor has been known to upgrade 15 MW of vapour. However, its capital cost is high and it is uneconomic unless the temperature lift is small. The vapour should be at or above atmospheric pressure, because the volume of gas is inversely proportional to its pressure and vacuum systems therefore require large and expensive compressors; moreover, their mechanical efficiency is lower.
3. Use thermal vapour recompression (TVR). This is another kind of heat pump, but with significant differences. Cheap steam can be used instead of expensive mechanical or electrical power, and capital cost is less, but the percentage of heat which can be upgraded is much lower; 1 MW of driver steam will upgrade 0.1–0.5 MW vapour, depending on the conditions. Quite high temperature lifts can be obtained; TVR may be used to upgrade vapour in the middle of a multi-effect evaporator train.

All these methods have been extensively used in industry. Evaporators with up to seven effects have been used in the dairy and sugar processing industries. However, the law of diminishing returns applies; in an evaporator system with a total vaporisation load of 1 MW, going from a one-effect to a two-effect system saves 0.5 MW but adding an extra effect to a four-effect system saves only (0.25–0.20) MW or 0.05 MW – one-tenth as much. Also, the large temperature drops required in a multi-effect system demand a high initial steam temperature and pressure, or a low final vapour temperature and pressure (necessitating vacuum pumps or ejectors), or low temperature differences within effects and hence large areas of heating surface. So three or four effects are the usual maximum, even on large bulk chemicals plants. Likewise, MVR around such a system will involve a very large temperature lift, the thermal efficiency will be low and the capital cost very high.

6.5.1.1 *Analysis by pinch methods*

The basic method of analysis, as with distillation columns, is to split the GCC. The reboiler and condenser loads of each effect are taken out of the background process and plotted as a box. Generally the boiling and condensing loads for each effect are almost equal, especially for backward-feed evaporators, and it plots as a rectangle. A multi-effect system will plot as a "stack" of boxes; the ΔT_{min} contribution of the evaporating and condensing streams will often be lower than that for streams elsewhere in the process.

The position of the evaporator system is checked against the Appropriate Placement principle; if it lies partially or wholly to the side of the background GCC, modifications can be considered. A major benefit is that altering the conditions in evaporators has very little effect on the separation because it is sharp. This contrasts sharply with distillation columns, where changing temperatures and pressures or adding intermediate reboilers affects the separation efficiency and reboiler load in a complex manner. As a result, there is considerable scope for tailoring an evaporator system to fit ideally with the remainder of the process. Effects can be added – or removed; their working temperatures can be changed; the heat loads on different effects need not be equal if heat is exchanged directly with other process streams; and inappropriately placed effects can be altered without significantly affecting the rest of the system.

As an example, Figure 6.17 shows the split GCC for a system containing a four-effect evaporator; the total evaporation load is 12 MW. Steam is supplied at 170°C (a shifted temperature of 160°C), the ΔT_{min} for each effect is 20°C and the final vapour has a condensing temperature of 90°C (shifted temperature 80°C). The

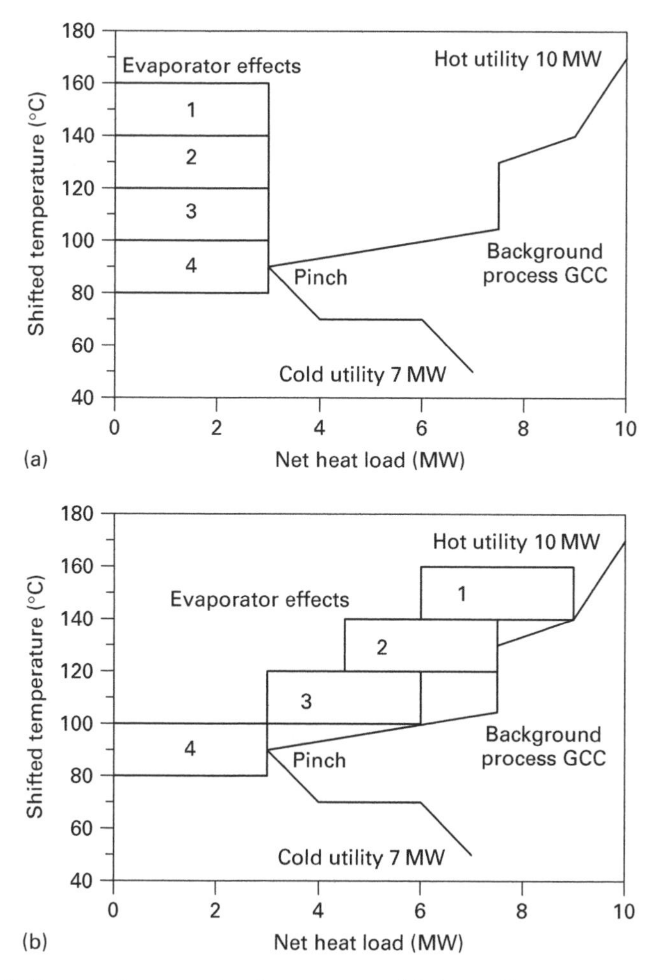

Figure 6.17 Split grand composite for process with four-effect evaporator

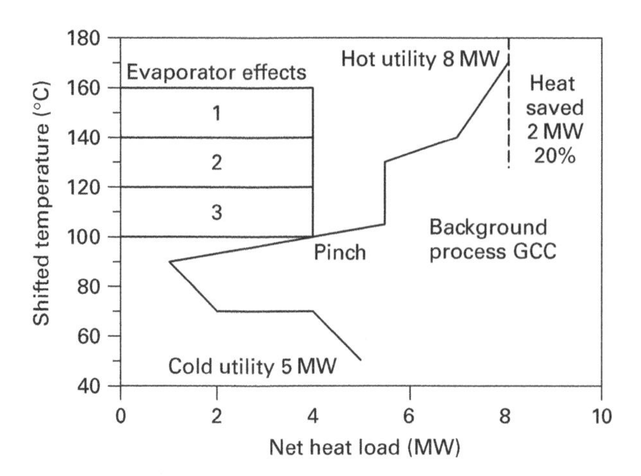

Figure 6.18 Evaporator system with three effects

latent heat loads of each effect plot as horizontal lines, each of 3 MW. We can see that the use of multiple effects has considerably reduced the process heating requirement, as it is only 10 MW in total and most of this is clearly due to the background process.

Can we do better? Treating the evaporator as a unit operation suggests only one possibility – add another effect. We can calculate the current energy consumption as (12/4) or 3 MW and the new consumption as (12/5) or 2.4 MW, a saving of only 0.6 MW. A 6% saving looks unlikely to justify the cost of additional pipework, tanks and heating surface. Moreover, assuming the steam supply temperature to the first effect is fixed, the additional effect would have to go below the existing ones. The final vapour temperature would then be only 70°C and a high vacuum pump would be needed as this corresponds to a pressure of 0.31 bar (absolute).

For the split grand composite, the evaporator effects can either be plotted in a vertical line (Figure 6.17(a) or touching the background GCC (Figure 6.17(b)). Both methods show clearly that the evaporator system straddles the pinch, thus incurring a 3 MW energy penalty, but it is only effect 4 which is responsible.

With the culprit clearly identified, we can suggest ways of removing the energy penalty. The first possibility is to remove effect 4 and redistribute the heat load to the other effects, so that they each evaporate 4 MW instead of 3 MW. Figure 6.18 shows that this cuts the energy penalty to 1 MW. The pinch has now moved and is caused by the condensing vapour from effect 3.

We could remove effect 3 by the same principles, but at those temperatures the background GCC would be unable to absorb the additional heat which would be added to effects 1 and 2, and there would be no net saving. But there is no need to remove effect 3 completely – we simply need to reduce the load on it by 1 MW. This can be done by transferring the load to the other effects and taking some of the vapour from effect 2 and use it not to heat effect 3 but to heat the process directly. The result is shown in Figure 6.19. The effects have now been given unequal evaporation duties; this is known as load shifting.

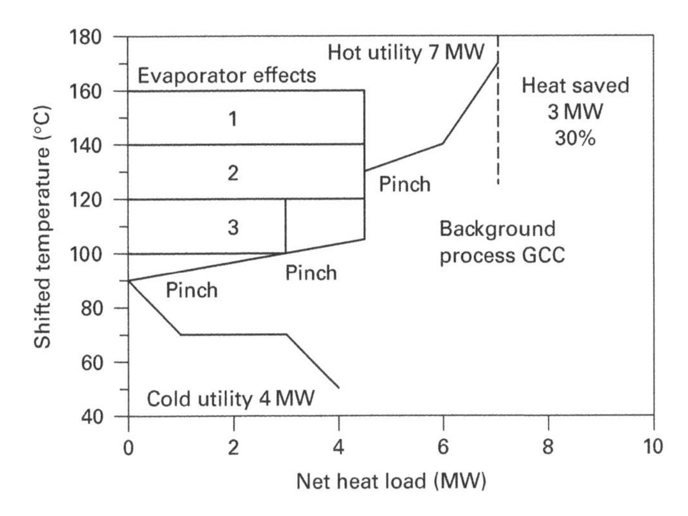

Figure 6.19 Evaporator system with three effects and load shifting

Let us compare Figures 6.17 and 6.19. The advantages of the new system are clear:

1. There is a 30% energy saving.
2. There is one less effect, so fabrication and piping costs are lower.
3. There is no need for a vacuum pump as previously used on effect 4.

Thus, pinch technology has given both capital and energy savings compared with conventional design. Moreover, the energy was saved by reducing the number of effects, a move which would seem ridiculous by unit operations principles. Again, we see the importance of process integration – that it gives an overview of the whole process and shows how one part affects another.

This, however, may seem little comfort to the engineer who already has a system like Figure 6.17 installed. The costs of enlarging all three of the remaining effects would be considerable. For a retrofit situation, therefore, we look for other alternatives.

Effect 4 is the only one which violates the pinch; we wish to move it to a position where its 3 MW heat load can be released to the process. Careful examination of Figure 6.17 shows that this is the case above 140°C. So we repipe effect 4 so that it is in parallel with the rest of the system rather than in series, and feed it with live steam like effect 1. The extra steam used is compensated for by the heat released from effect 4 into the process. The result is shown in Figure 6.20. With only one repiping and no additional surface area, we have removed the penalty.

We should note in passing that a few auxiliary changes will be needed with this retrofit system. Twice as much feed liquor as before must be heated to the highest temperature, so either the energy saving will be slightly less than shown or additional heat exchange between feed and product will be required. However, sensible heating loads are small compared with latent heat ones, so the difference is minor.

MVR can be considered as an alternative – or in addition – to changing the configuration of effects. MVR around the entire evaporator would involve a temperature

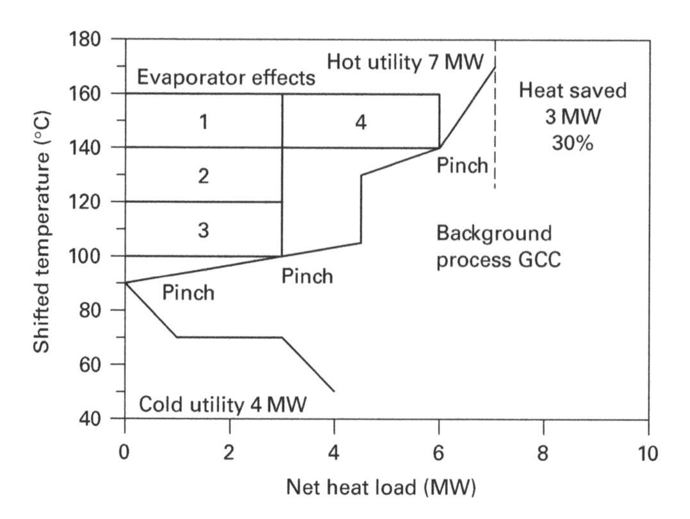

Figure 6.20 Retrofit to evaporator system with two effects in parallel

lift of 80°C, which will almost certainly be uneconomic. However, Figure 6.17 showed us that only effect 4 is across the pinch, so we only need to pump round this effect. This will save 3 MW of heat at a temperature lift of only 20°C.

For a retrofit, then, we have the option of shifting effect 4 or installing MVR. Both will save 3 MW of heat, although the MVR will require a small power input to the compressor. The choice between the options will depend on the comparative costs of repiping, moving tanks, power consumption and a steam compressor. It should be noted that MVR generally produces superheated steam – especially if compressing steam well below atmospheric pressure – and some form of desuperheating will be needed before it is re-introduced to the calandria.

If we are installing an evaporator in a new process, even more possibilities may be explored. From the process flowsheet we can find the amount of water to be evaporated and hence the total latent heat load. If we choose which temperatures the system will work at, we can plot the resulting box on the GCC of the rest of the process and see how it relates to it. We can see whether the Appropriate Placement principle has been obeyed, and develop alternative systems. If we change the number of effects, this can either alter the ΔT in each effect or the operating temperature range of the entire system.

Let us take an example. We have a system for which the background GCC is shown in Figure 6.21 and we wish to evaporate 5 kg/s of water. At a latent heat of 2,400 kJ/kg, this corresponds to a heat load of 12 MW. Low-pressure steam is available at 140°C and we can pull a slight vacuum to condense vapour at 95°C. The ΔT_{min} for the process is 10°C, so that the supply steam has a shifted temperature of 135°C and the condensing vapour shifted temperature is 90°C. We can thus plot a "feasibility box" for the system which is the large box in Figure 6.21. However, a single-effect system working over this temperature range would not be very efficient. Although it is above the pinch, it fits partly above and partly below the GCC. No less than 7.5 MW of heat cannot be absorbed by the process at 90°C shifted temperature and would be transferred across the pinch.

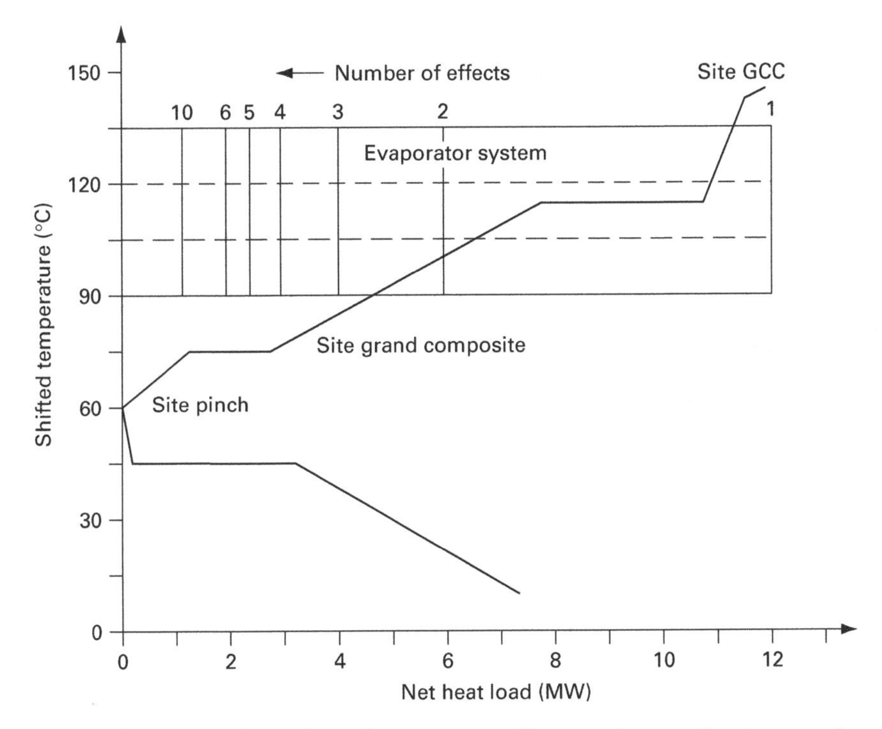

Figure 6.21 Selection of number of evaporator effects and operating temperature

If we use a multiple-effect system with n effects, the external heat load required will fall by a factor $1/n$. So a two-effect system will only require 6 MW of external heating and a three-effect system will need 4 MW. Figure 6.21 shows that the three-effect system working at the original temperatures is completely appropriately placed; all the heat released by condensing the vapour from the final effect can now be used to heat the rest of the process. So there is a substantial energy saving – 7.5 MW. But the ΔT across each effect is lower than for the single-effect system; 15°C as against 45°C. So the heat transfer surface area will have to be larger and the capital cost will be greater. Also, three sets of equipment will be needed instead of one.

There are further alternatives. Maintaining the same ΔT of 15°C across each individual effect, we could run a two-effect system working between 135°C and 105°C shifted temperature. Only two sets of equipment are needed and there is no need for a vacuum pump. Or one could even use a single-effect system working between 135°C and 120°C, which would incur an energy penalty of only 1 MW and give further simplifications.

In an existing system, similar trade-offs occur. However, because effects of a given size already exist, the biggest benefits again tend to come from shifting the temperature of effects rather than adding new ones. In the example above, the three-effect system could be transformed to the one-effect system simply by linking the effects in parallel rather than in series. We must also note that for an existing effect of given area, if we change the ΔT across the effect we will also change the heat load in inverse proportion.

Figure 6.22 shows some alternative layouts for an evaporator system of a given total area. System (a) can be achieved with six small effects, or three large effects, or six small effects in parallel pairs. If the upper temperatures in this system are judged to be too high, system (b) is split across the pinch. This could require a high vacuum to generate the vapour from the final effect, so (c) shows a similar system which explots the pocket in the GCC and has a higher vapour outlet temperature. System (d) shows the opportunity for using effects with unequal heat loads. If the effects must all be the same size (e.g. because an existing six-effect system is being reconfigured) then (e) shows a way of doing this, with the first two effects in parallel.

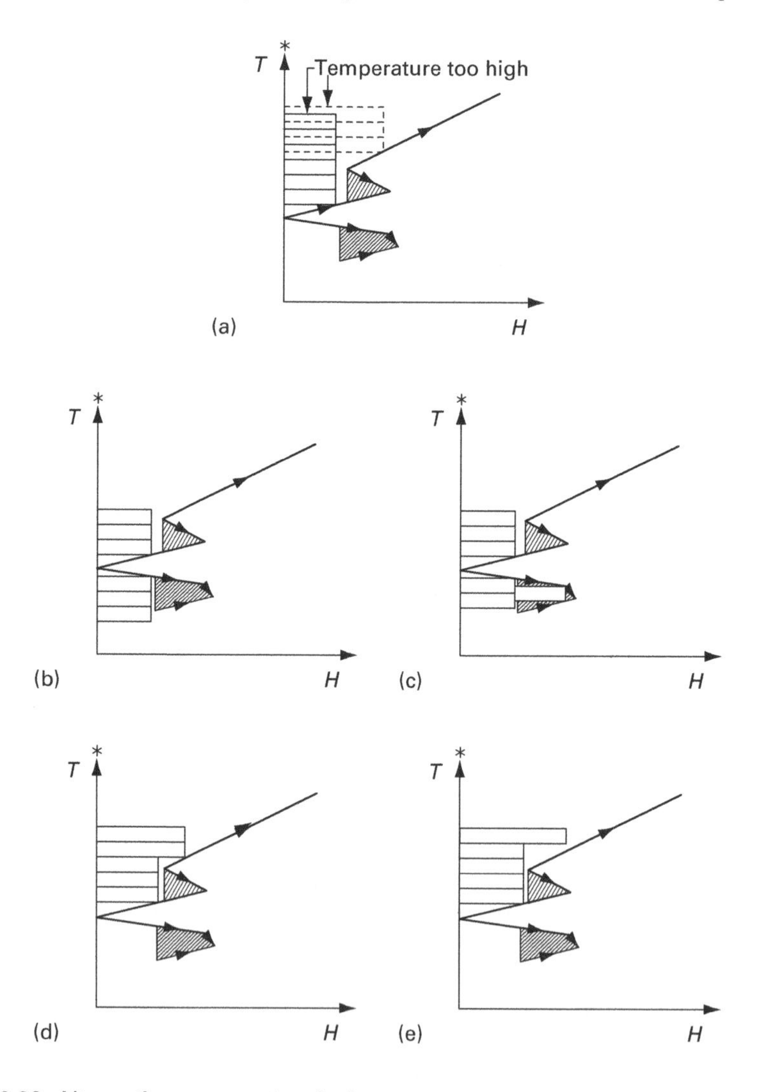

Figure 6.22 Alternative evaporator designs

An important point to bear in mind in all cases is that there are sensible heat streams associated with the evaporator as well. Their stream data will change as the effect temperatures and loads are changed, but this is not allowed for by the analysis. Thus we should not get carried away with the elegance of "packing boxes" ever more tightly around the GCC. It is important to recalculate the stream data rigorously and obtain new targets, especially if there are two or more alternative designs which are very different in structure but similar in targets. After retargeting, the calculated relative benefit of these schemes could reverse.

An evaporator system for a gelatine plant is described in the Case Studies section. Evaporators have been covered in numerous papers including Smith and Linnhoff (1988) and Smith and Jones (1990).

6.5.2 Flash systems

Flash systems are frequently used as a cheap cooling method, with the additional advantage of achieving some separation. A hot liquid is passed into a vessel at lower pressure; the temperature drops and the sensible heat evaporates some of the liquid, which is drawn off at the top of the tank.

Heat recovery from a slurry can be difficult in conventional exchangers because the solids can accumulate on the heat exchange surfaces. Fouling is worse than with pure liquids and blockages are possible. In many cases, a flash system provides the best solution to the problem. The clean top vapour is condensed and the heat recovered to the rest of the process. Such a scheme was successfully demonstrated at William Grant's distillery in Girvan, Scotland, with the aid of a grant from the Department of Energy. The recovered heat was used to heat water used in the cooking process, reducing the amount of steam heating required.

However, in a flash system the energy is all recovered at a single, low temperature level. In contrast, if it were recovered in a heat exchanger, the heat will be released over a range of temperatures. The question is, how does this affect the overall energy consumption of the process?

Process change analysis can be used here as well to find the answer. Look at the GCC in Figure 6.23(a), for a process including a hot slurry stream between 140°C and 80°C (shifted temperatures). This stream is taken out of the analysis to yield the background process, and this gives another form of split GCC as shown in Figure 6.23(b). It can be seen that all the heat from this stream can be recovered into the process. The hot utility requirement is 0.4 MW.

Now let us suppose we flash the stream instead. The same amount of heat will be recovered from the condensing vapour, but at a single temperature; it can thus be represented by the horizontal line in Figure 6.23(b). It is clear that this does not lie fully above the GCC and that some of the heat will therefore be wasted. So the flash has caused an energy penalty of 0.4 MW and the overall hot utility use will be doubled to 0.8 MW, as can be seen in Figure 6.23(c).

If the flash were carried out at a higher temperature, the horizontal line would lie above the background process GCC. The remainder of the heat could then be

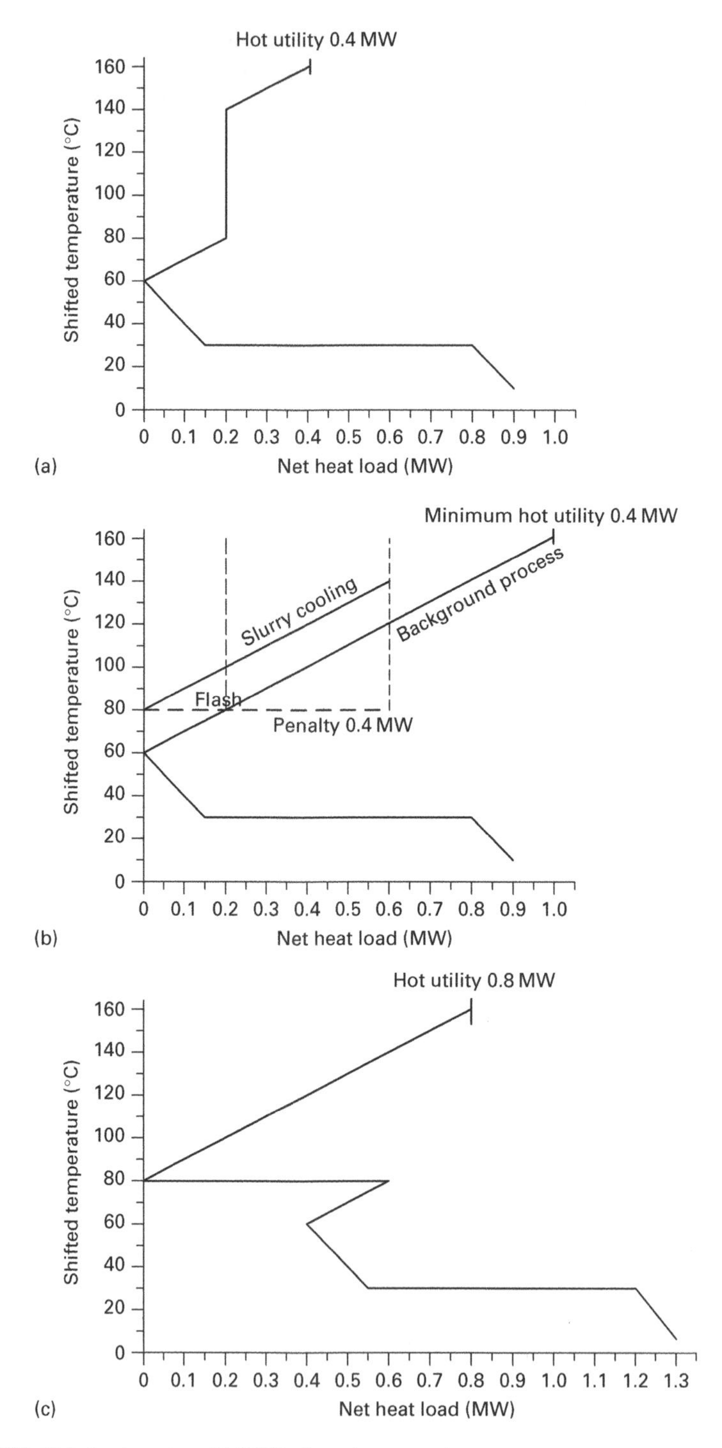

Figure 6.23 Original and split GCCs for slurry system

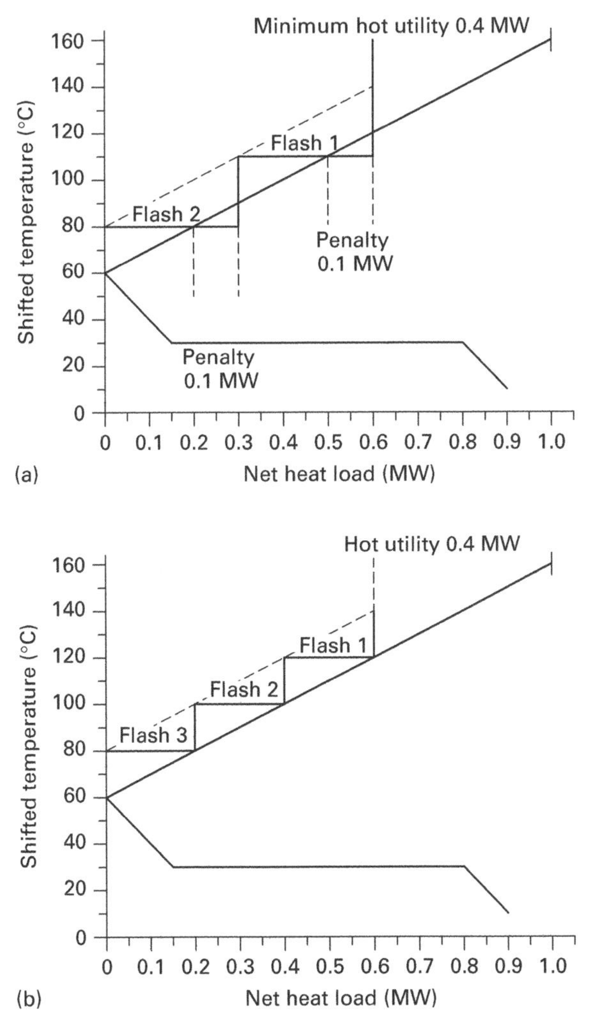

Figure 6.24 Optimisation of a flash system for heat recovery

recovered by heat exchange. This has two disadvantages. Firstly, a heat exchanger containing a slurry is still required. Secondly, less liquid is removed from the slurry by flashing, and this is often undesirable for process reasons. Downstream flows will be greater, and in many cases a flash is being used deliberately to remove water from a slurry; otherwise, an evaporator would be needed.

The solution is to substitute a multi-stage flash. This will achieve the same total separation as the single flash but recover the heat over a range of temperatures. Moreover, a simple construction on the GCC will show how many stages are required in the system. This is illustrated in Figure 6.24. It can be seen that a two-stage system will still incur an 0.2 MW energy penalty (giving 0.6 MW hot utility overall), but a three-stage system will fit entirely above the background process GCC.

What is the result? As with the evaporator, we have achieved precisely the required amount of separation. We have also ensured there was no energy penalty compared with using conventional heat exchange. The analysis has told us simply and rapidly how many stages we need to achieve this.

Although the flash system is shown plotted as a triangle, it is really an unbalanced heat load and thus resembles a reactor system or a utility.

6.5.3 Solids drying

In dryers, moisture in a solid is evaporated; the latent heat load has to be provided. This makes drying a highly energy-intensive unit operation, typically accounting for 10–20% of the total industrial energy consumption for many developed countries.

Virtually all dryers use air as the carrier gas and water as the solvent to be evaporated. The high heat requirement of dryers is almost entirely due to the latent heat of evaporation of the water. Much of the heat supplied to the dryer emerges as the latent heat of the vapour in the exhaust gas, which can only be recovered by condensing the water vapour from the exhaust. However, as saturation humidity increases almost exponentially with temperature, the dewpoint of exhaust air is generally 50°C or lower. It is very rare for this to be above the process pinch and the heat must thus be wasted. Hence it is usual to simply vent the dryer exhaust from a stack, possibly recovering a small proportion of the heat as sensible heat.

Kemp (2005) lists seven ways of reducing energy consumption of dryers, compatible with the insights of pinch analysis:

1. Reducing the inherent energy requirement for drying, for example by dewatering the feed.
2. Increasing the efficiency of the dryer, by reducing heat losses, total air flow or batch times.
3. Heat recovery within the dryer system, between hot and cold streams.
4. Heat exchange between the dryer and surrounding processes.
5. Use of low-grade, lower-cost heat sources to supply the heat requirement.
6. Combined heat and power; co-generate power while supplying the heat requirement to the dryer.
7. Use of heat pumps to recover waste heat to provide dryer heating.

Methods 1 and 2 can be categorised as ways of directly reducing the dryer heat duty (i.e. process change), methods 3 and 4 use heat recovery to reduce the energy targets and methods 5, 6 and 7 reduce the cost of the utilities or the primary energy requirement.

Heat supply to most dryers is in the form of hot air. Ambient air is drawn in and heated in either a direct-fired or indirect-fired furnace. The heat load of the dryer therefore plots as a sloping line. Composite curves for a typical dryer are shown in Figure 6.25. It is clear that the scope for heat recovery in the basic system is limited. Some heat can be recovered from the dryer exhaust to the cold feed air. Usually this is only a small proportion of that available, but nevertheless the actual cost savings can be considerable because dryers are so energy intensive.

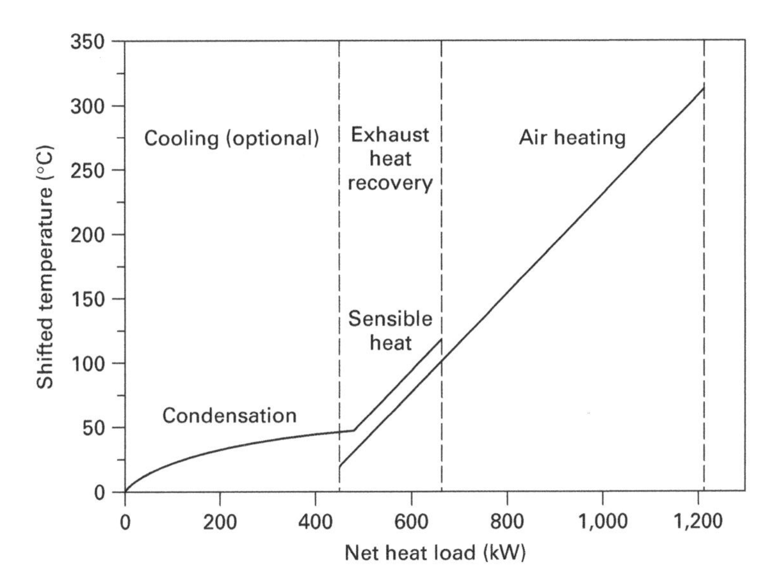

Figure 6.25 Composite curves for a typical dryer

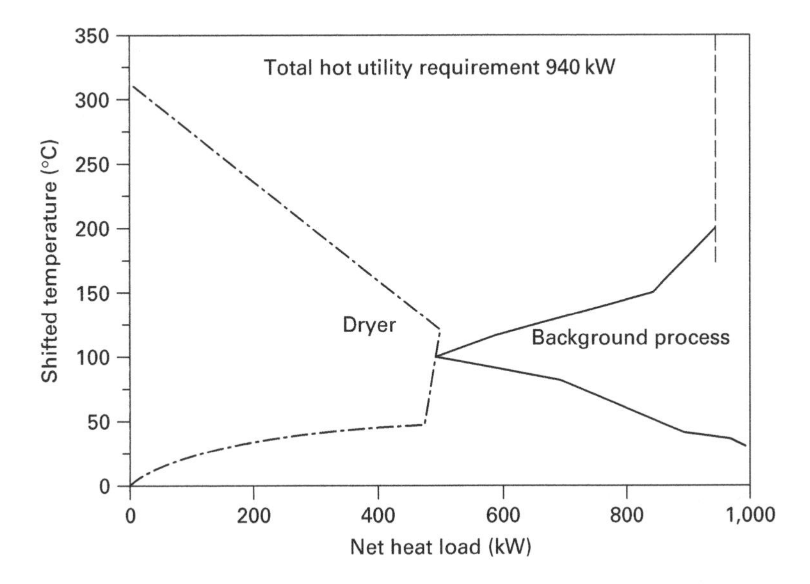

Figure 6.26 Split GCC for drying process

The split GCC is again an effective means of visualising the system. Figure 6.26 shows the split GCC for a process containing a dryer; the main cold stream is air being heated to 320°C and the hot stream is the heat available from the exhaust gas, mainly as vapour condensation below the dewpoint of 47°C. This is matched with a typical background process with a pinch at 100°C.

It is clear that the dryer is working across the pinch; it is not so clear what can be done about it. The only possibilities are:

(i) Reduce the temperature at which the dryer requires heat.
(ii) Raise the temperature at which the dryer exhaust stream releases heat.

The first can be achieved by using a low-temperature dryer extracting heat below the pinch as warm air or warm water. The reduced temperature driving forces would normally cause a huge increase in the size and capital cost of the dryer. However, if a dispersion dryer (e.g. a fluidised bed or cascading rotary dryer) can be substituted for a layer dryer (e.g. an oven or tray unit), the much enhanced heat transfer coefficients may allow low-temperature drying with a small unit. Warm air can be fed directly to the dryer; warm water can heat it indirectly via internal coils. Alternatively, some preheating of the wet feed solids may be carried out in a predryer working on below-pinch waste heat. This has particular advantages for sticky or temperature-sensitive materials.

The second option is virtually impossible with conventional air dryers as the dewpoint cannot be altered significantly. Recycling exhaust gases and raising the humidity will raise the dewpoint; however, it may adversely affect drying. In any case no heat can be recovered above the boiling point, 100°C, unless the entire system is placed under high pressure – an extremely expensive option. However, in some cases a heat transformer (see Chapter 5) has been used to absorb moisture from the exhaust gas and recover some of its heat.

If instead, the superheated form of the solvent being evaporated is used as the carrier gas instead of air, a very different picture emerges. The recovered vapour can then be condensed at high temperature, above the pinch. The commonest case is superheated steam drying, which also has the advantage of a better heat transfer coefficient between vapour and solids than for air. The steam is recirculated and reheated; a bleed equal to the evaporation rate is required, and this steam can be condensed significantly above 100°C to yield useful heat. Superheated steam drying has previously been advocated for heat transfer or safety reasons, but it clearly has energy advantages too. The main drawback is that a large fan is required to recirculate the steam, and the power consumption of this can cancel out the savings from heat recovery. An interesting solution to this problem is the airless dryer (Stubbing 1993), where no gas recirculation is used; the water driven off from the solids in the early stages of drying forces the air out of the system to create the superheated steam atmosphere.

Figure 6.27 shows the split GCC for both low-temperature and superheated steam dryers.

If these options are not possible, it is very difficult to reduce the energy consumption of a dryer significantly using pinch technology. However, the net cost of supplying the heat can be substantially reduced by using a co-generation system; the exhaust from either a gas turbine or a reciprocating engine is hot enough to supply almost any hot gas dryer. The exact inlet temperature is easily controlled by adding a varying amount of cool dilution air. The main limitations on such a system are the capital cost and the cleanness of the exhaust; gas turbines and gas engines are more acceptable in the latter respect than diesels.

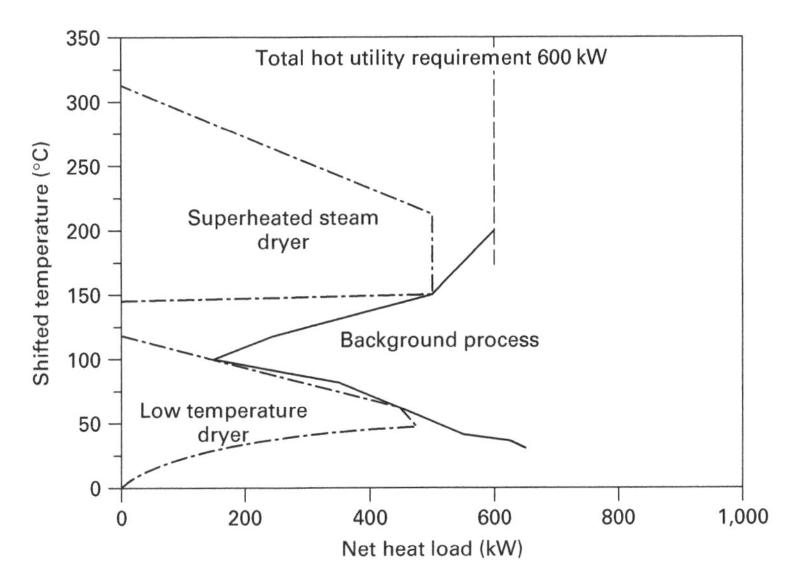

Figure 6.27 Split grand composite for energy-efficient dryer configurations

Heat pumps can be an option. For many dryers, the temperature lift is too high. However, for dryers using a large air recycle with a low temperature lift, including many food and agricultural dryers, heat pumping may be economic.

6.5.4 Other separation methods

Other separation systems rely on mass separating agents, for example solvent extraction. Process integration does not usually affect the design of these systems. However, if the mass separating agent is regenerated thermally, for example by a solvent recovery system, this can be integrated with the rest of the process.

6.6 Application to the organics distillation process case study

6.6.1 Identifying potential process changes

Having described the range of possible process changes, can we apply any of them to our case study? It seems a simple process, with just a few streams, so surely nothing is likely to be possible? Nevertheless, if we do a "brainstorm" for possibilities, we may be surprised at the number that come up. The reader may like to try this himself before proceeding further.

In brainstorming techniques we initially list all possibilities which come to mind, whether or not they seem feasible or even remotely sensible. "Off-the-wall" ideas should not be squashed; they may provide the stimulus for other, more realistic ones. When a good range of ideas has been collected, we then narrow them down to a sensible range, evaluate and rank them.

Some of the possibilities which could be considered here are:

(a) Run down some hot streams to lower temperatures to recover more heat, especially above the pinch.
(b) Change distillation column temperatures and pressures to improve integration.
(c) Add intermediate reboilers and condensers to the distillation columns, or consider divided wall columns.
(d) Flash off some vapour from the crude streams during heating, giving a new cold stream at a lower temperature.
(e) Avoid some heating and cooling tasks altogether, for example look for streams which are being cooled and reheated, or vice versa.

Under (a), we could recover heat from the bitumen and wax streams on the vacuum distillation unit. They are well above the pinch and the GCC, so this would directly reduce the hot utility targets; cooling by 50°C would recover 62.5 kW from the wax and 187.5 kW from the bitumen. However, the final product would then be more viscous and difficult to pump, and the heat transfer coefficients would be very poor, giving large heat exchangers. Fouling would also be a severe problem, and regular cleaning would be required. It is unlikely that heat recovery will prove economic under these circumstances, especially for these very modest amounts.

We have seen the benefits of changes of types (b) and (c) earlier in the chapter. However, here, with the two columns working at similar temperatures, it will be difficult to shift one above the other. An intermediate condenser or pumparound might be considered on the atmospheric column (in effect, the heavy oil recycle already provides one on the vacuum column), shifting some of the overheads load above the pinch. This looks likely to be a capital-intensive solution involving major column modifications and affecting the separation.

Option (d) would involve replacing the flash of the vacuum crude feedstock after the furnace with a low-temperature flash, followed by heating lower-temperature liquid under vacuum. The feedstock would then undergo partial vapourisation at lower temperatures, although larger diameter tubes would be needed to carry the vapour, and two-phase vapour/liquid flows present some transport problems. Is there any worthwhile gain? To find out, we do a flash calculation, which shows that the vacuum crude would drop from 155°C to 125°C, corresponding to the column pressure, and then estimate that we would gradually heat the new feed to 290°C; the overall heat load must be the same as before. Now we retarget the overall crude distillation unit (CDU) and the vacuum distillation unit (VDU) individually, and compare with before. To our dismay we find that the energy saving is only 60 kW for the VDU and 0 for the combined plants – the vacuum feed is still above the overall pinch temperature, and only a small amount of the heavy oil stream is at a high enough temperature to heat it. Therefore, there is no incentive to replace the current system by a flash before the furnace. Similar considerations arise for modifying the flash on the atmospheric unit. The opposite possibility would be to replace the flash systems by conventional reboilers; this should not incur any energy penalty in this case, but there is no obvious gain from replacing the current system.

Finally, in (e), we look for situations like streams being cooled, separated and reheated, where a separation at a different temperature may save energy. The overheads and light oil recycle fit this configuration, so either a change in separator temperature or a partial reheat of the light oil with below-pinch heat could be considered, although again this would affect the internal column conditions and separation. However, a further opportunity of this type can be seen when the atmospheric and crude units are operating together. The bottoms are cooled down and then reheated as vacuum crude. Eliminating this would remove a major cooling and heating task, and does not even involve changing any separations! The 1,030 kW of bottoms heat which is currently thrown away to cooling water, and which we had considered recovering to heat the crude feed, would be retained and give a direct load reduction on the vacuum unit furnace. This would not even need a new heat exchanger, just some new pipework!

Of the various process change possibilities, this final one is clearly the most promising and deserves further detailed assessment.

6.6.2 Eliminating bottoms rundown: detailed analysis

We need to produce targets and develop a feasible heat exchanger network for this option. The process change alters the stream data by removing the bottoms stream (258–155°C) and changing the supply temperature of the vacuum crude. Both have a CP of 1 kW/K. We assume the same 3°C temperature drop through heat loss between residue and vacuum crude as at present, so the vacuum crude enters the furnace at 255°C instead of 152°C and leaves at 319°C. Thus we have removed a hot stream of 1,030 kW and reduced a cold stream by 1,030 kW. Retargeting shows that the pinch remains at 123°C, as before, and the hot and cold utility targets are unchanged compared with the combined plants. This is logical; the eliminated hot and cold streams were both fully above the pinch. However, the heat exchange falls by 1,030 kW, as does the existing energy consumption (Table 6.1).

Next, we consider a possible heat exchanger network. There are two hot streams above the pinch – middle oil and heavy oil – and only one cold stream, the atmospheric column crude feed. So a maximum energy recovery (MER) design can only be achieved if the cold stream is split above the pinch, giving the network in Figure 6.28(a).

Table 6.1 Targets with process change, eliminating bottoms rundown

Situation	*Current hot utility (kW)*	*Current heat recovery (kW)*	*Hot utility target (kW)*	*Potential heat saving (%)*	*Cold utility target (kW)*	*Target heat recovery (kW)*
Current units separately	8,500	1,640	6,435	24.3	2,345	3,705
Current units combined	8,500	1,640	6,085	28.4	1,995	4,055
Combined with process change	7,470	1,640	6,085	18.5	1,995	3,025

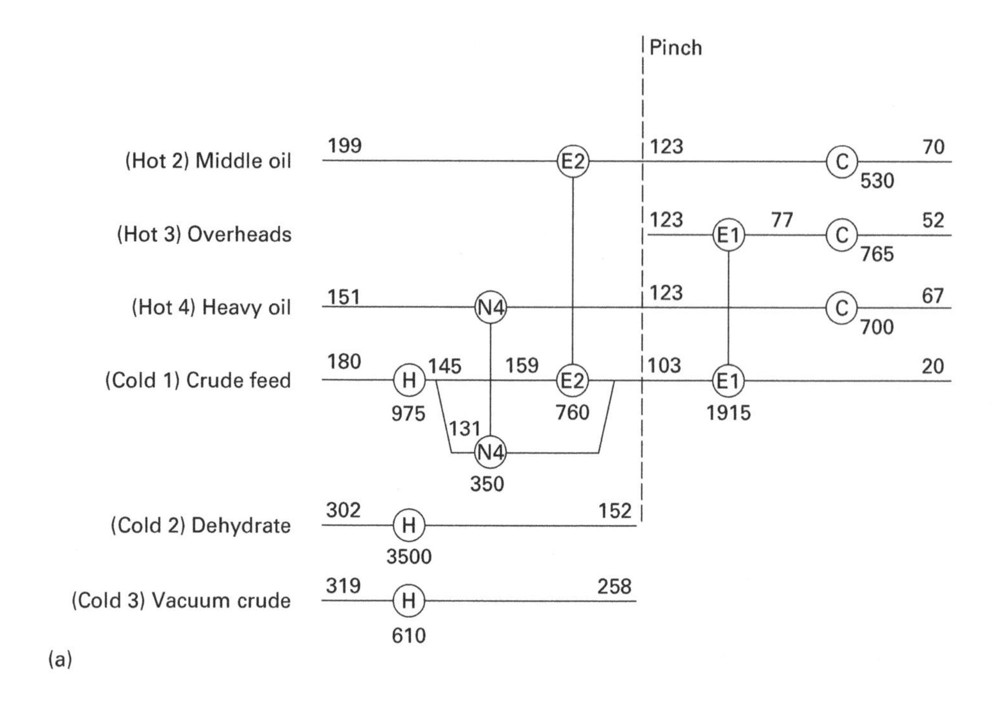

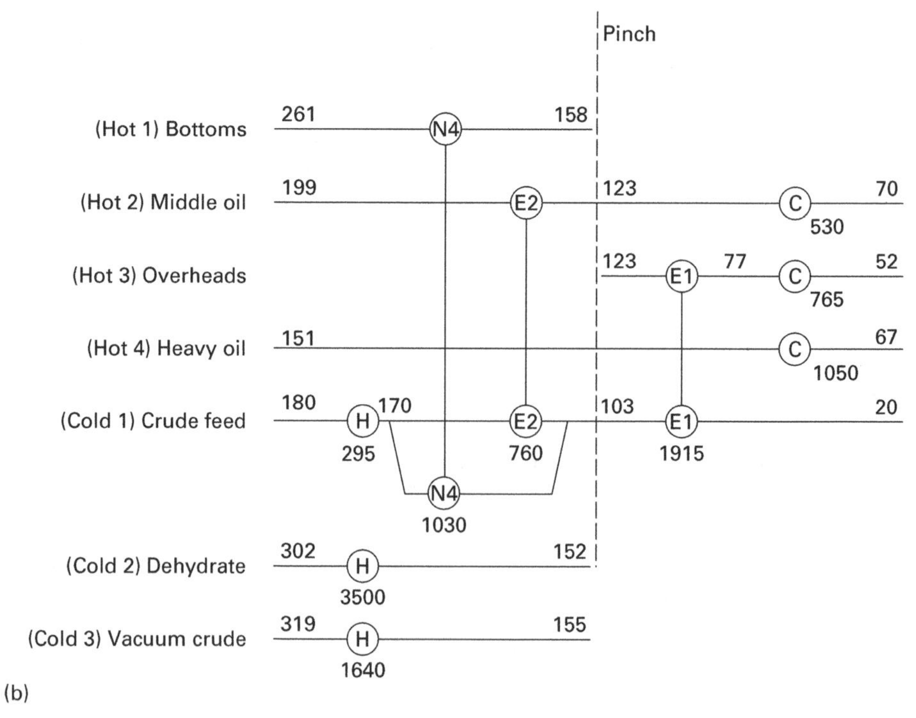

Figure 6.28 MER networks with process change for combined and separate plants

When the plants are operating separately, however, the MER network will be the one in Figure 6.28(b), based on Figure 4.41. We therefore have a dual base case.

Below the pinch the networks are identical, with the large overheads–crude feed exchanger E1. The middle oil–crude feed exchanger E2 is also common to both options. The difference is that the crude feed is also matched with the heavy oil in the combined units, saving 350 kW, and with the bottoms residue in the separate units, saving 1,030 kW. These options are not mutually exclusive, but their economics will depend heavily on the planned operating pattern, as each will only operate for a proportion of the plant's total operating hours; the projected energy savings will be reduced in proportion to the time which the exchanger is unused, and the payback time will increase. In particular, if the plants normally operate together, the justification for the exchanger between the residue and the crude feed will virtually disappear.

Compared to the existing process, the avoidance of bottoms rundown and reheat saves 1,030 kW of heat without the need for an additional exchanger. This provides a much more compelling argument for linking the two plants than the 350 kW recovered by the heavy oil–crude feed heat exchanger identified in Section 4.9.2 (although the two projects can be additive). Since the cost of the extra pipework is small, the project will give a very short payback time, maybe only a month or two – better than any of the heat recovery projects outlined in Section 4.9.

We have to allow for the alternative cases where the plants run together or either one is operating in isolation; all three situations must be operable. Provision of separate exchangers, heaters and coolers for each situation would be expensive. However, with a little ingenuity, we can consider multi-purpose exchangers, used for different duties in different situations. There are two alternatives:

(a) Use a process–process heat exchanger as a heater or cooler when plants are operating separately.
(b) Use a process–process exchanger with a different process stream when the plants are operating separately.

(a) may run into compatibility problems. Cooling water and steam can often alternate with aqueous streams, but if the process fluid is a hydrocarbon, cross-contamination occurs. Conversely, with (b), if the alternative process fluids are incompatible, multi-purpose use is not an option; but if, as here, they are simply different fractions of the same mixture, there should be no problem.

So we can envisage the following. Add a new stream split to the crude feed stream, with a new exchanger N4 on it. When the two plants are operating together, the hot stream to N4 is the heavy oil from the vacuum crude plant. When the atmospheric unit is operating on its own, the bottoms is used as the hot stream. The two hot-side streams are compatible. Obviously the two matches have different duties and the exchanger size can either be a compromise or sized for one duty, in which case it will over- or underperform on the other duty. A quick $Q/\Delta T$ comparison shows that the bottoms match is the more onerous duty (20.0 kW/K as against 16.9 kW/K). Assuming the plants normally operate together, if we size N4 for the normal, smaller heavy oil duty, we find that this will recover 950 kW on the bottoms duty, not far below the maximum figure of 1,030 kW. The resulting flowsheet is shown in Figure 6.29.

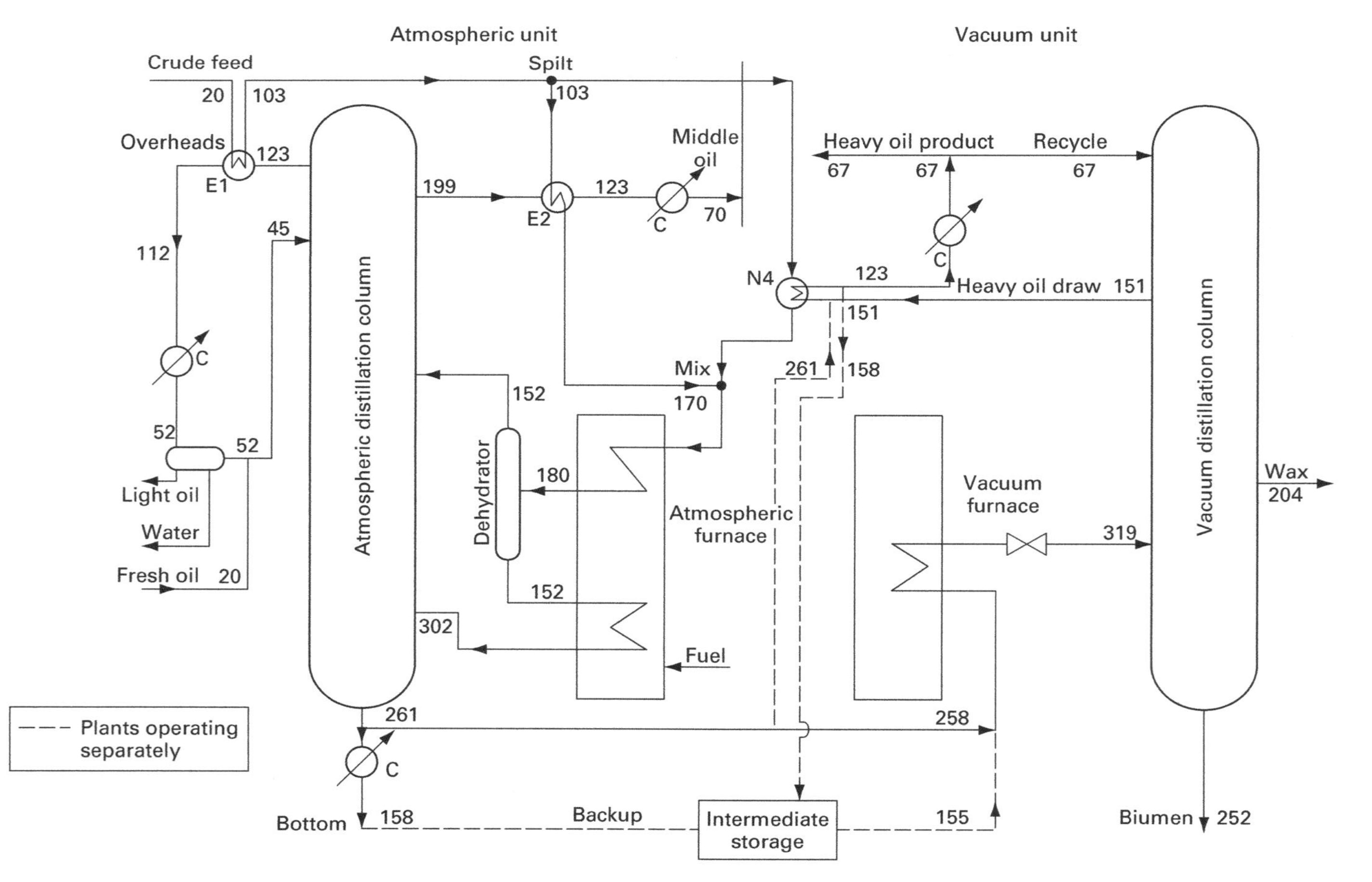

Figure 6.29 Flowsheet with process change and allowance for separate operation of plants

The heavy oil cooler and vacuum furnace are left at their existing sizes and will be oversized for their new duties when the plants are operating together, but will still be needed to perform as before during separate operation. The bottoms cooler will not be needed for simultaneous operation and, if N4 is installed as a dual-purpose exchanger, will be oversized for separate operation. Theoretically it could be re-used for the new duty, if the cold-side fluid is compatible with the materials (which were previously chosen for a cooling water duty). However it is probably preferable to retain it as an emergency cooler and purchase a new purpose-designed heat exchanger.

6.6.3 Economic assessment

Various possibilities obtained from the network design and optimisation in Section 4.9 as well as here can be combined for an overall assessment. One way of doing this is to list all the possible project alternatives, as in Table 6.2. However, with several different options for heat exchanger networks, each of which could be considered with or without the process change, this is somewhat unwieldy. It can be seen however that the projects in the bottom section of the table, including the process change, give consistently lower paybacks than the corresponding projects without the process change (which usually involve the new exchanger N3). The repiping required for the process change is assumed to cost £10 K.

An alternative is to list the effect of individual changes, that is adding a particular exchanger. Care is needed with this approach, as different items are not always purely additive. Here, for example, the required size of the new exchanger N3 depends on whether E1 and E2 have been enlarged, as this alters the temperature driving forces throughout the network. However, in this case the network configuration does allow this. The result is shown in Table 6.3. Options 3a and 3b are alternatives and the savings cannot be combined.

In both Tables 6.2 and 6.3, the process change options have been evaluated assuming that the plants operate together for the full working year. Obviously, if this is not the case, the savings will be reduced pro-rata; even so, the process change is clearly by far the best single project, with the enlargement of E1 as next best. The evaluation of the dual-purpose option for N4 is more tricky. It has been assumed that it incurs an additional repiping cost of £10 K, and that the plants work together 75% of the time, so the calculated saving is for 75% of 350 kW plus 25% of 950 kW. If the processes work separately for much less than 25% of the year, the savings from the bottoms–crude feed match will fall and may not repay the repiping cost.

The economics of the process change are far better than those of any of the other heat recovery schemes. Therefore, it is a “no-brainer” and should form part of any overall project. It does not harm operability, as the existing configuration can still be used when the plants operate separately. The final choice of projects depends on available capital and the payback criteria of the company at a given time, but a combination of the process change and enlarging exchanger E1 gives over 75% of the total available energy and cost savings (1,835 kW out of the 2,415 kW target) with a payback time only just over 6 months.

Table 6.2 Economics of various heat recovery options, with and without process change

Description	*Hot utility (kW)*			*Heat recovery (kW)*	*Exchanger area (m^2)*	*Capital cost £*	*Energy saving (kW)*	*Cost saving (£/yr)*	*Payback time (yr)*
	Atmos	*Vacuum*	*Combined*						
Current	6,860	1,640	8,500	1,640	130.7	0	0	0	–
E1 enlarged	6,055	1,640	7,695	2,445	342.8	58.7 K	805	48.3 K	1.21
E1/E2 enlarged	5,825	1,640	7,465	2,675	426.8	88.9 K	1,035	62.1 K	1.43
E1, New N3	5,025	1,640	6,665	3,475	477.3	100.3 K	1,835	110.1 K	0.91
N3, E1, E2	4,795	1,640	6,435	3,705	586.4	136.0 K	2,065	123.9 K	1.1
N3, N4, E1, E2	4,445	1,640	6,085	4,055	716.4	176.8 K	2,415	144.9 K	1.22
New N3 only	5,830	1,640	7,470	2,670	218.4	31.0 K	1,030	61.8 K	0.5
Process change	6,860	610	7,470	1,640	130.7	10.0 K	1,030	61.8 K	0.16
PC+E1 enlarged	6,055	610	6,665	2,445	342.8	68.7 K	1,835	110.1 K	0.62
PC+E1, E2	5,825	610	6,435	2,675	426.8	98.9 K	2,065	123.9 K	0.80
PC+E1, E2, N4	5,475	610	6,085	3,025	556.8	139.7 K	2,415	144.9 K	0.96

Table 6.3 Economics of specific heat exchanger installation projects

Project	*Description*	*Energy saving (kW)*	*Additional area (m²)*	*Capital cost (£)*	*Cost saving (£)*	*Payback time (yr)*
1	E1 enlarged	805	212.1	58.7 K	48.3 K	1.2
2	E2 also enlarged	230	84.0	30.2 K	14.0 K	2.2
3a	New N3	1,030	159.6	47.1 K	61.8 K	0.76
3b	Process change	1,030	–	10.0 K	61.8 K*	0.16
4a	New N4, heavy oil only	350	130.0	40.8 K	21.0 K*	1.9
4b	New N4, dual purpose	350, 950	130.0	50.8 K	31.2 K*	1.6

*Depends on number of hours when plants run simultaneously.

In effect, the work we did to optimise networks for the current layout in Section 4.9 was redundant! They were a useful teaching aid, but will never be the optimal economic solution (except in the unlikely situation that it is decided that the plants will never operate simultaneously). Fortunately, most of the sizing calculations also applied to the networks with process change, but often this will not be the case. This illustrates very clearly why process change opportunities should always be investigated during the initial targeting phase, and *before* heat exchanger network design!

Looking at the overall case study, we see that even this fairly simple plant has given a wide range of options in terms of heat recovery, economics and operability. However, we have not needed complex software to evaluate this. Simple targeting, network design by hand and spreadsheet analysis have been sufficient. Outside the oil refining and large-scale bulk chemicals industries, this is almost always the case.

6.7 Summary and conclusions

Process changes can be very worthwhile, and frequently save more energy at a shorter payback than additional heat exchangers. They often involve separation unit operations such as distillation and evaporation, but may apply to any process stream. Ingenuity and lateral thinking is often needed to identify possibilities, and the "plus–minus principle" helps to show whether they are likely to be worthwhile. Process changes need to be investigated and checked in the early stages of targeting analysis, as they alter the stream data set and the resulting heat exchanger network.

Exercises

E1.1 For the organics distillation unit, recalculate the targets, balanced composite and balanced GCCs for the system with process change and also including utility streams. Draw the balanced network grid and construct a heat exchanger network including utility streams. Compare with the networks obtained in Section 4.9.3. What are the implications?

References

Dhole, V. R. and Linnhoff, B. (1993). Distillation column targets, *Comp and Chem Eng*, 17(5/6): 549–560. (Paper originally presented at *Europ Symp on Computer Applications in Process Engineering ESCAPE-I, Elsinore, Denmark, May 24–28 1992.*)

Fonyo, Z. (1974). Thermodynamic analysis of rectification 1. Reversible model of rectification, *Int Chem Eng*, 14: 18–27.

Fraga, E. S. and McKinnon, K. I. M. (1994). CHiPS: A process synthesis package, *IChem E Symp Series*, 133: 239–255.

Glavic, P., Kravanja, Z. and Homsak, M. (1988). Heat integration of reactors; 1. Criteria for the placement of reactors into process flowsheet, *Chem Eng Sci*, 43: 593.

Hindmarsh, E. and Townsend, D. W. (1984). Heat integration of distillation systems into total flowsheets – a complete approach. *AIChE Annual Meeting*, San Francisco, November.

Kaibel, G. (1987). *PhD Thesis*, Technical University, Munich, Germany.

Kemp, I. C. (1986). Analysis of separation systems by process integration, *J Separ Proc Technol*, 7: 9–23.

Kemp, I. C. (2005). Reducing dryer energy use by process integration and pinch analysis, *Drying Technol*, 23(9–11): 2089–2104. Paper originally presented at *14th International Drying Symposium (IDS 2004)*, Sao Paulo, Proceedings Volume B, pp. 1029–1036.

Linnhoff, B., Dunford, H. and Smith, R. (1983). Heat integration of distillation columns into overall processes, *Chem Eng Sci*, 38(8): 1175, August.

Linnhoff, B. and Vredeveld, D. R. (1984). Pinch technology has come of age, *Chem Eng Progr*: 33–40, July.

Porter, M. E. and Momoh, S. O. (1991). Finding the optimum sequence of distillation columns – an equation to replace rules of thumb (heuristics), *Chem Engg J*, 46: 97.

Rajah, W. and Polley, G. T. (1995). Synthesis of practical distillation schemes, *Chem Eng Res Des* (*TransIChemE* Part A), 73A: 953–966.

Smith, R. (1995). Distillation sequencing. Chapter 5 of *Chemical Process Design*, McGraw Hill. Also Chapter 11 of *Chemical Process Design and Integration*, Wiley 2002.

Smith, R. and Linnhoff, B. (1988). The design of separators in the context of overall processes, *ChERD*, 66(3): 195–228.

Smith, R. and Jones, P. S. (1990). The optimal design of integrated evaporation systems, *J Heat Recov Syst CHP*, 10(4): 341–368, July.

Smith, R. and Omidkhah Nasrin, M. (1993). Trade-offs and interactions in reaction and separation systems. Part 1: Reactors with no selectivity losses. Part 2: Reactors with selectivity losses, *ChERD* (*TransIChemE* Part A), 71(A5): 467–473 and 474–478.

Stubbing, T. J. (1993). Airless drying: its invention, method and application, *Trans Inst Chem Eng*, 71(A5): 488–495.

7 Batch and time-dependent processes

7.1 Introduction

So far, we have dealt exclusively with continuous processes. However, pinch analysis can also be applied to batch processes, with suitable modifications. This type of analysis has been largely neglected, because batch processes are in general less energy intensive than continuous bulk systems. However, the techniques are worth studying for several reasons:

- Existing batch processes often have little or no heat recovery and there can be easy energy saving opportunities.
- The analysis often gives useful non-energy operational benefits, such as debottlenecking – identifying rate-limiting steps and reducing cycle times.
- Many continuous processes incorporate semi-batch sections, which must be included in the analysis to give the overall picture.
- The methods can be applied to other situations where the stream data varies with time, for example, during start-up and shutdown, or day/night variations.

Although continuous processes reign supreme for the production of bulk chemicals, there are many batch plants in existence. Indeed, their use is increasing in many countries because of the trend towards high-value, low-tonnage products such as specialty chemicals, fine chemicals and pharmaceuticals. However, the throughput and energy use of these processes is low, and therefore energy and cost savings from heat recovery are also limited.

For batch systems, both pinch analysis and practical heat recovery projects are much more difficult than for continuous processes, for a number of reasons. Many streams are present for only certain time periods, which restricts the possibilities for heat exchange. Also, they may not run between constant temperatures with constant heat capacity flowrates; for example, much heating is done *in situ* in vessels. Heating or cooling are supplied by an external jacket or internal coils, and the vessel contents gradually change in temperature. Often products of high value and low production rates are involved, and energy use has therefore been neglected and considered to be unimportant.

Nevertheless, despite these difficulties, pinch analysis has a part to play. Because heat recovery has been neglected in the past, significant energy savings are possible on many plants, although the project giving an optimal rate of economic return may be very different to the "theoretical best" project. Far more important in many cases, however, are the operability benefits which we can identify due to our greater understanding of the process, particularly debottlenecking (Section 7.8).

Pinch analysis of a batch process allows calculation of some or all of the following:

- targets for maximum heat exchange (MHX) within a batch;
- possibilities for heat storage within and between batches;
- rescheduling process operations to increase heat exchange;
- debottlenecking by finding which operations and equipment are rate-limiting;
- designing heat exchanger networks to achieve targets;
- identifying "principal matches" that achieve most heat recovery;
- modelling vessels which gradually heat or cool;
- analysis of the utility systems and reduction of peak loads.

The time-dependent nature of the analysis methods, in contrast to the techniques described so far which have assumed continuous, steady-state situations, is also exploited in the following situations:

- start-up and shutdown of continuous or batch processes
- buildings and other non-process operations
- multi-plant sites where different sections operate at different times.

There is designated and there is multi-purpose batch production, and there are cyclic and there are random batch processes. Designated and cyclic production is often found in the food and drink industry. Heat integration has achieved considerable savings on batch processes in the brewing industry, mainly via the water system. Multi-purpose and random production is found, for example in the manufacture of pharmaceuticals, of glue and resins, and of other low volume specialist products. Although effective process integration might seem unlikely in multi-purpose and random production in particular, experience suggests that it can be highly worthwhile even in the least likely environment. In a summary report issued by the Energy Technology Support Unit (ETSU) in Britain (Brown 1989) the results were published of 26 practical applications of pinch analysis across various sectors of industry. Eighteen of these studies involved some aspect of batch or partial batch production, and worthwhile projects were identified in nearly all cases. These usually came from better energy management but were often not primarily aimed at energy cost reduction. Cost savings came from:

- capacity increase (debottlenecking),
- improved yield and product quality,
- energy savings

in that order.

The gains also included reduced product cycle times and less reworking of off-specification product. Surprisingly, cost savings as a percentage of total operating costs were often greater than those found in continuous processes. A major reason is that, in many batch process environments, there has simply never before been a *methodical* search for process integration opportunities.

Further evidence of savings was provided and analysis techniques were further developed in an EU-sponsored project under the Joule program involving research and application groups from several European countries (Ashton *et al.* 1993).

7.2 Concepts

The main features of the batch analysis can be summarised relatively briefly and will then be fleshed out in Sections 7.3–7.7. To illustrate the concepts, we will initially use the simplest possible form of batch process, a reactor with a single feed and single product (just as we began illustrating continuous processes in Chapter 2 with a two-stream example). The feed is heated from 20°C to 120°C; a mildly exothermic reaction then occurs and the temperature gradually rises to 130°C. Finally the product is cooled to 30°C.

The batch cycle is 1 h, divided up as follows:

Filling and heating: 0.5 h (time t = 0.0–0.5 h);
Holding at constant temperature: 0.3 h (t = 0.5–0.8 h);
Cooling and discharge: 0.2 h (t = 0.8–1.0 h).

The mass of reactants and product is 250 kg and the specific heat capacity is 4 kJ/kgK, so 1,000 kJ (1 MJ) is required to heat or cool the liquids by 1°C. Hence the heating and cooling both have a heat load of 100 MJ and the heat released by the exothermic reaction is 10 MJ. The existing process, with no heat recovery, requires 100 MJ of hot utility and 100 MJ of cold utility. An acceptable ΔT_{min} is 10°C. How do we analyse this process?

Firstly, we can identify streams, which are as follows:

Cold stream A: 0.0–0.5 h, 20°C–120°C; total heat load 100 MJ, instantaneous heat flow 200 MJ/h.
Hot stream B: 0.8–1.0 h, 130°C–30°C; total heat load 100 MJ, instantaneous heat flow 500 MJ/h.

We could now draw composite curves, using the total heat load for the batch period for the horizontal axis. Figure 7.1(a) shows that the hot composite lies completely above the cold composite and, apparently, 100% heat recovery can be achieved and the hot and cold utility targets are zero. In effect, we have averaged the heat flows over the period of the batch and this method is therefore known as the **time average model (TAM)**.

However, there is an obvious snag. Common sense tells us that heat released at the end of the batch, between 0.8 and 1 h, cannot be used to heat up a cold stream which only existed half-an-hour earlier. Just as heat will not flow up a temperature gradient, it cannot flow backwards in time. We have something like a **"time pinch"**. Nevertheless, the TAM is useful as it gives us a limiting "best case" which shows what would happen if all time constraints were removed.

To allow for the effect of time, we can split the process into **time intervals**. The boundaries can be chosen as times when streams begin or end. So here, we split the batch into three periods, 0.0–0.5 h, 0.5–0.8 h and 0.8–1.0 h. A quick targeting

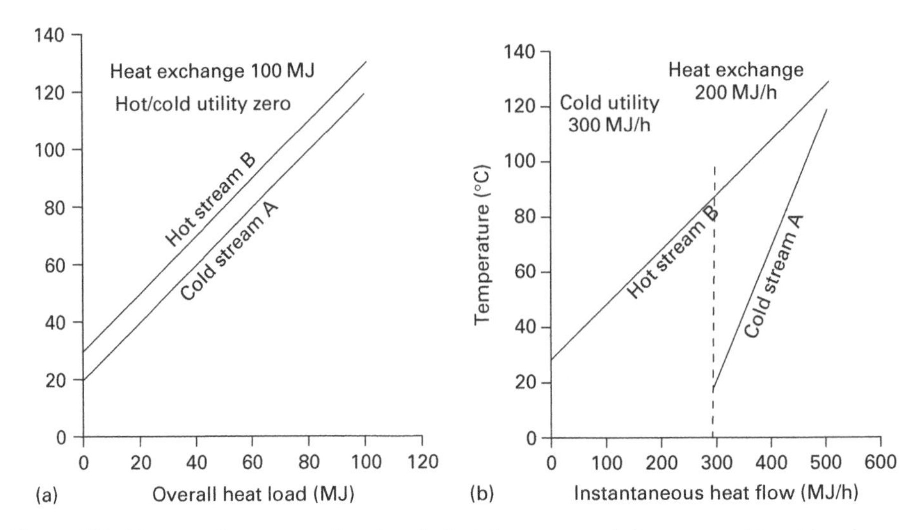

Figure 7.1 Composite curves for simple batch process: (a) averaged heat load, (b) instantaneous heat flow

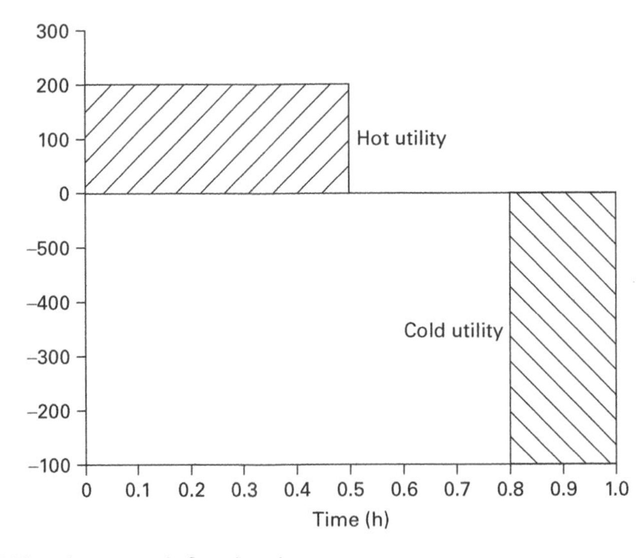

Figure 7.2 Utility-time graph for simple process

calculation tells us that 100 MJ of hot utility is required in the first time interval and 100 MJ of cold utility in the third. This is the **time slice model (TSM)**.

A useful by-product is that the change in utility requirements with time can be plotted as a **utility-time graph** (Figure 7.2). Here, this shows that hot and cold utility are never required simultaneously, and both could be supplied alternately to the vessel jacket, which doubles as both a heater and a cooler. The area under the graph corresponds to the total hot and cold utility use over the batch period.

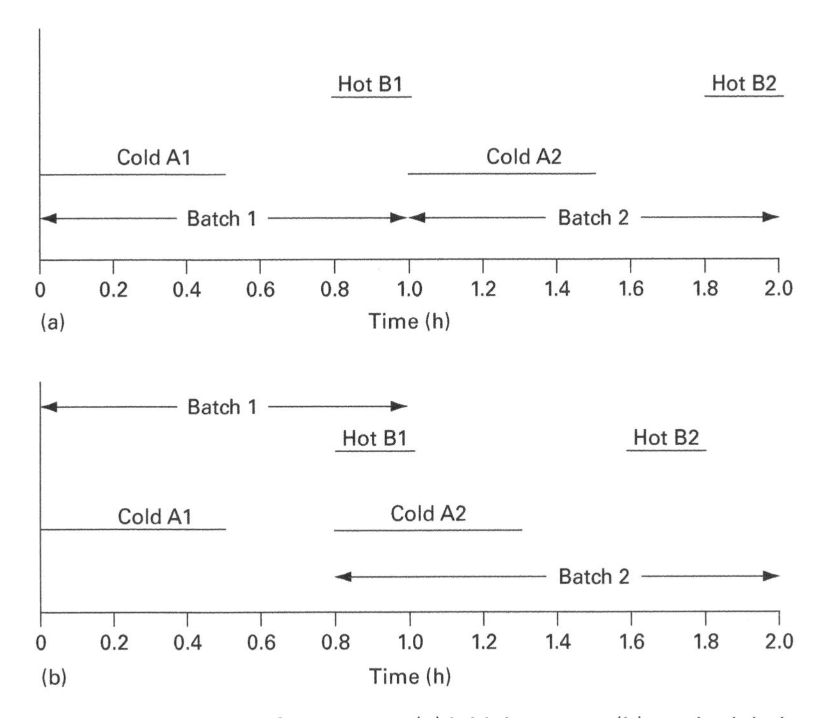

Figure 7.3 Time event charts for streams: (a) initial process, (b) rescheduled

Clearly, with the process as it stands, no heat recovery is possible. However, the batch process may be repeated so that heat may be recovered into the next batch, or it may be possible to change the operating times of streams. We can represent the streams on a **time event chart** to explore these possibilities. Figure 7.3(a) shows the streams for the batch and an identical second batch, which could be a following batch in the same equipment or a batch in a parallel processing line. If we can move batch 2 forward in time by 0.2 h, its cold stream A2 will overlap with the hot stream B1 in batch 1 and they can exchange heat directly as shown in Figure 7.3(b). Changing the timing of processes in this way is known as **rescheduling**.

However, several factors may limit the heat recovery which can be achieved by rescheduling:

Load limitations: The instantaneous heat flows of the streams may be mismatched. It is important to retarget for the rescheduled process using the TSM. Figure 7.1(b) shows that only 200 MJ/h of the hot stream heat flow can be used by the cold stream; the remaining 300 MJ/h goes to waste. Likewise, the cold stream is not matched for the period 0.5–0.8 h and must be heated by hot utility. Multiplying by the lengths of the time periods, we find that 40 MJ of heat is recovered and 60 MJ of hot and cold utility are still required during the batch period.

If, however, we could also change the *duration* of the hot or cold stream, and hence match the heat flows better, further recovery could be achieved. At the limit, if the cold stream could be heated in 0.2 h at 500 MJ/h, complete heat recovery of all 100 MJ could be achieved. Alternatively, the hot stream could be slowed down to operate over 0.5 h at 200 MJ/h. All these rescheduling options are discussed in depth in Section 7.7.

Types of stream: The liquid may not be flowing continuously through a heat exchanger. It is common to heat and cool the contents of a vessel *in situ*. At time 0, the vessel contents will be at 20°C and all the heat from a hot stream running between 130°C and 30°C can be used. In contrast, when the vessel has reached 120°C, no heat can be recovered from the hot stream (allowing for the ΔT_{min} of 10°C). So heat recovery targets will be significantly different for *in-situ* heating and cooling. Four separate stream types A–D can be defined, as explained in Section 7.3.

Equipment occupancy: If heating and cooling are done in the same vessel, it is not possible to exchange heat directly with the next batch. On the other hand, if the hot liquid is discharged to a separate storage tank (after 0.8 h) and circulated through the reaction vessel jacket (or a separate heat exchanger), it can heat up the next batch. Such opportunities can be identified using another kind of time event chart, based not on streams but on equipment occupancy – the Gantt chart. Figure 7.4 compares the two systems, assuming that filling and discharge of the reaction vessel each take 0.1 h, so that initial cycle time is 1.2 h. With hot liquid discharge, heat recovery is now possible, but there are also production benefits. At first sight this may seem surprising as, although the batches overlap in time, the processing period per batch is slightly longer (1.4 h), as the hot liquid must be held to await

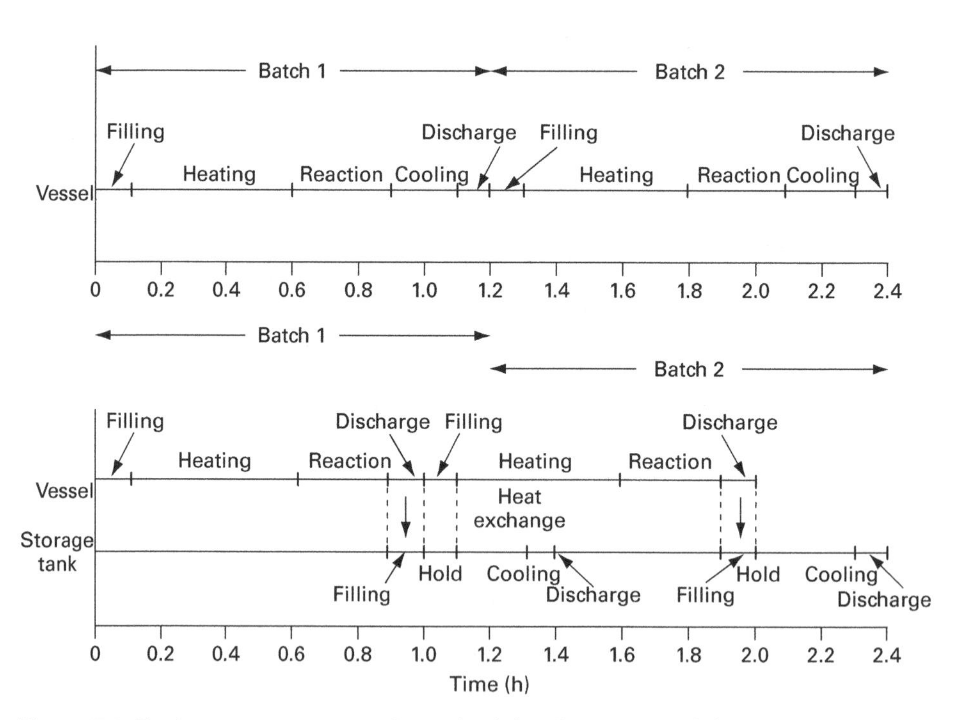

Figure 7.4 Equipment occupancy charts for (a) jacket heating, (b) system with separate storage tank and coils

the filling of the vessel and the storage tank must also be discharged at the end. But the important point is that the batch only occupies the reaction vessel for 1 h instead of 1.2 h. Since the batch reactor is the rate-limiting step in this process, an increase in throughput of 20% is obtained – a third batch could have been introduced after 2 h instead of having to wait till 2.4 h. Thus the equipment occupancy chart is the key tool in identifying **debottlenecking** opportunities.

Finally, an alternative to rescheduling is to store heat and re-use it later. Thus, some or all of the 100 MJ released at the end of the batch could be stored at suitable temperatures and used to heat the next batch. Both rescheduling and heat storage opportunities can be identified rigorously by the Cascade analysis described in Section 7.5.

Note that the targets for even this simple batch process can vary all the way from zero heat recovery (TSM for non-rescheduled process) to 100% (TAM), with a number of intermediate options.

7.3 Types of streams in batch processes

In a continuous process, a stream flows from a constant supply temperature to a constant target temperature at a constant flowrate and heat load, and is present at all times. In batch processes and other time-dependent situations, one or more of these assumptions will generally not hold. Instead, four basic types of stream can be identified:

Type A: A stream which operates between fixed supply and target temperatures and has a fixed heat load, but only exists for a certain time period. All streams on continuous plants in steady-state are Type A; they may be known as **flowing streams**.

Type B: A stream which gradually changes in heat load over a time period although it remains at set temperatures (e.g. a volatile product being boiled off from a batch reaction).

Type C: A stream which gradually changes in temperature but whose heat load is constant (e.g. liquid being heated in a reaction vessel by electric resistance coils at constant power).

Type D: A stream which changes in both temperature and heat load with time. The classic example is a jacketed reaction vessel heated by steam or cooled by water. The heat transfer rate depends on the temperature difference between utility and vessel contents, so it constantly changes. These are probably the most common type of stream in batch processes.

For a Type C stream, the heat load is constant and the vessel contents heat up at a constant rate (ignoring the thermal capacity of the vessel itself). For a Type D stream, the heat transfer rate from jacket to vessel contents Q_J is regulated by the following equation:

$$Q_J = AU(T_J - T_L) = m_L C_{PL}\, dT_L/dt \tag{7.1}$$

where

T_J = temperature of heating medium in jacket;
T_L = temperature of liquid in vessel;

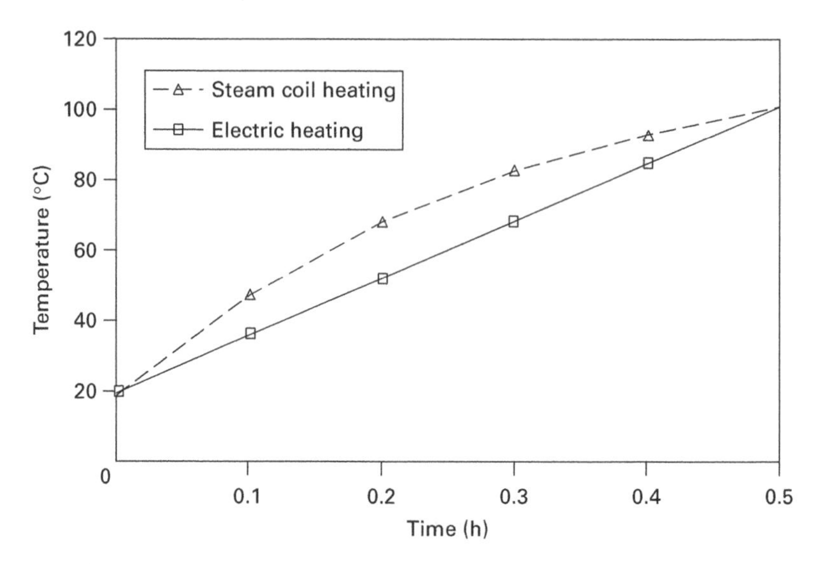

Figure 7.5 Plot of vessel temperature against time

m_L = mass of liquid in vessel;
C_{PL} = specific heat capacity of liquid;
dT_L/dt = rate of rise of liquid temperature.

Rearranging gives:

$$dT_L/dt = (AU/m_L C_{PL})(T_J - T_L) \tag{7.2}$$

Equation (7.2) can be integrated to give the following relationships:

At any instant,
$$T_L = T_J - (T_J - T_{L1})e^{-kt} \tag{7.3}$$

where
T_{L1} = initial temperature of liquid in vessel;
t = time from start of batch;
$k = (AU/m_L C_{PL})$.

And over the batch period,

$$t = 1/k \ln((T_J - T_{L1})/(T_J - T_{L2})) \tag{7.4}$$

$$H = m_L C_{PL}(T_{L2} - T_{L1}) \tag{7.5}$$

where
T_{L2} = final temperature of liquid in vessel;
H = total heat transferred to liquid.

Thus, graphs of vessel temperature rise and heat transfer rate against time are of logarithmic form for a Type D stream. Figure 7.5 compares how vessel temperature T_L varies with time for electric and jacket heating. Figure 7.6 shows how the rate of heat supply Q to the vessel changes.

A similar set of calculations apply if heat is being exchanged between a hot vessel which is cooling down and a cold vessel which is heating up. Both T_J and T_L vary,

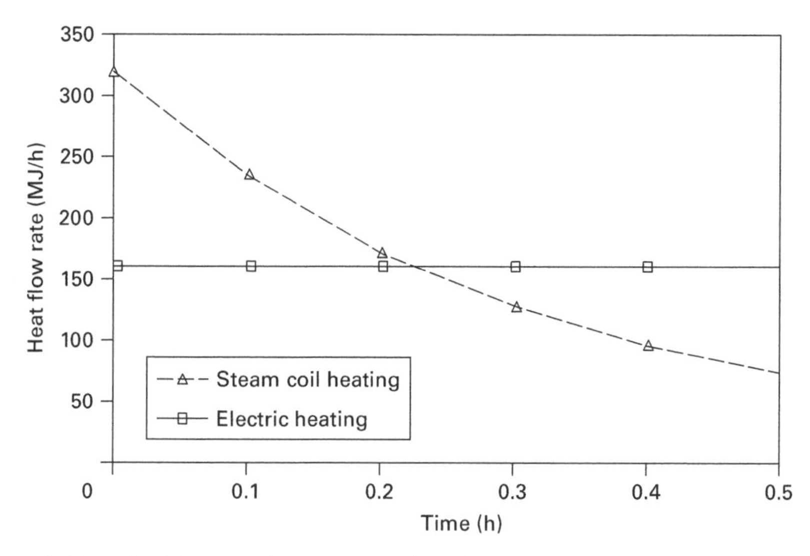

Figure 7.6 Plot of utility heat flow rate against time

and the vessels will eventually equilibrate at an intermediate temperature unless separate hot or cold utility are supplied.

7.4 Time intervals

Time intervals can be defined initially by stating that the boundaries are times when streams begin, end or change heat capacity flowrate significantly. This is precisely equivalent to the definition of temperature intervals, and is adequate for Type A streams. However, this definition meets problems where the temperature or heat load vary continuously with time, as for streams of Types B, C and D. None of these types of stream are easy to handle rigorously by the time-interval method. To model them precisely, one would need an infinite number of time intervals. However, for practical purposes, good approximations can be obtained by taking an approximation to the stream conditions during the time interval or by using "snapshots" at various times, and treating the stream as if it were Type A. The stream may be split into several time intervals with different temperatures in each. Care has to be taken not to generate an over-optimistic target. The detailed methodology is covered by Kemp (1990).

A continuous process at steady-state can in fact be regarded as a special case of a batch process. All the streams are of Type A, and for each stream the time period is the whole of the processing cycle.

7.5 Calculating energy targets

Energy targets for batch processes are calculated by the Cascade analysis (Kemp and Deakin 1989). The overall methodology gives targets for the TAM, TSM and heat storage opportunities between time intervals, from which rescheduling opportunities (Section 7.7) can also be deduced.

7.5.1 Formation of stream data

The TAM, TSM and utility-time graph have already been briefly defined. They will now be illustrated by application to a batch process directly derived from the four-stream example in Chapters 2–4, whose stream data were given in Table 2.2. This continuous process is transformed to an equivalent semi-batch process by assuming that each stream only exists for a limited time period. The cycle time is taken as 1 h and the streams exist during the following time periods:

Cold stream 1: 0.5–0.7 h
Hot stream 2: 0.25–1.0 h
Cold stream 3: 0–0.5 h
Hot stream 4: 0.3–0.8 h

The hot streams predominate in the latter part of the cycle and the cold streams in the earlier part. This is again typical of many batch processes, where a cold feed is heated and reacted at an elevated temperature and the product is then cooled before discharge. All four streams are Type A.

The stream heat loads over the 1 h batch period (in kWh) are taken to be equal to those during 1 h of the original continuous process in Table 2.2; the two processes would thus be equivalent in terms of material processed overall. The intermittent heat flows (in kW) are significantly higher than for the continuous process. The resulting set of stream data is shown in Table 7.1.

7.5.2 Time average model

The first stage in energy targeting is to use the TAM, averaging the heat loads in kWh over the batch period of 1 h. This yields the same heat flows as for the original continuous process. Hence, the same problem table and grand composite curve will apply, and the targets will be 20 kW hot utility and 60 kW cold utility, as calculated in Section 2.1.4. Likewise, for a 1-h batch cycle, 20 kWh hot utility and 60 kWh cold utility are required (Table 7.2).

7.5.3 Time slice model

We now divide the process into time intervals. The boundaries occur when streams start or finish, and are therefore at times of 0, 0.25, 0.3, 0.5, 0.7, 0.8 and 1.0 h, giving a total of six time intervals.

During each time interval, a certain combination of streams will exist and their conditions can be taken as constant. The time event diagram (Figure 7.7) provides a useful visualisation of which streams exist in which periods. Hence a heat cascade can be set up for that time interval and the hot and cold utility targets are calculated

Table 7.1. Stream data for semi-batch process

		Stream temperatures		*Shifted temperatures*				*Op. times*			
Stream Number	*Type*	*Supply °C*	*Target °C*	*Supply °C*	*Target °C*	*CP kW/K*	*Heat flow kW*	*Start h*	*End h*	*CP. Dt kW/K*	*Heat load kWh*
1	C	20	135	25	140	−10	−1,150	0.5	0.7	−2	−230
2	H	170	60	165	55	4	440	0.25	1	3	330
3	C	80	140	85	145	−8	−480	0	0.5	−4	−240
4	H	150	30	145	25	3	360	0.3	0.8	1.5	180

Table 7.2 Infeasible and feasible heat cascades for time-average model (TAM)

T (°C)	t (h)	*Infeasible cascade*	*Feasible cascade*
		⇩	⇩
165		0	20
	ΔH	*60*	*60*
145		60	80
	ΔH	*2.5*	*2.5*
140		62.5	82.5
	ΔH	*−82.5*	*−82.5*
85		−20	0
	ΔH	*75*	*75*
55		55	75
	ΔH	*−15*	*−15*
25		40	60
		⇩	⇩

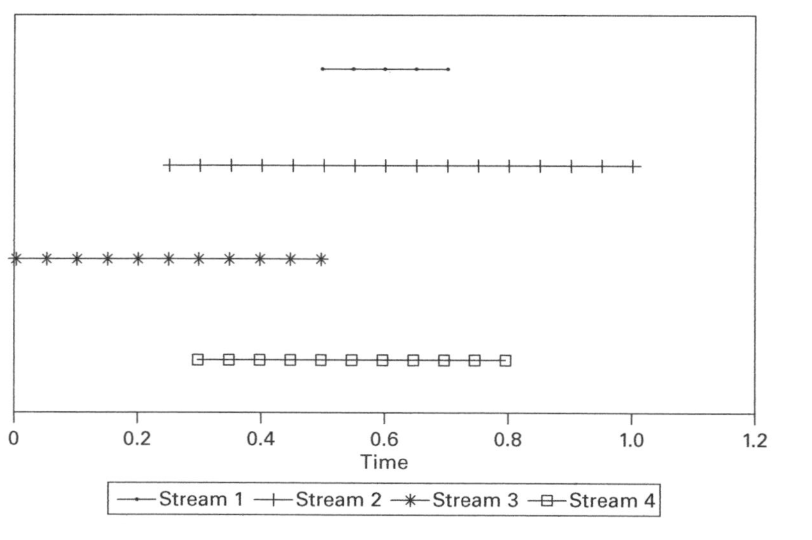

Figure 7.7 Time event diagram for main example

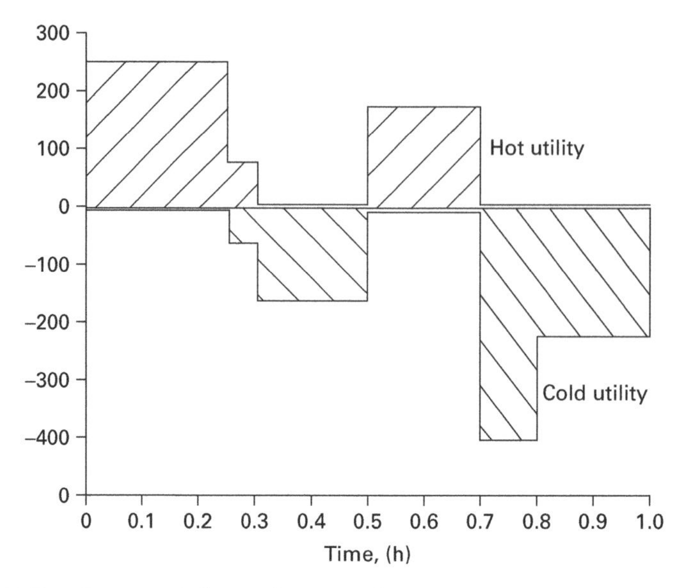

Figure 7.8 Utility-time graph for main example

by the problem table method, in just the same way as for a continuous process. An infeasible cascade is generated, the maximum negative value is found and that amount of hot utility is supplied. The targets obtained correspond to the maximum heat recovery (MHR) within that time interval by direct heat exchange. They may be calculated either as heat flows (in kW) or, by multiplying by the length of the time interval, as heat loads (in kWh). The heat flow graph is helpful for constructing the utility-time graph, by simply reading off the hot and cold utility loads in each time interval, giving Figure 7.8. However, heat loads are needed to calculate the overall hot and cold utility targets, as the heat flows in each cascade must be weighted by the duration of the time interval. Adding up the hot and cold utility heat loads over the batch period gives the TSM targets. These correspond to the situation where all possible heat is recovered by direct heat exchange within each time interval. They are therefore also known as MHX targets.

The heat cascades for each time interval may be placed side by side to allow heat flows in the whole batch to be visualised, as in Tables 7.3–7.6.

Only three time intervals require hot utility and four require cold utility. The TSM targets are 198 kWh hot utility and 238 kWh cold utility; the corresponding heat recovery is 272 kWh. Note that the utility requirements are much higher than the TAM targets of 20 kWh and 60 kWh. However, there is a considerable saving (over 55%) compared with the requirements with no heat recovery (470 kWh and 510 kWh).

It is also clear from the utility-time graph, Figure 7.8, that the heat loads are very sharply peaked. The main hot utility requirement comes at the beginning of the process and the cold utility requirement at the end.

Table 7.3 Infeasible heat cascades in terms of heat flows

T (°C)	t (h)	0.0–0.25	0.25–0.3	0.3–0.5	0.5–0.7	0.7–0.8	0.8–1.0
		⇩	⇩	⇩	⇩	⇩	⇩
165		0	0	0	0	0	0
	ΔH	0	80	80	80	80	80
145		0	80	80	80	80	80
	ΔH	−40	−20	−5	35	35	20
140		−40	60	75	115	115	100
	ΔH	−440	−220	−55	−165	385	220
85		−480	−160	20	−50	500	320
	ΔH	0	120	210	−90	210	120
55		−480	−40	230	−140	710	440
	ΔH	0	0	90	−210	90	0
25		−480	−40	320	−350	800	440
		⇩	⇩	⇩	⇩	⇩	⇩

Table 7.4 Feasible heat cascades in terms of heat flows

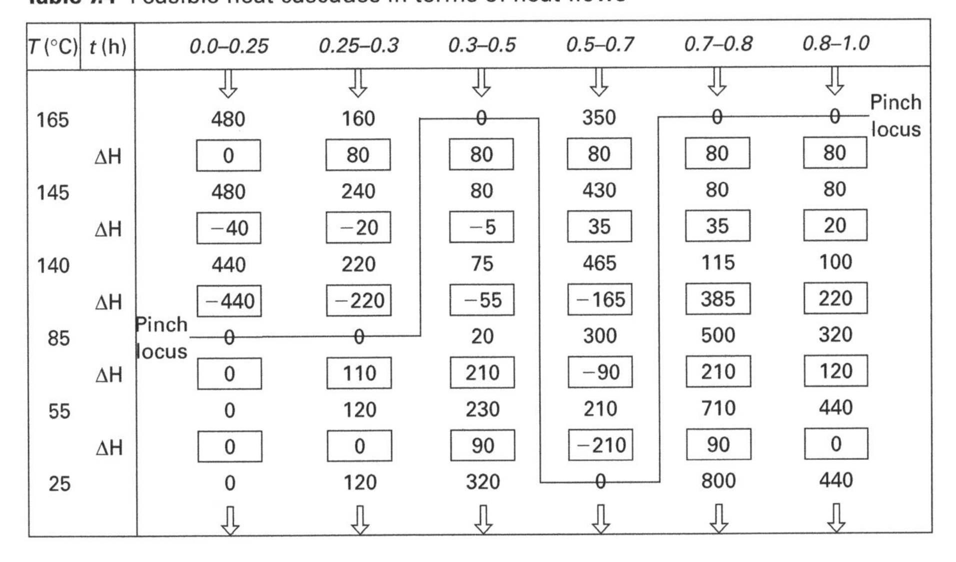

T (°C)	t (h)	0.0–0.25	0.25–0.3	0.3–0.5	0.5–0.7	0.7–0.8	0.8–1.0
		⇩	⇩	⇩	⇩	⇩	⇩
165		480	160	0	350	0	0
	ΔH	0	80	80	80	80	80
145		480	240	80	430	80	80
	ΔH	−40	−20	−5	35	35	20
140		440	220	75	465	115	100
	ΔH	−440	−220	−55	−165	385	220
85		0	0	20	300	500	320
	ΔH	0	110	210	−90	210	120
55		0	120	230	210	710	440
	ΔH	0	0	90	−210	90	0
25		0	120	320	0	800	440
		⇩	⇩	⇩	⇩	⇩	⇩

7.5.4 Heat storage possibilities

Next, heat storage may be considered. Although this may prove to be impractical, the insights gained are also useful in identifying possibilities for rescheduling.

In Section 3.5.1, we saw how zonal targeting could be used to evaluate heat recovery possibilities between separate process plants with different pinch temperatures. The stream data sets are combined, or the problem tables or grand composite curves

Table 7.5 Infeasible heat cascades in terms of heat loads

T (°C)	t (h)	0.0–0.25	0.25–0.3	0.3–0.5	0.5–0.7	0.7–0.8	0.8–1.0
		⇩	⇩	⇩	⇩	⇩	⇩
165		0	0	0	0	0	0
	ΔH	0	4	16	16	8	16
145		0	4	16	16	8	16
	ΔH	−10	−1	−1	7	3.5	4
140		−10	3	15	23	11.5	20
	ΔH	−110	−11	−11	−33	38.5	44
85		−120	−8	4	−10	50	64
	ΔH	0	6	42	−18	21	24
55		−120	−2	46	−28	71	88
	ΔH	0	0	18	−42	9	0
25		−120	−2	64	−70	80	88
		⇩	⇩	⇩	⇩	⇩	⇩

Table 7.6 Feasible heat cascades in terms of heat loads

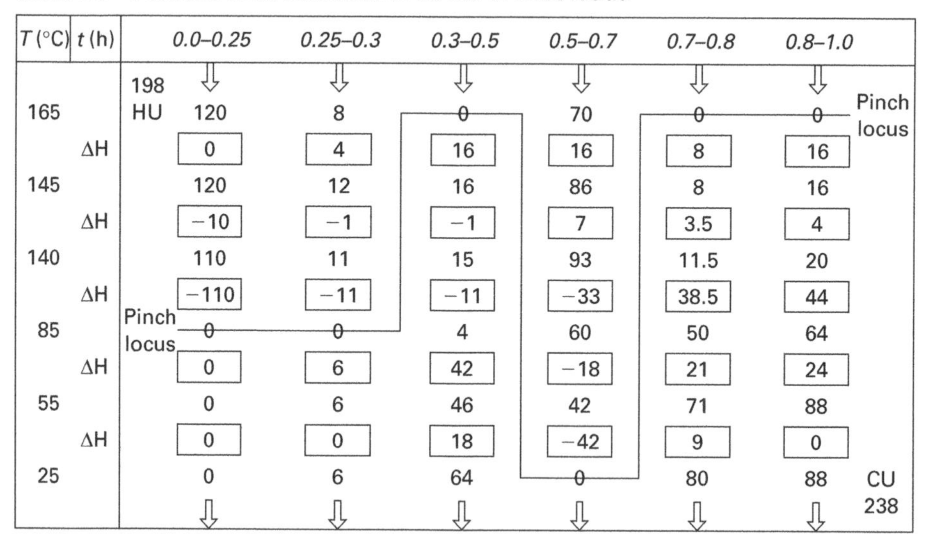

T (°C)	t (h)		0.0–0.25	0.25–0.3	0.3–0.5	0.5–0.7	0.7–0.8	0.8–1.0	
		198	⇩	⇩	⇩	⇩	⇩	⇩	
165		HU	120	8	0	70	0	0	Pinch locus
	ΔH		0	4	16	16	8	16	
145			120	12	16	86	8	16	
	ΔH		−10	−1	−1	7	3.5	4	
140			110	11	15	93	11.5	20	
	ΔH		−110	−11	−11	−33	38.5	44	
85		Pinch locus	0	0	4	60	50	64	
	ΔH		0	6	42	−18	21	24	
55			0	6	46	42	71	88	
	ΔH		0	0	18	−42	9	0	
25			0	6	64	0	80	88	CU 238
			⇩	⇩	⇩	⇩	⇩	⇩	

can be superimposed. This identifies how much below-pinch heat from one process can be used to provide above-pinch heat for the other, so that more heat is recovered and overall utility requirements fall.

The same principle can be used for two time intervals. If an early time interval has a pinch at high temperature, this heat could be used in a later time interval with a low pinch temperature if some way can be found to store it. Again, heat recovery will increase and utility targets will fall.

Table 7.7 Feasible heat cascades with heat storage within and between batches

T(°C)	t(h)		0.0–0.25	0.25–0.3	0.3–0.5	0.5–0.7	0.7–0.8	0.8–1.0	
165		20 HU	6	8	0	6	0	0	
	ΔH		0	4	16	16	8	16	
145			6	12	16	22	8	16	
	ΔH	4 ⇒	−10	−1	−1	7	3.5	4	⇒ 4
140			0	11	15	29	11.5	16	
	ΔH	110 ⇒	−110	−11	−11 ⇒ 4	−33	38.5 ⇒ 50	44	⇒ 110
85		Temperature pinch	0	0	0	0	0	0	
	ΔH		0	6	42 ⇒ 18	−18	21	24	
55			0	6	24	0	21	24	
	ΔH		0	0	18 ⇒ 42	−42	9	0	
25			0	6	0	0	30	24	CU 60

Looking at Table 7.6, the third time interval (0.3–0.5 h) has a "threshold" pinch at the top temperature of 165°C and releases 64 kWh of heat below that temperature, while the fourth time interval (0.5–0.7 h) has the opposite kind of threshold pinch with 70 kWh of cooling required. A similar deduction might be made from the utility-time graph. So there is potential for heat transfer between these two time intervals. The easiest way to calculate how much heat can be recovered is to combine the stream data from the two time intervals and calculate the heat cascade. The resulting cascade shows that only 6 kW of heat now needs to be supplied, and no cooling; there is a double pinch at 85°C and 25°C (the latter being a threshold). So the potential saving available from storing heat between these two intervals is 64 kWh, corresponding to a hot utility target of 134 kWh and a cold utility target of 174 kWh.

Likewise, if the batches are repeated at regular intervals, it is clear that high-temperature heat from the end of one batch can be used to provide heat at the beginning of the next. If it is assumed that heat storage matches, like heat exchange, can take place with a ΔT_{min} of 10°C, then all the cascades can be combined into one. The targets will then fall to the values given by the TAM, 20 kW hot utility and 60 kW cold utility. We can deduce that 178 kWh of heat storage has been used in total, but it is not so easy to determine what temperatures it is required at. Logical deduction based on the TSM cascades can be used. Kemp and Deakin (1989a) gave a rigorous analysis method, the available heat cascade, which shows what storage times and temperatures are needed to reach the TAM target. For information, the results are given in Table 7.7. However, the algorithm is complicated and no computer program using it is commercially available.

In practice, heat recovery in batch processes using heat storage is rare and likely to remain so. It is difficult to store heat for the relatively long periods required for batch processing – tens of minutes or hours – without unacceptable heat losses. Also, most practical heat storage methods will require a much larger driving force than ΔT_{min} or store heat at a fixed temperature level. Hence, heat storage tends to be much more costly than heat exchange. However, it is sometimes used in large-scale continuous systems, for example for storing hot water and steam in a total site system, and the analysis methods above can be used in this situation.

Possible heat storage systems include:

(a) Hotwells – large storage tanks of liquid maintained at roughly constant temperature. All the heat supplied is degraded to the mean temperature of the hotwell.
(b) Stratified storage tanks – similar to hotwells, but mixing and turbulence is kept to a minimum so that the liquid remains in layers with the hottest layers at the top of the tank. A variety of storage temperatures is thus obtained and some degree of countercurrent heat exchange can be achieved.
(c) Thermal regenerators – for example Cowper stoves, where hot and cold gas is successively passed through a brick chamber.
(d) Heat wheels – more commonly used as heat exchangers as their storage ability is very limited.
(e) Systems using latent heat of evaporation – for example steam drums and accumulators, balls filled with a pressurised water/inert gas mixture.
(f) Systems using latent heat of fusion – plastic balls containing a eutectic, ice (in low temperature systems).

Heat storage methods and equipment are comprehensively reviewed in the book by Dincer and Rosen (2002).

The Cascade analysis can be adapted to give targets for the configuration of the actual heat storage system, allowing for the additional temperature penalty, as described by Kemp and Deakin (1989a) and Kemp (1990). The concepts are directly equivalent to those for heat recovery in total site systems via the site steam system or a recirculating heat transfer fluid (Section 5.4.3), which also involve a temperature degradation by the intermediate heat transfer system.

Instead of using heat storage to recover heat between two streams in different time intervals, it may be possible to reschedule the process operations so that the streams fall in the same time interval. This will affect the ratio of heat exchange to heat storage by altering the data for the TSM, but leaves the TAM target unchanged. The overall cascade shows where heat storage is possible and thus indicates where rescheduling could be employed to turn this into heat exchange. The methods are described further in Section 7.7.

An alternative way of evaluating the TSM was proposed by Golwelker (1994). Instead of producing heat cascades for each individual time interval, his method starts with the heat cascades for the individual streams. This can reduce the amount of calculation required, especially where there are significantly fewer streams than time intervals. It can also help to clarify the amount of heat available for rescheduling purposes in a given temperature–time interval. The method and its application are discussed by Shenoy (1995), who also gives numerical examples.

Summarising, targeting for batch processes involves the following steps:

- Collect data on times of operations, temperatures and heat loads.
- Use the TAM to calculate MHR by heat exchange and storage.

- Divide the process into time intervals.
- Carry out approximation for streams with variable temperature or heat load.
- Add further time intervals if necessary.
- Use the TSM to find MHX targets for each time interval.
- Note variation in hot and cold utility requirements over batch period, using the utility-time graph.
- Check pinch locations and deduce opportunities for heat storage or rescheduling.
- Refine calculations near the pinch temperatures if necessary.
- Allow for practical heat storage systems with fixed working temperatures or increased ΔT_{min}, if desired.

7.6 Heat exchanger network design

7.6.1 Networks based on continuous or averaged process

In the design procedure for heat exchanger networks in continuous processes described in Chapter 2, the basic principle was to start at the pinch and work outwards, which would ensure that the targets were achieved. Early studies on batch processes, using the TAM, postulated that since the MHR target was equal to that for the continuous process, the optimal network design would also be the same as for the continuous process. The MER network derived in Section 2.3.3 (Figure 2.18) for the continuous process can be applied to the batch situation, as shown in Figure 7.9. Note the distinction between heat recoverable by direct exchange and storage on each match.

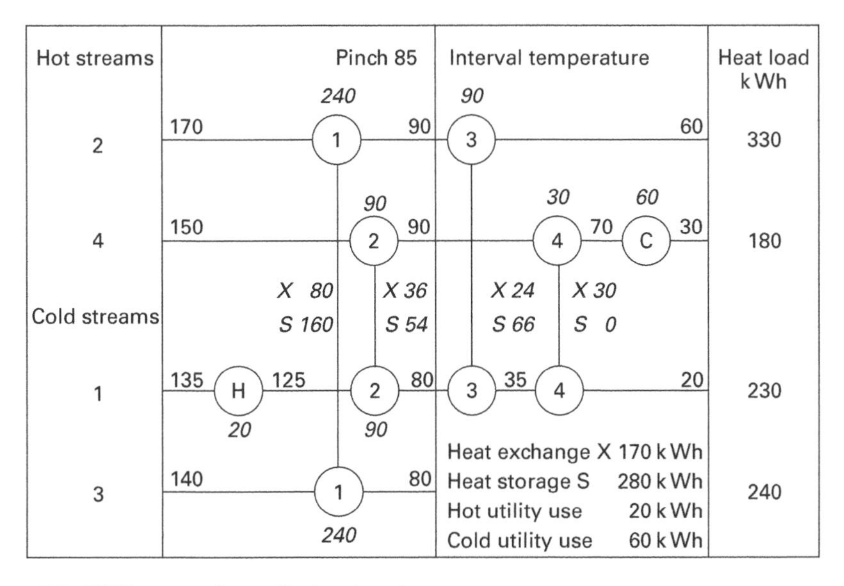

Figure 7.9 MER network applied to batch process

Now direct heat exchange is limited to periods when both streams actually exist, and even then is limited by the instantaneous heat flow rate Q_{inst} available from each stream. For a Type A stream this is defined by:

$$Q_{inst} = C_P(T_2 - T_1)$$

That is the product of the heat capacity flowrate and the temperature change of the stream. The heat recovered by the match is given by multiplying the lower of the two values of Q_{inst} and the period of time when the two matched streams coexist.

Applying this to the network in Figure 7.9 gives values for the heat exchange (X) possible from each match and the heat storage (S) required to make up the required heat load; these values are shown in italics by the side of the match to which they refer. Although the network can indeed achieve the TAM energy target for the batch process, less than half the heat recovery can be carried out by direct exchange (170 kWh out of 450 kWh). Moreover, a complicated storage system with closely spaced temperature levels and long storage times would be needed to achieve the required ΔT_{min} of 10°C. Using this network to obtain MHR by a combination of heat exchange and storage would certainly be uneconomic.

Often, a heat exchanger has an acceptable payback but a heat storage system does not, because a considerable amount of extra equipment is needed. The network of Figure 7.9 could therefore be adopted with only the direct heat exchange being carried out. Heat exchange is 170 kWh. However, the TSM (Section 7.5) showed that 272 kWh can be recovered by direct heat exchange, so the network based on the continuous process is certainly not optimal.

7.6.2 Networks based on individual time intervals

For each time interval, a network can be designed which achieves MHX. Hence, superimposing all six networks will give an overall network structure which achieves the MHX target.

The first time interval requires only utility heating and the last two only require cooling. The networks for the other three time intervals are shown in Figure 7.10. In each case they have been developed by starting from the pinch temperature for that time interval (obtained from Table 5.6) and working outwards; none require more than two matches. Matches 1, 3 and 4 were in the MER network, but match N is new. A stream split is required in the (0.3–0.5 h) and (0.5–0.7 h) intervals where there are two hot streams but only one cold stream.

The network achieved by combining all the matches identified in the individual time intervals, Figure 7.11, achieves the MHX target of 198 kWh of hot utility, using four exchangers recovering 272 kWh in total. Stream splits are required on both cold streams. Figure 7.11 is the simplest form of **maximum heat exchange network** – defined as one which achieves the MHX/TSM targets for direct heat exchange.

Note that both exchangers 3 and 4 have to transfer heat across the overall pinch at 85°C. This apparently contradicts a basic principle of process integration, and does mean that the network cannot achieve the TAM target. It is necessary because the pinch temperature for each time interval is different; in most time intervals, heat is

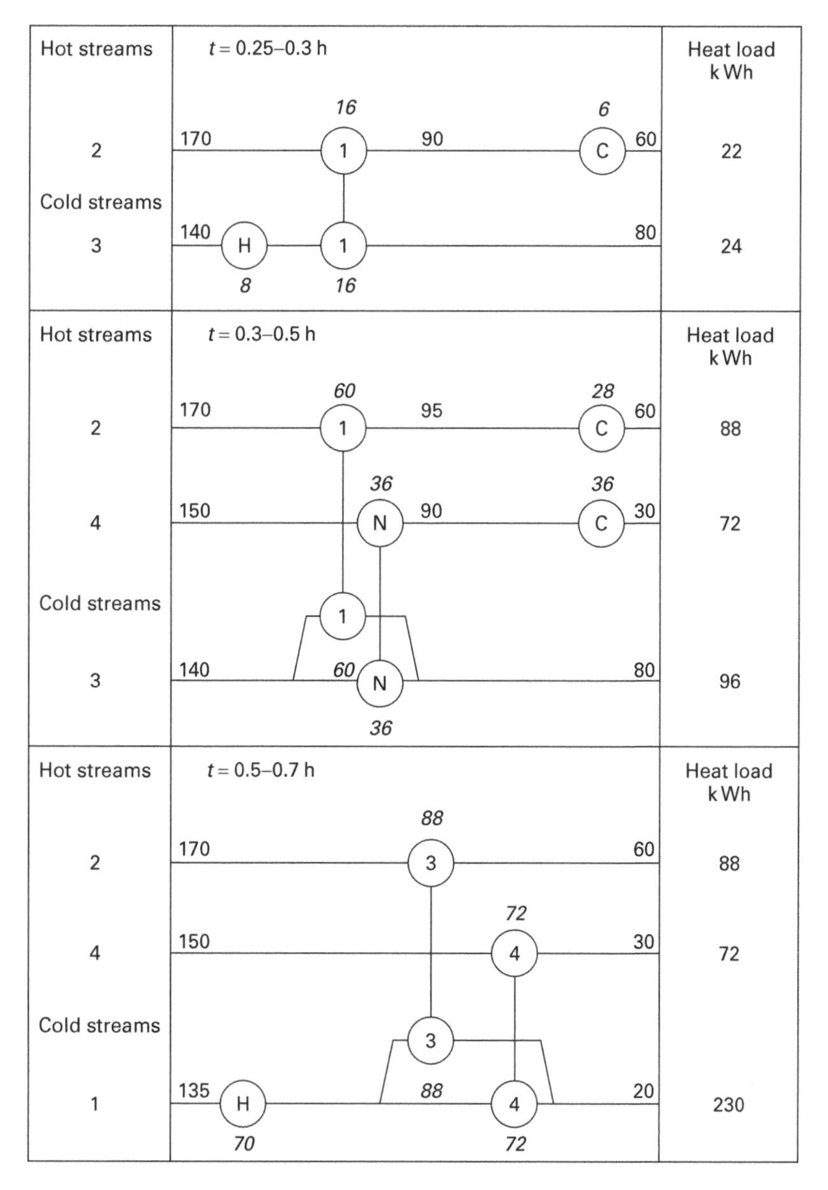

Figure 7.10 Heat exchanger networks for individual time intervals

either rejected above the overall pinch or required below the overall pinch. It is therefore difficult or impossible to design a network which achieves both the MHX and TAM targets. In practice, an MHX network is far more likely to be cost-effective.

The network of Figure 7.11 may seem uneconomic because four exchangers are still required for what could be a relatively small plant. However, use of special heat exchangers can reduce costs considerably.

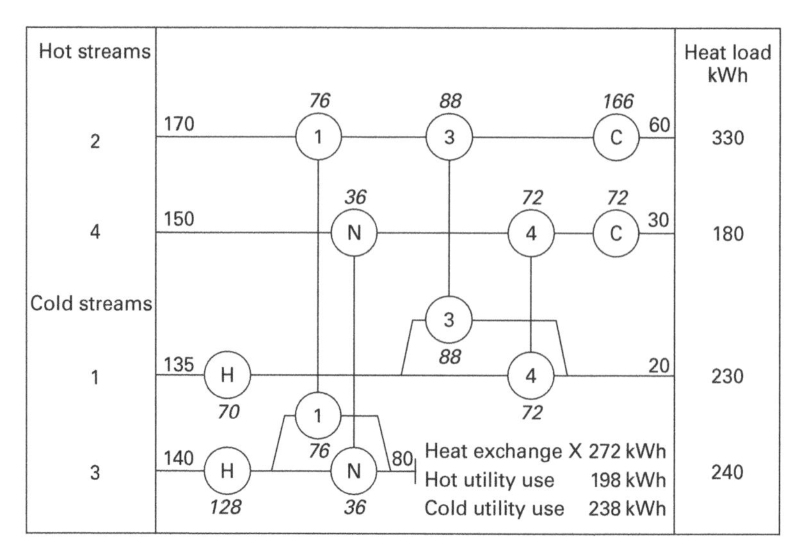

Figure 7.11 MHX network

Instead of splitting streams 1 and 3, multi-stream heat exchangers could be used, for example a gasketed or welded plate exchanger or a plate-fin exchanger, as described in Section 4.2.6. In each case, the "hot" side would be divided into a section through which hot stream 2 flows and a separate section in which stream 4 flows, thus matching both simultaneously against the cold stream. Use of multi-stream heat exchangers would reduce the number of units required in Figure 7.11 from four to two. Moreover, since streams 1 and 3 do not exist at the same time, it is theoretically possible to use the same exchanger for both, so that all four matches could be achieved with a single multi-purpose exchanger. Such units are quite common in speciality chemicals plants. The main limitation on multi-purpose exchangers is that cross-contamination will occur between the liquids used in succession (here streams 1 and 3). This may be unacceptable because of safety requirements, side reactions, unacceptable impurity levels or other product quality reasons. The problem may be overcome by a brief purge (e.g. with steam) before the second fluid is passed through; this is particularly effective for gaseous streams.

Similar opportunities should always be looked for on batch plants, which are often small so that the economic return from heat recovery is limited by the need for many exchangers with small heat loads and operating only intermittently. Combining the matches into a small number of multi-stream or multi-purpose heat exchangers can make an apparently uneconomic heat recovery project viable. With plate exchangers, heat recovery may be achievable without using any extra exchangers, as a single frame can accommodate both heat exchange and a heating or cooling section.

The MHX network can be relaxed to remove small or undesirable matches. The techniques, including removal of stream splits and loop breaking, are the same as for continuous processes. Possibilities for this four-stream example are described in depth by Kemp and Deakin (1989b). For example, if multi-stream and multi-purpose heat exchangers cannot be used, and the stream splitting in Figure 7.11 is judged to be undesirable, matches 1 and N can be placed in series, as can matches 3 and 4.

If it is intended to use heat storage, these matches can be included in the network. The heat delivered to storage from the first time interval is added to the network for that time interval as a cold stream. Likewise, the heat released from storage in the later time interval is added to that network as a hot stream. Again, this method is covered in depth by Kemp and Deakin (1989b).

7.7 Rescheduling

7.7.1 Definition

Rescheduling is the alteration of the timing of certain process operations. The result is that streams stay in the same temperature range as before, but move into different time intervals. This means that the overall heat recovery target, as predicted by the TAM, is not affected by rescheduling. However, the individual time cascades will be different and the MHX target obtained from the TSM will change. Rescheduling thus allows some heat to be recovered by direct heat exchange where it was previously recovered via heat storage.

Rescheduling can give a number of benefits:

1. It can increase the amount of direct heat exchange and reduce the MHX energy target.
2. High intermittent loads on hot and cold utilities can be reduced ("load smoothing" or "peak lopping").
3. Operability may be improved or capital cost reduced by removing the need for some heaters and coolers.

Rescheduling is clearly a form of process change. However, instead of changing stream temperatures, we are altering stream times. The sequence of operations is as follows:

1. Obtain targets for the existing schedule by the TSM, Cascade analysis or utility-time curves.
2. Identify beneficial scheduling changes.
3. Obtain new heat exchange and storage targets with the new schedule.

7.7.2 Classification of rescheduling types

Rescheduling opportunities can be broadly divided into four classes. These will be illustrated by referring to our simple two-stream example. The streams will be assumed to be of Type A so that heat exchange is always feasible in temperature terms.

Stream	*Heat load (MJ)*	*Operating time (h)*	*Heat flow rate (MJ/h)*
Cold A	100	0.0–0.5	200
Hot B	100	0.8–1.0	500

The current operating times of the two streams are effectively represented on a time event diagram, Figure 7.12(a).

The four different types of rescheduling to allow streams A and B to exchange heat are as follows:

Type 1: If there are two parallel processing lines, their hours of operation could be regulated so that A and B can exchange heat, with no change to times within

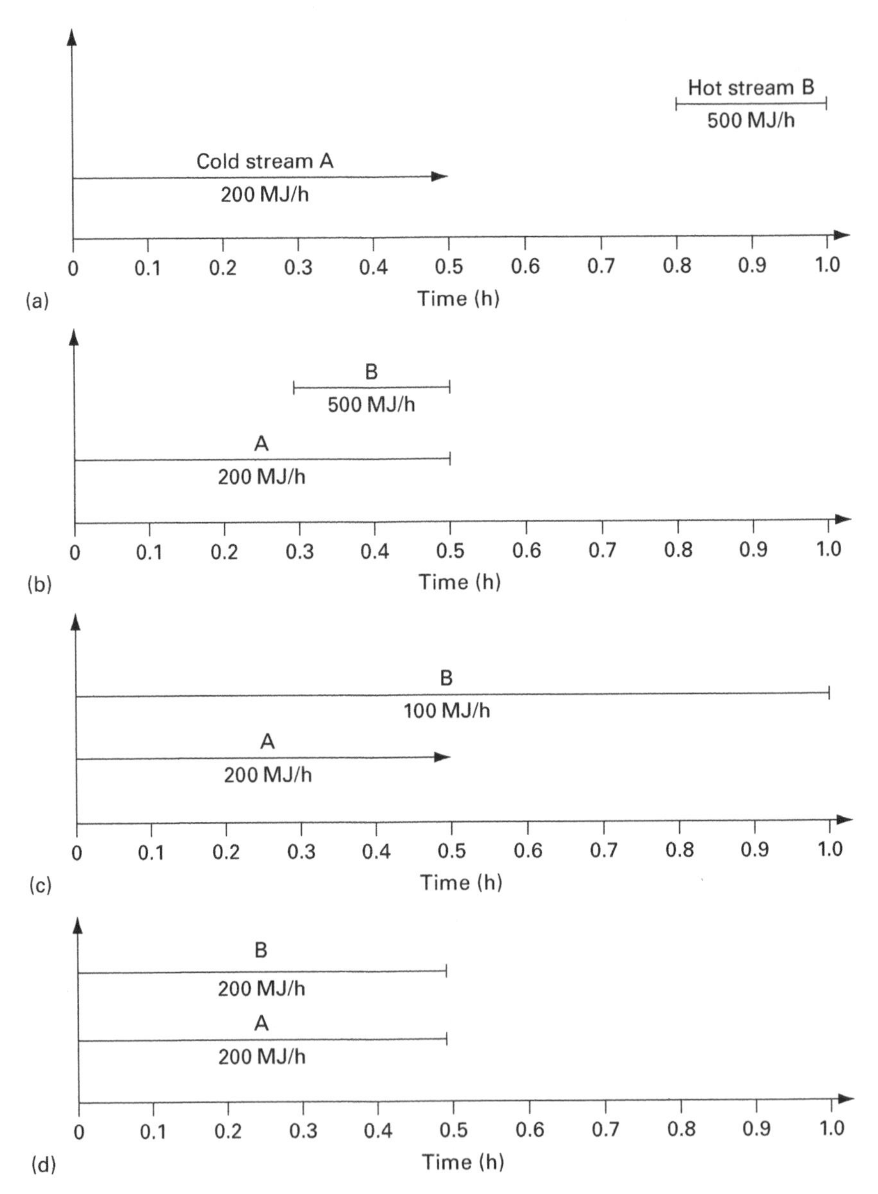

Figure 7.12 Time event charts for rescheduling of two-stream problem

the batch. This could be achieved by running plant A 0.5–0.8 h later or plant B 0.5–0.8 h earlier (see Figure 7.12(b)). Heat recoverable is (200 × 0.2) or 40 MJ.

Type 2: Altering the time of a stream within a batch, but not its duration and flowrate. This is done by delaying stream A or moving B forward by 0.5–0.8 h, thus changing the internal scheduling of the plant. Again Figure 7.12(b) is the result. Heat recoverable is again (100 × 0.4) = 40 MJ.

Type 3: Altering the duration of a stream by changing its flowrate, retaining the previous start or finish time. Stream B can be extended to run from 0 to 1 h, when its heat flowrate will fall from 500 to 100 MJ/h. The heat recoverable by this method is (100 × 0.5) or 50 MJ (see Figure 7.12(c)).

Type 4: Altering both the timing and duration of a stream. Stream B could run from 0.0 to 0.5 h and its heat flowrate would then be 200 MJ/h. Since this matches stream A exactly, heat recovery between the streams reaches its maximum value, (200 × 0.5) = 100 kWh (see Figure 7.12(d)). Note that we have now reached the TAM target.

Type 1 rescheduling may incur little or no disruption to operations and can yield projects which are operationally acceptable and give a good economic return. However, it links together two separate processes which then have to be run in conjunction to achieve the energy savings; hence there can be some loss of operating flexibility.

In contrast, to carry out rescheduling opportunities of Types 2, 3 and 4, major changes to the plant will often be required and these may be costly or impracticable. Examples are:

- rerating of pumps or altering their speed to achieve a change in flowrate,
- enlargement of heat exchangers to handle higher transient loads,
- provision of additional holding capacity.

Indirect heat storage can be used instead; this imposes less constraints but requires greater temperature differences and more costly equipment. In effect, there is a three-way trade-off between reduced flexibility (with rescheduling and direct heat exchange), increased cost and complexity (with heat storage) or increased energy use (with neither).

7.7.3 Methodology

As with most sections of pinch analysis for batch processes, identification of scheduling changes is a difficult task. The time event diagram is inadequate for most cases as it does not show the temperature range or heat loads of the streams unambiguously, but it does give an important representation of when streams coexist. The TSM and the utility-time graph indicate when there are major peaks in hot and cold utility use and can thus suggest where it would be valuable to increase or reduce the hot or cold stream loads within a time interval. The equipment occupancy time event chart (Gantt chart) can again be a valuable tool.

If heat storage possibilities have been identified using the Cascade analysis, these can be converted to direct heat exchange by suitable rescheduling of streams, as shown by Kemp and Deakin (1989b) and Kemp (1990). For the main example, the overall cascade in Table 7.7 gives both the storage required and the temperature range (e.g. 64 kW is being stored between periods 0.3–0.5 h and 0.5–0.7 h). Consulting the time event diagram, Figure 7.7, and the networks obtained for the individual time intervals, Figure 7.10, the streams which are exchanging heat via storage can be deduced. The heater load in period 7 falls on cold stream 1 and is reduced from 70 to 6 kWh, so this is the stream which has received heat and either stream 2 or 4 could have supplied it. If one or more of these streams can be rescheduled, it should be possible to recover some of the heat by direct exchange. Alternatively, if different streams in these time intervals can be rescheduled, new networks can be developed.

However, rescheduling may also give opportunities which do *not* have a corresponding heat storage possibility, because heat storage still has a one-way time constraint. A hot stream can only store heat for use in a later time interval, not an earlier one. However, if a cold stream from an earlier interval is rescheduled to operate later, it can receive this heat. So the utility-time plot and time event chart can give additional information. Looking at Figure 7.8, the period 0.5–0.7 h requires net heating, and heat is available not only from the time interval before (as identified above) but also from the following interval, 0.7–0.8 h. The time event diagram, Figure 7.7, shows that only hot streams exist at this time, suggesting that we might want to increase the operating period of cold stream 1 or reduce that of cold stream 4.

The analysis initially aims to find all rescheduling possibilities based on the stream data alone, without reference to how they can be applied in practice. This ensures that no opportunities are overlooked initially; some will be eliminated later when the physical significance of changing certain streams is considered. Also, because rescheduling moves streams in time so that they can exchange heat directly, the ideal heat storage analysis with ΔT_{min} = 10°C can be used as a basis; there is no need here to impose additional temperature penalties.

Rescheduling should always be treated rigorously as a process change. After making any alterations to time periods of streams, the stream data should be recalculated, and then the energy targets using the TSM. A useful tool at this stage is the batch utility curves (Gremouti 1991), which show the temperature ranges of heat not being directly exchanged.

Almost invariably, there is some element of trial and error. The analysis has suggested that it may be beneficial to extend stream 1 forwards or backwards in time, or both, but it is not clear what option will be best until the TSM targets are calculated. Table 7.8 shows the results of various scenarios.

Moving stream 4 or stream 1 forwards gives only small benefits. Option (d) saves 64 kWh but requires both streams 1 and 3 to be moved by 0.2 h, whereas moving stream 1 back by just 0.1 h removes the whole of the hot utility requirement in the middle of the process. Thus (e) will be by far the easiest rescheduling possibility if it is allowed by process constraints.

Table 7.9 shows the heat cascades for the rescheduled process, option (e). Note that one time interval has disappeared as no streams now start or stop after 0.7 h. The network will have the same structure as the MHX network in Figure 7.11, but the heat loads on the matches to stream 1 will have to be increased.

Table 7.8 Rescheduling options for four-stream example

Situation	*Stream 1 operating times (h)*	*Hot utility (TSM) kWh*	*Heat exchange (MHX) kWh*	*Additional heat recovered kWh*
(a) Current	0.5–0.7	198	272	0
(b) (a) +Stream 4 0.3–0.7 h	0.5–0.7	180	290	18
(c) Stream 1 moved forward	0.3–0.7	179	291	19
(d) (c) +Stream 3 0.0–0.7 h	0.3–0.5	134	336	64
(e) Stream 1 moved back	0.5–0.8	128	342	70

Table 7.9 Feasible heat cascades (as heat loads) for rescheduled process

T(°C)	t(h)		*0.0–0.25*	*0.25–0.3*	*0.3–0.5*	*0.5–0.8*	*0.8–1.0*	
		128	⇩	⇩	⇩	⇩	⇩	
165		HU	120	8	0	0	0	Pinch locus
	ΔH		0	4	16	24	16	
145			120	12	16	24	16	
	ΔH		−10	−1	−1	10.5	4	
140			110	11	15	34.5	20	
	ΔH		−110	−11	−11	5.5	44	
85		Pinch locus	0	0	4	40	64	
	ΔH		0	6	42	3	24	
55			0	6	46	43	88	
	ΔH		0	0	18	−33	0	
25			0	6	64	10	88	CU 168
			⇩	⇩	⇩	⇩	⇩	

Rescheduling may also be used to improve operability without increasing heat recovery, for example by minimising the number of changeovers between heating and cooling on a single jacketed vessel, or to reduce peak heating and cooling loads. Similar opportunities may also appear during network design (e.g. if a large heater or cooler is needed on a stream but is only required for a short period, it may be possible to reschedule some streams to eliminate it or reduce it in size). Again Kemp (1990) gives examples.

7.8 Debottlenecking

As pointed out earlier, energy is often a relatively minor factor in batch processes. Far more important are opportunities to increase throughput (production rate) and operability (which may lead to improved product quality and yield). In particular, debottlenecking to give increased production can give economic benefits which far exceed the total plant energy bill.

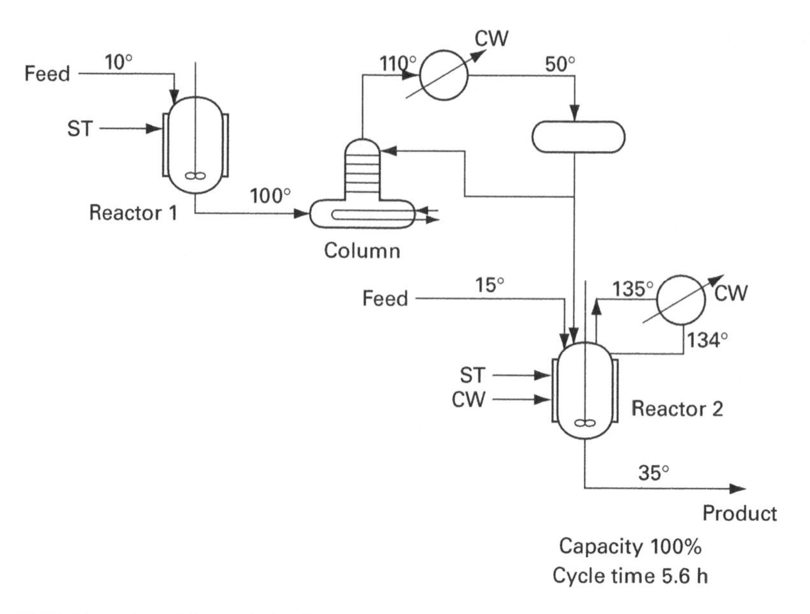

Figure 7.13 Flowsheet for original process

Batch processes are usually limited by one or more of the following four parameters:

- material flow (e.g. waiting for next charge);
- heat flow (e.g. waiting to reach temperature);
- equipment capacity (e.g. waiting for next empty vessel);
- labour (e.g. waiting for next shift).

These parameters are typically interlinked, and heat flow plays a dominant role for technical feasibility. Material flow may be controlled by the rate of heating the charge or cooling the product. Equipment capacity may be controlled by reactor residence time, which in turn is controlled by heat transfer. The quality of separation may be limited by cooling temperature limits, etc.

In most processes, it is only one or two sections which limit the production rate of the overall plant. These "rate-limiting steps" can be considered to be "productivity pinches". As in traditional energy-based pinch analysis, the greatest benefits come from identifying these pinches and improving that section of the process. Probably the most useful single tool in doing this is the **equipment occupancy chart**, or **Gantt chart**, which can be considered to be a development of the time event chart. This chart shows which items of equipment are in use all the time and, conversely, which ones have spare capacity.

A highly instructive example of successful batch process integration by this means was reported by Gremouti (1991) and is described below, in slightly modified form. Figure 7.13 shows the flowsheet of a speciality process operated in cycles but in the context of overall random production and in multi-purpose reactor vessels. Two reactions take place in succession and there is an intermediate distillation column to separate the products from the first reaction. The reactions take place within

Table 7.10 Operational schedule for original process (h)

Vessel	*Filling*	*Heating*	*Reaction/Reflux*	*Cooling*	*Discharge*	*Total*
Reactor 1	0.7	1.0	1.5	–	0.3	3.5
Column	0.3	–	2.0	–	0.5	2.8
Reactor 2	0.5	1.1	2.5	1.25	0.25	5.6

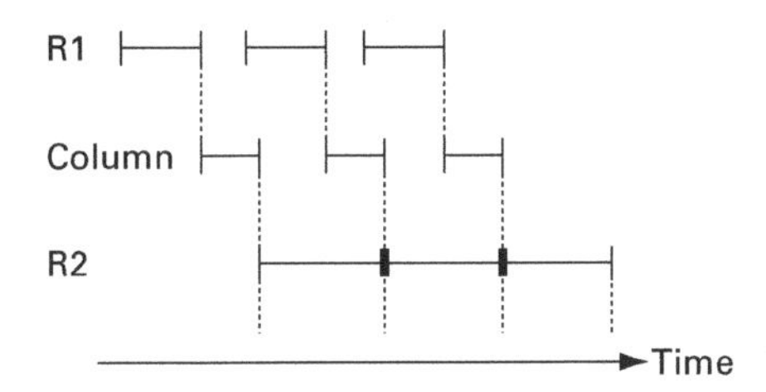

Figure 7.14 Equipment occupancy (Gantt) chart for original process

jacketed vessels and there is an initial heating period and, for Reactor 2 only, the product is cooled. In addition the time required to fill and discharge each vessel must be allowed for. Table 7.10 shows the schedule of operations and Figure 7.14 shows the equipment occupancy chart for the process. The cycle time is 5.6 h.

It is obvious from the chart that reactor No. 2 is limiting. In order to debottleneck the plant we have the following options:

(a) Accelerate some of the operations taking place in Reactor 2 to reduce the cycle time.
(b) Transfer some of the operations currently performed in Reactor 2 to another vessel.

It was not acceptable to reduce the reaction time, as this would necessitate changing the reaction conditions with possibly deleterious effects on yield and selectivity. The filling and discharge processes might be speeded up, but new, larger pumps would be needed and the gain would be small. However, the feed heating and product cooling processes could be performed outside the reactor, and this is a much more promising possibility.

There are no spare vessels nearby which can be reused. One option is to add a heater to heat the Reactor 2 feed and a cooler to cool the Reactor 2 discharge. It would be difficult to use the same item for both tasks as it would need to be cleaned twice per cycle to prevent cross-contamination.

Another possibility is to use the hot product itself to heat the incoming feed. Applying the Cascade analysis shows that the total heat loads (kWh) and temperatures are compatible, but the streams do not exist at the same time and the heat flowrates (kW) do not match. A Type 3 or 4 rescheduling would therefore be needed on at least one of the streams. Options are as follows:

(i) Discharge the product through a heat exchanger and store the heated feed (at 78°C) until it is required.
(ii) Store the hot product (at 135°C) until the feed is required.

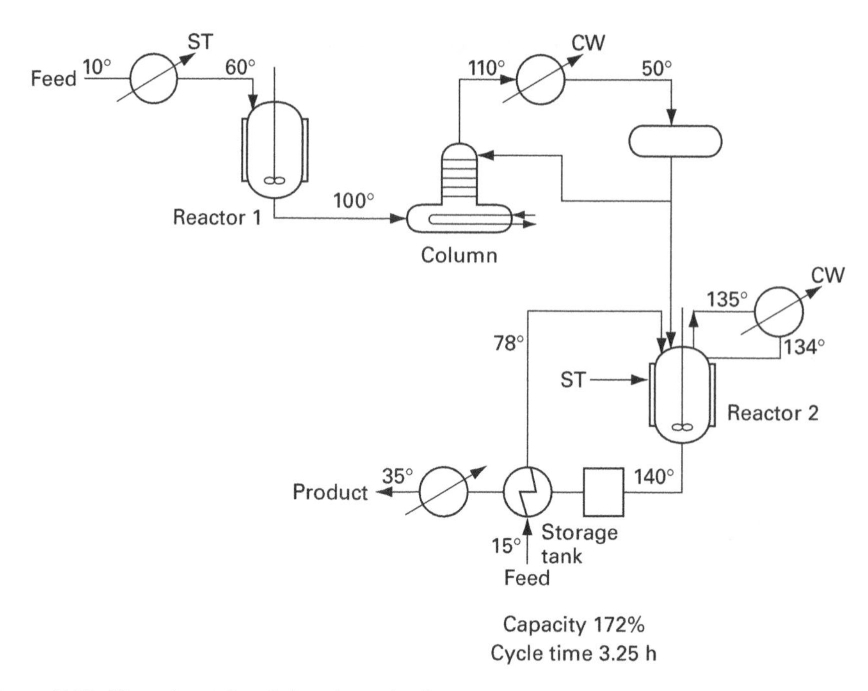

Figure 7.15 Flowsheet for debottlenecked process

The choice between them depends on process-specific factors. Normally the lower-temperature storage option would be preferred to reduce heat losses; but in this case, option (ii) was preferred because the feed was less stable than the product and also because the feed is currently added over a longer period (0.5 h as against 0.25 h for discharge) so the heat transfer rate and required exchanger area are halved. The required heating and cooling rates are actually considerably faster than those in the current reaction vessel, but this does not require an excessively large exchanger as its surface area to volume ratio is much better than that of a jacketed vessel. The storage time is relatively short, as the new feed can begin to enter the vessel as soon as the previous product has been discharged. For start-up and shutdown, steam or cooling water can be fed to the heat exchanger.

There is an energy saving due to this heat recovery, but the cost benefits are negligible compared to those of increasing the production and saving the capital cost of a second heater/cooler.

This would reduce the equipment occupancy time for Reactor 2–3.25 h. Looking at the Gantt chart in Figure 7.14 or at Table 7.10, we see that Reactor 1 is now the rate-limiting step as it is used for 3.5 h. It can be debottlenecked in the same way through external feed heating while charging occurs.

Figure 7.15 shows the flowsheet for the final project proposal.

Table 7.11 is the revised schedule and Figure 7.16 is the Gantt chart. Cycle time is reduced from 5.6 h to 3.25 h. Capacity (throughput) is increased by no less than 72% without any modifications to the reactors. All that is required are two inexpensive exchangers and an insulated storage tank.

Table 7.11 Operational schedule for modified process (h)

Vessel	*Filling*	*Heating*	*Reaction/Reflux*	*Cooling*	*Discharge*	*Total*
Heater 1	–	1.0	–	–	–	1.0
Reactor 1	0.7	–	1.5	–	0.3	2.5
Column	0.3	–	.0	–	0.5	2.8
Reactor 2	0.5	–	2.5	–	0.25	3.25
Storage	0.25	–	–	–	0.5	0.75

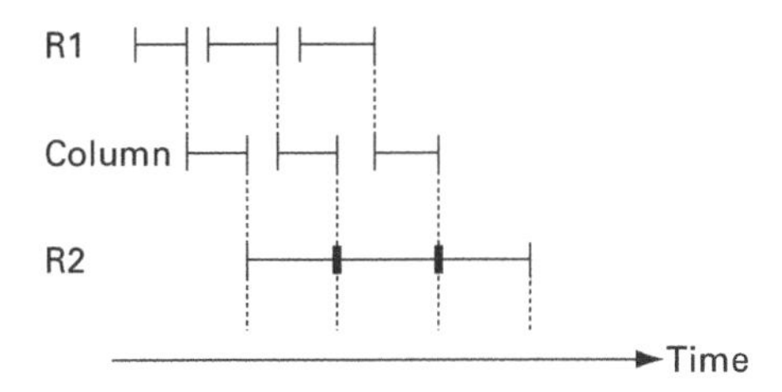

Figure 7.16 Equipment occupancy (Gantt) chart for debottlenecked process

Summarising, the debottlenecking method for batch processes involves the following steps:

- Collect data on times of operations, temperatures and heat loads, and vessel occupancy.
- Generate an equipment occupancy diagram.
- Find which item(s) of equipment are rate-limiting.
- Consider ways to speed up these operations or transfer them to other vessels.
- Check that another section of the process has not become the limiting step.
- Use pinch analysis to check for any heat recovery and rescheduling opportunities.
- Re-draw the equipment occupancy chart and implement the chosen solution.

7.9 Other time-dependent applications

7.9.1. Start-up and shutdown

A continuous process being started up or shut down can be handled effectively by the time-dependent analysis. This may be particularly useful to identify high transient hot and cold utility loads during this period, e.g. when a feed or discharge stream is suddenly started or stopped. The peak can then be minimised, and the plant operators can be warned to prepare for it. An example has been published by Kemp (1990, 1991). Likewise, the time-dependent analysis could be used for control, by predicting plant behaviour if a stream suddenly disappears, and may provide a simple alternative to rigorous sensitivity analysis. Since there are often fears that highly integrated plants will be difficult to start up, shut down and control, this is potentially a major benefit.

7.9.2 Day/night variations

On sites where there are a number of operations taking place at certain periods of the day, the 24 h can be divided into time intervals. The methods above can then be used to identify any opportunities for linking together two sections of the site and rescheduling their operating hours, in order to recover heat or to remove peaks in the utility-time graph. In the case study on the hospital site in Section 9.6, heat recovery from an incinerator combined with extending its operating time by 2 h meant that only two site boilers were required instead of three.

7.10 Conclusions

Heat recovery in batch processes is possible, but generally gives lower absolute savings than for continuous processes, because energy use is generally lower and there are major constraints on whether hot and cold streams coexist at the same time. However, pinch analysis can identify substantial benefits on batch processes, and also on other time-dependent situations. Indeed, it could be considered that these are the general case; a continuous process is really just a special case of a batch process with all streams of Type A, and existing at all times in steady-state!

The two most commonly useful techniques for batch and time-dependent processes are the TSM and the time event (Gantt) chart. These are the key weapons in evaluating heat exchange and debottlenecking possibilities. The utility-time graph is also useful as it helps to identify possibilities for rescheduling. However, heat storage is rarely feasible or economic. Multi-stream exchangers often considerably increase the potential heat recovery and economic benefit.

On batch processes, the low energy use generally means that the biggest benefits come from debottlenecking or capital cost reductions. For continuous operations, the time-dependent analysis is useful for start-up and shutdown, and systematic variations during the day.

References

Ashton, G. J. *et al.* (1993). Design and operation of energy efficient batch processes. Report by Linnhoff March Limited, *Contract No. JOU-0043 (SMA), JOULE Programme, Rational Use of Energy* (The Commission of the European Communities).

Brown, K. J. (1989). Process Integration Initiative. A review of the process integration initiatives funded under the Energy Efficiency R&D programme. Energy Technology Support Unit (ETSU), Harwell, Oxon, UK.

Dincer, I. and Rosen, M. A. (2002). *Thermal Energy Storage: Systems and Applications.* John Wiley & Sons, Chichester, UK. ISBN: 0471495735.

Golwelker, S. (1994). *Energy Integration of Batch Processes: Pinch Technology Approach*, M.Tech. Thesis, Indian University of Technology, Bombay.

Gremouti, I. D. (1991). *Integration of Batch Processes for Energy Savings and Debottlenecking*, M.Sc Thesis, University of Manchester (UMIST), U.K.

Kemp, I. C. and Deakin, A. W. (1989). The cascade analysis for energy and process integration of batch processes, *Chem Eng Res Des*, 67(5): 495–525, September. (a) Part 1 – Calculation of energy targets, pp. 495–509. (b) Part 2 – Network design and process scheduling, pp. 510–516. (c) Part3 – A case study, pp. 517–525.

Kemp, I. C. (1990). Process integration: process change and batch processes. ESDU Data Item 90033. Available by subscription from ESDU International plc, 27 Corsham Street, LONDON N1 6UA.

Kemp, I. C. (1991). Some aspects of the practical application of pinch technology methods, *ChERD (Trans I Chem E)*, 69(A6): 471–479, November.

Shenoy, U. V. (1995). *Heat Exchanger Network Synthesis: Process Optimization by Energy and Resource Analysis*, Chapter 9.1. Gulf Publishing Company, Houston, USA.

8 Applying the technology in practice

8.1 Introduction

As can be seen from the previous chapters, pinch analysis or process integration has developed into a large and complex subject in the last quarter-century. Even the details given are only a summary of the large number of papers which have been published developing the original procedures. Some of the new theoretical developments are complex and difficult to use in practical situations. The question which the average process engineer may ask is, "Which techniques are the most useful? How can I apply them on my plant?"

We need to reinforce the original goals of pinch technology:

- Obtain a rapid understanding of the important factors regulating the energy consumption of a process.
- Allowing approximate but meaningful energy targets to be set using short-cut calculations.
- Pre-optimisation to identify the most promising schemes before embarking on the costly and time-consuming detailed design phase.

8.2 How to do a pinch study

The stages in a pinch analysis of a real process plant or site were outlined in Section 2.5.2. Summarising, the key steps are:

- Produce the heat and mass balance from the process flowsheet.
- Extract the stream data for the process integration analysis.
- Select a ΔT_{min}, using supertargeting if helpful.
- Calculate energy targets and pinch temperature for the current process.
- Investigate possibilities for process change.
- Look at total site aspects including utility levels, combined heat and power (CHP) and heat pumping.
- Design the heat exchanger network, starting with the maximum energy recovery (MER) design and relaxing.

There is significant overlap between some of these stages. For example, a CHP system has a major effect on the costs of utilities and the differential price between different levels, and this substantially affects the economics of process changes and heat recovery projects. Note also that the network design stage comes at the very end; process change and total site aspects should not simply be added as an afterthought.

The following sections look in more detail at practical aspects of some of these stages, particularly the heat and mass balance and data extraction, in the light of the techniques developed earlier.

8.3 Heat and mass balance

The first essential step in a process integration study is to form a consistent heat and mass balance, with all heat losses and leaks accounted for, but it is very rare to find that one is available. The data need not always be precise, but it is essential that they are self-consistent. Generally, the information available from recorded plant data is inadequate, conflicting or both. Temperature readings may sometimes be trusted, assuming that they have been calibrated reasonably recently; flow measurements can often be seriously in error. The engineer must allow a significant time to prepare the heat and mass balance and be prepared to make significant changes to reconcile the data.

For a new plant in the design stage, there will usually be some sort of balance available from the flowsheet. Even this may not always be self-consistent and careful checking is required. Frequently, the extent of likely heat losses is unknown. We should also remember that during commissioning, the process conditions may be changed significantly from the original design values in order to obtain reliable operation.

Likewise, for an existing plant, one should never rely on the original design flowsheet. It can be used as an initial guideline, but it is rare to find that it represents current operating conditions accurately. If possible, it is always best to form a heat and mass balance of the plant as it is operating now; a suitable method is as follows.

- Collect data on mass flows and compositions.
- Form the mass balance for the significant streams in the process flowsheet.
- Cross-check against annual production rate and annual working hours (optional).
- Collect data on temperatures, steam, fuel and water flows, specific and latent heats.
- Form the heat balance, allowing as necessary for heat losses.
- Cross-check against annual fuel consumption and energy bill (optional).
- Alter mass and heat flows as necessary to reconcile the balances, varying the least reliable data first.

Discrepancies of up to 50% in the mass balance and even more in the heat balance are not uncommon, especially in older, poorly instrumented plants. To resolve the conflict, it is often necessary to assume errors in instrument readings or estimated mass flows that have been accepted as gospel for many years. The engineer must be prepared to take these difficult decisions and defend them to plant and management personnel. The reliability of instrument readings generally decreases in the following order:

- Temperatures of liquids, gases and solids.
- Mass flowrates of liquids and solids.
- Annual production rates and fuel consumption.
- Mass flowrates of gases, including steam meters.
- Ancillary data (e.g. annual working hours, heat losses, solids moisture content).

It is important to use a set of data which was collected at roughly the same time; plant conditions can vary considerably for different days or shifts.

Modern plant instrumentation and centralised computer control systems can make the task of data collection and recording very much easier. However, it is still useful and educational to form the heat and mass balance yourself (e.g. using a spreadsheet) and make sure that it is indeed self-consistent. The data recording system may only have limited cross-checking facilities.

An accurate balance is frequently a very valuable by-product of a pinch study. It helps both plant management and operators to understand better what is happening on their plant and may in itself reveal cost savings. In one study, the balance revealed that plant throughput was one-third lower than the original design values but that the steam supply to the distillation columns had not been reduced accordingly. By turning down the heating, annual savings of £190,000 were made at zero capital cost.

Data for performance of existing heat exchangers is also valuable. When we know the heat loads and the temperatures on both hot and cold sides, since the heat exchange surface area is known, we can work out the effective heat transfer coefficient across the exchanger. This is often far more accurate than trying to estimate it from correlations. However, the correlations can be useful in suggesting how the heat transfer resistance is split between the hot and cold sides, thus allowing us to deduce the film heat transfer coefficients. These values can then be used in cost targeting calculations to select the optimum ΔT_{min}.

8.4 Stream data extraction

Extraction of stream data from a plant flowsheet is rarely discussed in research literature, yet it is probably the biggest source of difficulties in a process integration study. Once a consistent heat and mass balance has been obtained, data extraction is reasonably straightforward if all the process flows are separate liquid streams which do not mix together or change in composition, and with all heating and cooling carried out indirectly. Unfortunately, such processes are rare, especially in the "non-traditional" industries such as food processing. Mass exchange makes data extraction particularly awkward; it is difficult to know where to draw the boundary conditions and which sections to treat as an undisturbed "black box". We should always apply the key criterion for a stream, that it should change in heat load but not in composition. Even so, special care will be needed in handling the following:

- How to subdivide streams passing through intermediate vessels.
- Streams with variable specific heat capacity, partial vaporisation or partial condensation.
- Mixing points.
- Separation processes (distillation, evaporation, absorption, flash systems, drying, etc.).
- Chemical reactions within the process system.
- Significant heat losses.
- Heating by direct firing or direct steam injection.
- Cooling by cold water injection or sprays.

In the last two cases, any change in utility load will affect the composition of some process streams and alter the heat and mass balance and hence the stream data, making exact targeting an iterative process.

Several aspects have already been covered in Section 3.1, such as choosing streams, handling latent heat streams and using heat load data. Some (e.g. mixing), have been mentioned briefly but will be covered in more detail below. The UMIST (University of Manchester Institute of Science and Technology) course on process integration (UMIST 1996) has been a valuable source of material for this section.

It is also vital that streams to and from a vessel in which a significant change in composition takes place (e.g. a distillation column, evaporator or reactor) are kept separate. Again, the temperature of the unit operation can be optimised by the methods given in Chapter 6 on Process Evolution, but it cannot be ignored, it is a vital processing step.

8.4.1 Mixing and splitting junctions

Mixing of streams was briefly mentioned in Chapter 3, where it was recommended that for the analysis we run down streams to the same temperature and mix them isothermally whenever possible, to achieve the best energy targets. However, there are other effects, and splitting junctions must also be considered.

The top drawing of Figure 8.1(a) shows schematically two cold streams leaving separate units at different supply temperatures, mixing and then requiring heating to a common target temperature. In terms of capital targeting, the system is really only one stream as shown in the lower drawing of Figure 8.1(a), because it can be satisfied by only one unit. However, Figure 8.1(b) shows what may happen if the system is regarded as only one stream for energy targeting. If the mixing temperature lies below pinch temperature, then the "cooling ability" of the cold stream below the pinch is degraded. More heat must therefore be put to utility cooling, and by enthalpy balance, heat must be transferred across the pinch increasing hot utility usage. To ensure the best energy performance at the targeting stage, the mixing must be assumed isothermal, as shown in Figure 8.1(c). If $T = 120°C$ then the system is regarded as two streams. If $T < 120°C$, then it is three streams. Hence in

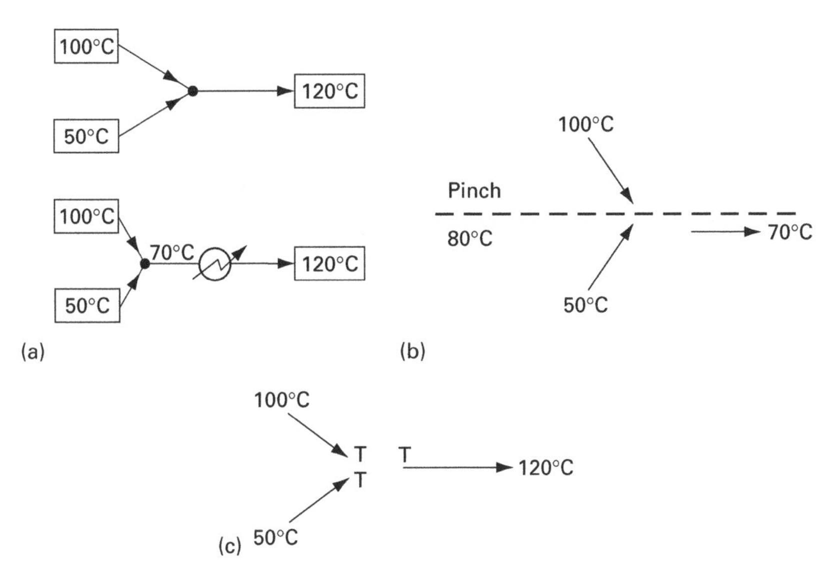

Figure 8.1 Mixing streams

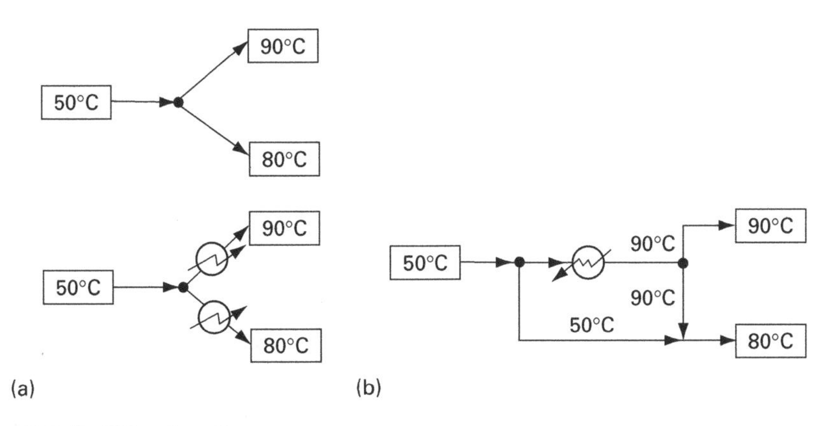

Figure 8.2 Splitting junctions

stream mixing, the data for units targeting is incompatible with the data for energy targeting. However, this should not cause confusion at the design stage if the above principles are thoroughly understood. It merely means that the designer might require one more unit than minimum if non-isothermal mixing cannot be allowed in an MER design.

Stream splitting (where the branches have different target temperatures and are not re-joined) is illustrated in Figure 8.2(a). In this case, two units are needed because of the different target temperatures, and hence for capital targeting the system is represented by two streams. Similarly for energy targeting, the two target temperatures means two streams. Figure 8.2(b), however, shows that it might be possible to get away with one unit if bypassing and mixing can be used. Here, the second unit effectively is replaced by the bypass mixing junction which performs the heat transfer job.

Finally, a special case which should be mentioned is where a hot and cold stream are mixed. If these are treated as separate streams in the pinch analysis, a false result could be obtained. For once, the actual plant might beat the target! This is because the mixing, in effect, allows the temperature exchange to occur with a ΔT_{min} of 0, whereas the analysis would impose a finite ΔT_{min}. In this case, it may be necessary to combine the streams before performing the targeting analysis, assuming that it is intended to retain the match.

8.4.2 Effective process temperatures

Where some sort of heat exchange system already exists, should we include it in the stream data? Let us consider a flowsheet which includes a vessel in which a highly exothermic reaction is taking place. The reactor is cooled by a molten salt loop which rejects heat by raising 4 bar (60 psi) steam in a heat exchanger at 144°C. How do we extract the data?

One obvious possibility is to ignore this area completely – it is constrained in operation, it is self-balancing and we treat the steam produced as a "free gift" into the rest of the process. Another possibility is to consider the steam as a hot stream generated from this plant. Knowing the flowrate and the latent heat at the given temperature, we can assign a heat load to it which accounts for the heat it would release on condensing (and, if we wish, the sensible heat recoverable from cooling the condensate).

However, the most rigorous method is to include both the molten salt loop and the bottom product as hot streams. They can then be used to raise steam if necessary; however, there may be more worthwhile duties for them elsewhere on the site. In particular, there has been a very large temperature degradation in the heat exchangers and if there are some intermediate heating duties at say 200–300°C, the streams could be used for these. An energy saving will certainly result if the pinch falls in the range between about 140°C and 350°C.

An experienced engineer should raise questions of practicality at this point. The molten salt loop is carrying a high heat load and it is essential for the safe operation of the reactor that it can reject this heat. If a heat exchanger at lower temperature differences is used and it becomes fouled or the cold stream flowrate drops, a thermal runaway could occur. However, this can be handled by the analysis. Firstly, if a minimum temperature difference is required, this can be allowed for by increasing the ΔT_{min} contribution on the molten salt stream. Secondly, part of the heat load can be "ring-fenced" so that it is always used to raise steam, thus ensuring that the existing exchanger is retained as a back-up. Moreover, we still have the bottoms stream available at high temperature, and no such constraints exist on this as it has already left the vessel. This method is the correct way to handle similar situations such as "quench" streams from reactors or distillation columns, where all or some of the heat must be rapidly removed by cooling water (e.g. to prevent undesirable side reactions).

In summary, the extracted stream data might look like this:

Stream	*Heat load*	*Temperature range*	ΔT_{min} *contribution*
Molten salt cooling (H)	H1 or (H1–H3)	377–375	100
Bottoms product (H)	H2	375–180	10
Essential quench steam (H)	0 or H3	144–143	10
Condensate from quench steam (H)	0 or k(H3)	144–30	10
BFW heating of quench steam (C)	0 or k(H3)	30–144	10

(k is a fraction with $0 < k < 1$).

The other question we could ask ourselves is whether the reactor feed could be preheated. By analogy with the mixing vessel in Section 8.4.1, the feed is entering at 127°C to a reactor working at 375°C – definitely non-isothermal mixing! We could consider incorporating the feed preheat as an extra stream. However, this would alter the conditions in the reactor. To dissipate the additional heat, the circulation rate of the molten salt loop would have to rise or its temperature drop would have to increase. Again, there would be a significant danger of a thermal runaway! So the correct method here would be:

- Extract the stream data for the original process and calculate the targets and pinch temperature.
- Using the plus–minus principle and/or splitting the Grand Composite Curve (GCC), see whether it is worthwhile to preheat the reactor feed (i.e. is the pinch above 127°C? and can we use heat at 375°C instead of lower temperatures, to increase driving forces and reduce exchanger capital cost?).
- If preheating is worthwhile in energy terms, consider its effect on the reactor. If a new set of conditions can be safely achieved, extract the new stream data, retarget and evaluate the benefits compared to the base case.

8.4.3 Process steam and water

In many situations, heat is performed using a direct contact heater (steam injection), where the steam is mixed directly into the process fluid and changes its composition. Similarly, cooling by direct mixing with process water may occur. If we use heat exchange to reduce these heating and cooling loads, the flowrate and composition of all streams downstream of the injection point will change, and the heat loads on downstream processes and exchangers will alter. The same will occur if we replace the injected steam by indirect heating, whether by steam, hot oil or some other utility. Sometimes this may be beneficial, for example if the water is simply going to have to be separated out again later in a distillation column, evaporator or dryer. In other cases it may be positively harmful, for example if the extra water is providing the necessary dilution for effluent processing or a downstream reaction. Steam injection is frequently used where indirect heat exchange would be difficult (e.g. heavily fouling or viscous liquids), or slurries containing a high proportion of solids.

There is no way to deal with direct injection which is "best" in all situations, but the safest way for initial calculations is to assume that the flow is unavoidable and necessary for the process. Therefore, it should be included in the stream data, and this is done by treating it as a cold process stream with the appropriate mass flowrate. The associated amount of boiler feedwater preheat can also be added. Thus, if we know we are supplying 1 kg/s of steam at 125°C, this corresponds to a total enthalpy of 2713 kW being added to the system (from steam tables). If this steam is supplied in practice from boiler feedwater at 30°C, with an enthalpy of 126 kW, then the overall load on the stream is 399 kW sensible heat between 30°C and 125°C and 2188 kW latent heat at 125°C, giving a total of 2587 kW. Later on, if we decide that this steam flow can be reduced or replaced by indirect heating, the downstream process flows can be recalculated and retargeting can be performed with the new process flowsheet and heat load.

8.4.4 Soft data

Many streams on a plant have fixed supply and target temperatures, as they run between two unit operations working at definite temperatures. However, some temperatures may be variable (e.g. the temperature of final storage, or the temperature with which drainwater or hot air is rejected) to the surroundings. These temperatures are known as "soft" data and can potentially be changed for various reasons:

- To allow additional heat recovery (e.g. "running down" a stream) which is above the pinch temperature. The plus–minus principle will show whether the change is beneficial to the energy targets.
- To reduce the number of exchangers required in a network, by matching heat loads on hot and cold streams more exactly, or even eliminating an exchanger or cooler completely if there is no real need to cool a below-pinch stream.

In some cases entire streams are "soft" or optional. In the case study on the evaporator–dryer plant (Section 9.4), the warm, damp exhaust air from the dryer could be vented to atmosphere, or cooled in an exchanger. If the stream is entirely below the pinch, deliberately cooling it will simply increase the cold utility target without saving any hot utility. However, in a few cases, it may be worthwhile to use a below-pinch stream in this way if, in the network, it provides a more convenient way to heat a cold stream than other available below-pinch hot streams. This did not apply in the case study; the exhausts continued to be vented to atmosphere and were not included in the final stream data set.

The case study on the hospital (Section 9.6) shows another example of the use of soft data. The pinch was at ambient temperature, and heat could be extracted from the warm air and drainwater streams above the pinch, reducing the hot utility target. In fact, the temperature driving forces were so low in this case that heat recovery turned out to be uneconomic. However, the simple example in Figure 8.3 shows a case where it was worthwhile. Hot product was being delivered to storage at 80°C, but could have been run down as far as 30° without adversely affecting flow characteristics. In fact the pinch was at 65°C, so it was worthwhile to run

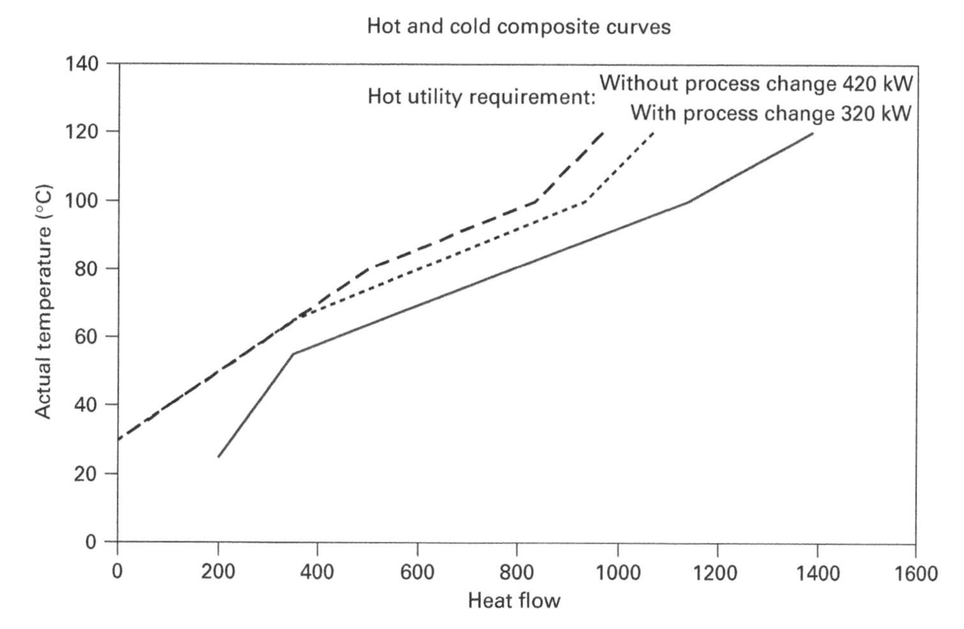

Figure 8.3 Heat recovery from a stream rundown to storage

the product down to that temperature. The broken line shows the hot composite curve before this process change. A saving of 100 kW was obtained.

One point to bear in mind is that if the product from storage will need to be reheated again in the next stage of the process, some or all of the extracted heat will need to be resupplied! However, at least the heat losses in storage should be less at a lower temperature.

8.4.5 Units

Throughout the book, we have worked in a set of units based on SI (metric) units adapted for the engineering scale and mutually compatible. The basis has been time in seconds (s), temperature in degrees Celsius (°C) or Kelvin (K), mass in kilograms (kg) and energy in kilojoules (kJ). (Joules are an impracticably small unit for engineering calculations.) This leads to mass flowrates in kg/s, heat capacity flowrates in kW/K and heat loads in kW. However, most plant data is recorded and reported in different units, such as tons per hour.

Other sets of units can be adopted, but either they must be mutually consistent or the correct conversion factors must be applied. Three compatible groups of units are shown in Table 8.1. The first set are as above; the second use hours as the time basis instead of seconds (more convenient for many flowrates) and the third use Imperial or US units. It can be seen from the table that assuring compatibility is not straightforward; it is all too easy to drop a factor of 1,000, especially in converting

Table 8.1 Compatible sets of units

Quantity	*SI-based unit*	*Metric hour-based units*	*US/Imperial units*
Time	s	h	h
Temperature	°C or K	°C or K	°F
Mass flowrate	kg/s	te/h (kg/h × 1,000)	lb/h × 1,000
Specific heat capacity	kJ/kgK	kJ/kgK	Btu/lb°F
Heat capacity flowrate	kW/K	MJ/h/K	MBtu/h/°F
Heat load	kW	MJ/h	MBtu/h
Heat transfer coefficient	kW/m^{2}K	MJ/m^{2}hK	MBtu/ft^{2}h°F
Surface area	m^2	m^2	ft^2

te = metric tonne, 1,000 kg; h = hour.

from mass flowrates to heat loads. Likewise, using hours instead of seconds with metric units not only affects heat loads, but also, more unexpectedly the required units for heat transfer coefficients; if this is overlooked, heat transfer area will be underestimated by a factor of 3.6.

Depending on the scale of the plant, it may also be best to change heat loads and CP to a set of units which are a factor of 1,000 higher or lower (e.g. between kW and MW, MJ/h and GJ/h, MBtu/h and MMBtu/h). Note that conventionally M (mega) is the prefix for 10^6 in SI/metric units, but in Imperial units M was used for 10^3 and MM for 10^6.

Many other units are in common use and can be used directly, as long as a mutually consistent set is adopted. For example, the aromatics case study in Section 9.3 has heat load data in thousands of tonne calories per hour (explained in the text).

8.4.6 Worked example

Some of the difficulties involved in stream data extraction are well illustrated by taking a direct-fired dryer. This involves mass exchange, heat losses and utility–process interaction. The flowsheet is shown in Figure 8.4 and the heat and mass balance in Table 8.2. The current unit dries 1 kg/s of solid from 15% moisture (dry basis) to zero. The dryer uses 1 MW of combustion heat and loses 0.134 MW in wall heat losses.

For such a simple flowsheet, one might expect data extraction to be easy; it is not. The first question is how to represent the heat that must be supplied to the dryer to evaporate the moisture and heat the materials. This can be done in three ways, depending on where one draws the control surface:

1. Treat the entire system as a black box requiring 1 MW of heating. The difficulty then is to assign a temperature range to this cold stream; here, the combustion temperature has been used. This method is not recommended.
2. Treat the dryer as a black box supplied with hot air; the cold stream is then the heating of the air. A further question arises as to whether one should consider (a) the heating of the small flow of combustion air or (b) the large mixed air flow which enters the dryer.

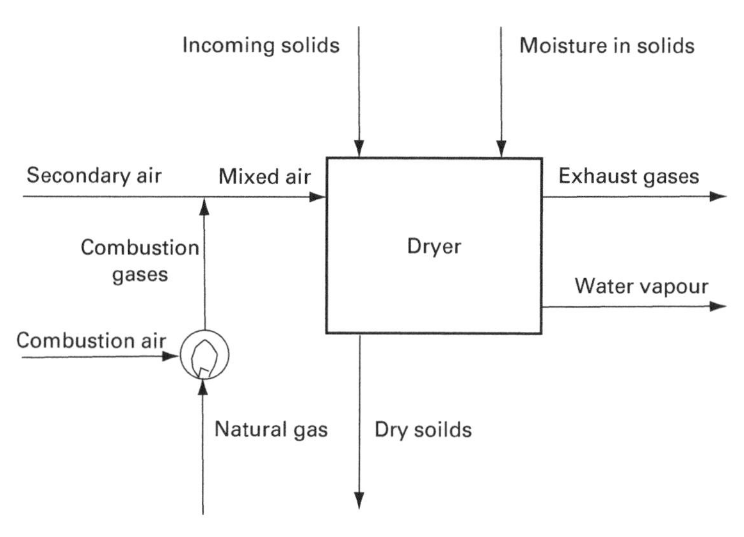

Figure 8.4 Flowsheet for dryer system

Table 8.2 Heat and mass balance for a direct-fired dryer

Process flow	*Mass flowrate (kg/s)*	*Temperature (°C)*	*Specific heat capacity (Cp kJ/kgK)*	*Enthalpy (h kJ/kg)*	*Heat flowrate (kW)*
Inputs					
Natural gas	0.0187	0		53,365	1,000
Combustion air	0.3853	20	1	20	7.7
Combustion gases	0.404	2,267	1.1	2,494	1007.7
Secondary air	2.226	20	1	20	44.3
Mixed air	2.63	400	1	400	1,052
Solids (dry basis)	1	20	1	20	20
Moisture	0.15	20	4.2	84	12.6
Total	3.78				1084.6
Outputs					
Exhaust gases	2.588	120	1	120	310.6
Water vapour	0.192	120		2,708	520
Dry solids	1	120	1	120	120
Total	3.78				950.6
Heat losses					134

3. Consider all the flows through the dryer individually and calculate the enthalpy change of the air, solids and water between inlet and outlet conditions. A temperature must then be assigned to the evaporation heat load.

Table 8.3 compares the stream data obtained from the various methods. It is not clear which of them is best; indeed, different methods are best in different situations.

Table 8.3 Stream data for direct-fired dryer

Stream	*Supply temperature (°C)*	*Target temperature (°C)*	*Heat load (MW)*	*Heat capacity flowrate (CP MW/K)*
Method 1: Black-box				
Dryer heat load	2,500	2,501	1.0	1.0
Method 2a: Air heating (combustion air only)				
Combustion air	20	2,501	1.0	
Method 2b: Air heating (mixed air input)				
Dryer inlet air	20	400	1.0	
Method 3: Detailed breakdown				
Dry air heating	20	120	0.26	0.0026
Solids heating	20	120	0.1	0.001
Evaporation load	20	120	0.51	0.0051
Heat/combustion losses	119	120	0.13	0.13
Optional hot stream for heat recovery				
Exhaust air only	120	20	0.76	0.0076

Method 1 is generally too crude, but could apply to a dryer using infra-red radiant heat where air preheating is inappropriate (or to an electrically heated dryer where it was desired to include this heat requirement in the targets).

Method 2(a) would apply if the dryer relies on radiant heat and thus requires a hot flame or high-temperature gas stream (~2,000°C). This applies, for example, to some types of rotary dryer.

Method 2(b) gives the correct results if the existing dryer is to be retained and the air is preheated by heat recovery. It also applies if hot exhaust gases from a CHP system (gas turbine or reciprocating engine) are substituted for the drying air.

Method 3 gives the correct targets where the solids themselves can be preheated (e.g. in a rotary predryer or pneumatic conveying duct), or where they can be heated within the dryer by heating coils, as in a fluidised bed.

Thus, 2(b) and 3 are the most generally useful methods. But even then, 2(b) will give an incorrect target if solids preheating is possible. Likewise 3 does not allow correctly for the inlet air temperature actually required to make an existing dryer operate. For some dryers, then, accurate targeting will be practically impossible.

Targeting with constraints (Section 4.5.2) will help in some cases, but not all. Moreover, the algorithm for constrained targeting is much more complex than the straightforward Problem Table.

In addition, heat can potentially be extracted from the exhaust by sensible cooling and condensation; should this be included in the stream data set? If it is, and the heat is not then used, the cold utility target will have a false high value (because the exhaust air could be discharged directly to atmosphere without cooling). Nevertheless it is a potential heat source and it has therefore been included at the bottom of Table 8.3.

The CP values have been calculated as averages over the temperature range. However, in method 3, the CP values are highly non-linear for the evaporation load and the exhaust air cooling. In the former case, it is best to calculate the latent heat load and assume it occurs at the top end of the temperature range, 119–120°C. Even this may not allow correctly for the temperature driving forces actually required in the dryer to achieve heat transfer at an acceptable rate. For the exhaust gas, condensation occurs below the dewpoint and a latent heat load must be added to the sensible cooling; a cooling curve can be constructed by calculating the dewpoint humidity at each temperature. The fact that the CP of a single stream can vary considerably over its temperature range has unfortunately been overlooked by several programmers who have produced simple targeting programs, making them difficult or impossible to use for systems involving phase changes.

The next question is whether to include heat losses in the dryer heat requirements. Energy consumption is 1,000 kW including them, 866 kW excluding them. Generally, it is best to include the heat losses as the targets will then be consistent with the actual current energy consumption of the plant. However, for method 3 a separate "heat losses" stream is then required and it must be allocated a temperature range. As with the evaporation load, the choice of temperatures is somewhat arbitrary and presents a further difficulty in targeting.

Thus, even for this simple unit operation, there is no clear-cut best method of data extraction, and different targets may be obtained using the different methods. This shows that it is difficult to make pinch technology techniques completely rigorous, and emphasises that accurate stream data extraction is a difficult task which requires care and thought and may need expert assistance. It is not simply a routine operation that can be done by "turning a handle".

Other illustrations of stream data extraction in practice may be found in the Case Studies in Chapter 9.

8.5 Targeting and network design

8.5.1 Targeting

The energy targeting calculation is still the cornerstone of pinch technology. The Problem Table method is well proven and is easy to implement given the set of stream data. Although hand calculation is generally too difficult, the algorithm is straightforward and can be transformed relatively easily into a simple computer program. The one practical difficulty is that a value is needed for ΔT_{min}. However, as pointed out, it is not necessary to find the precise optimum; as long as the value chosen is reasonably sensible, it will give adequate results and can be refined later.

8.5.2 Network design

The classic pinch-based network design method is to start by constructing the MER design and then relax the network. This is generally effective, but we must remember that there are some situations where the MER network is very different in structure to the most economic network, and the latter may be much more easily obtained by a sensible hand design. This mainly occurs where there are a large number of cross-pinch streams (Section 4.4.2). This does not invalidate pinch methods; but it does mean that care and thought are required when applying them.

For retrofits, in particular, it is often appropriate to start from the existing design and work towards the MER network, particularly for small plants where only a few exchangers will be economically justifiable. Ideally, one can also work outwards from the MER network and the two designs should converge in the middle. Working inwards from the existing design has in fact been practised since the early days of pinch technology, using the insights given to identify the pinch violation (inappropriately placed exchangers, heaters and coolers) which need to be changed. Section 4.7 pointed out that one should move, as quickly as possible, to a network simulation (network optimisation) using the actual sizes of existing heat exchangers in the network, rather than the indeterminate-size exchangers yielded by the original pinch design method. The results from these two different approaches can be compared directly in the Aromatics Plant case study (Section 9.3) where there is very little to choose between them.

8.6 Targeting software

8.6.1 Options available

It was originally hoped that pinch analysis techniques would allow energy targets to be found using only a pocket calculator. Sadly, this is not true except for the very simplest plants. Hand calculation is tedious and there is a high risk of mistakes. Computer software is therefore almost essential to perform a pinch analysis. Three broad types exist; dedicated high-level programs, general process simulators, and simplified targeting software.

Early examples such as TARGET (written at UMIST) and PROTAB (developed by ICI) were developed relatively quickly into highly sophisticated programs such as Supertarget (Linnhoff March) and Advent/Aspen Pinch (Aspen Tech). Later Hyprotech developed HX-NET® and this became the main Aspen specialised pinch software. These programs use complex targeting procedures and can perform automated network design (with a manual option to allow the user to design his own network). As a result, they are relatively costly to purchase or license.

An alternative is to use software only for the essential calculations of energy and cost targets and to generate the composite and GCC, without including network design. One program to take this approach was HERO (Chepro Ltd). Because of the limited functions of such programs, they are very much cheaper than full network design software. The spreadsheet supplied free with this book is of this type; its specification is given in Section 8.6.2.

Finally, a number of process simulators incorporate a pinch analysis section, early examples being HYSIM® and HEXTRAN®, and more recently Aspen Plus® and Hysys®. Simulators include a basic UA-type model of heat exchangers, and can be therefore used to perform network design if targets have been calculated by a spreadsheet (or if one does not wish to use pinch techniques at all). They are particularly useful for manually retrofitting existing networks.

8.6.2 Spreadsheet accompanying this book

To allow users to perform pinch analysis on their own plants as easily as possible, a free spreadsheet is supplied with this book. This performs the key targeting calculations and plots, as follows:

- Input of stream data (either as CP or heat load).
- Calculation of composite curve data, Problem Table, energy targets and pinch temperature.
- Plotting of composite curves and GCC.
- Plotting stream population over temperature range and basic grid diagram.
- Tables and graphical plots of variation of energy and pinch temperature over a range of ΔT_{min}.

Area and cost targeting is not performed because of the considerable extra complexity, the frequent lack of suitable data on heat exchanger coefficients, and the flat nature of most cost-ΔT_{min} plots. Topology traps can still be identified from the graphs of utility use and pinch temperature against ΔT_{min}. Most of the composite and GCC in this book have been generated using the spreadsheet. It was written for Microsoft Excel® but should run in similar compatible spreadsheets; however, fully correct operation cannot be guaranteed. A brief user guide is given in the Appendix.

The spreadsheet was the winning entry in a competition run by the Institution of Chemical Engineers for young process engineers, and was written by Gabriel Norwood.

8.7 Industrial experience

Some of the early results from pinch studies were described in Section 1.2. As mentioned, the sometimes startling gains obtained caused great interest and controversy, giving the impetus for widespread application of the techniques in industry.

In the UK, the Department of Energy funded a systematic programme of pinch studies in a carefully chosen range of representative industries. These were performed by consultants and companies' in-house teams, with the majority being done by Linnhoff March Ltd. The results were given in a series of published Energy Efficiency Office R&D reports and summarised by Brown (1988). Table 8.4 shows the savings identified. Overall, the average savings identified were 25% of the site energy bills. The majority of these came from heat recovery, with utility change coming second. However, process change analysis was then in its infancy and would probably account for a higher proportion nowadays; also, some projects then included under heat recovery or utilities could nowadays be classified as process change

Table 8.4 Results of EEO-funded process integration studies (Brown 1988)

		Energy cost savings identified			*Technology split*			
Company	*Site energy bill (£k/yr)*	*Saving (£k/yr)*	*Economic percentage of bill*	*Payback (months)*	*Process change (£k/yr)*	*Heat recovery (£k/yr)*	*Utilities change (£k/yr)*	*Other (£k/yr)*
Chemical								
Tioxide	5,000	1,240	25	36	200	380	600	
Procter & Gamble	2,000	1,437	72	54	200	234	980	23
Beecham	130	31	24	5		21		10
Cray Valley Products	1,000	281	28	29		80	177	24
William Blythe	230	96	42	11	20	8	68	
Staveley Chemicals	5,500	1,040	19	18	70	470	500	
Coalite	5,100	790	15	11	205	143	422	20
Bush Boake Allen	1,000	11	1	30		11		
May & Baker (1)	1,700	1,076	63	28		40	965	71
May & Baker (2)*	210	152	72	31	152			
Gulf Oil	12,000	1,420	12	7	800	620		
Shell Oil	12,000	1,600	13	19		1,600		
Subtotal	45,870	9,174	20		1,647	3,607	3,772	148

Food and drink								
Long John	1,500	358	24	27	148	90	120	
Tetley Walker	850	175	21	13		62	75	38
Cadbury Typhoo	500	84	17	6	42			42
Van den Berghs	5,000	500	10	24		500		
J. Lyons	3,500	748	21	16		191	517	40
Express (1)	1,600	103	6	11		16		87
Batchelors	600	105	18	7		38	67	
Express (2)*	270							
Subtotal	13,820	2,073	15		190	897	779	207
Textiles								
Courtaulds	13,000	334	3	11		257	77	
Edward Hall*	530	68	13	30				
Paper								
East Lancs Paper Mill	3,000	876	29	28	120	86	500	170
St. Regis	1,600	147	9	26	62	50	35	
Iron and steel								
Sheerness Steel	13,000	1,487	11	10		1,487		
British Steel	20,000	9,950	50			7,450	1,300	1,200
Subtotal	5,1130	12,862	25		182	9,330	1,912	1,370
Grand total	110,820	24,109	22		2,019	13,834	6,463	1,725

*New plant study.

Table 8.5 Summary of identified and implemented savings by sector (Brown 1988)

Sector	Energy bill (£M/yr)	*Savings identified* Process change (£M/yr)	Heat recovery (£M/yr)	Utilities changes (£M/yr)	Other (£M/yr)	Total savings (£M/yr)	*Implemented savings* (£M/yr)	(%)
Chemicals	160	4.9	10.7	18.2	3.0	36.8	15.6	42
Oil refining	81	0.6	11.5	–	–	12.1	6.8	56
Food and drink	29	1.5	3.9	5.1	1.7	12.2	4.3	35
Textiles	15	0.2	0.6	3.4	0.1	4.3	0.2	5
Paper and board	10	–	1.1	4.0	0.4	5.5	0.6	11
Iron and steel	34	–	9.0	1.3	1.2	11.5	2.9	25
Total (£M/yr)	329	7.2	36.8	32.0	6.4	82.4	30.4	37

(see Section 9.4 for an example). Also, for processes involving solids (covering many of the "non-traditional" process industries outside oil refining), direct heat recovery is more difficult and utilities and process change opportunities are correspondingly more important.

Because of the different payback criteria applied by different industries, and the reduction in energy prices in the late 1980s and 1990s, many of the identified projects were not implemented. However, 37% of the identified projects were implemented in practice, and led to cost savings of over £30 M/year at 1988 prices (Table 8.5).

One item not shown in these summary tables is the typical economic rate of return for each type of project. Heat recovery projects involve capital expenditure on heat exchangers and typically had paybacks of 1–2 years, although a few projects had paybacks down to 6 months. Paybacks for utility related changes were more wide-ranging, typically 1–5 years, with the longest paybacks for some of the large CHP projects (although total cost savings were also very large). A small number of cases had short paybacks of a few months, usually where a low cost utility was being substituted for a high cost one. However, some of the process changes and non-pinch projects identified by the studies involved little or no capital expenditure and therefore gave payback times as short as 1 month, or even zero. This re-emphasises the importance of considering process change possibilities, and the value of doing a systematic analysis to identify them.

The following sections describe how pinch analysis can typically be applied in a range of industrial sectors, using both the experience from the EEO studies and other analyses.

8.7.1 Oil refining

Oil refining is a major traditional major area for pinch studies, and of course the pioneering pinch study was on a crude unit, as described in Section 9.2. The ICI aromatics plant (Section 9.3) was another highly successful early application. The benefits there came from both heat recovery and process change. Since then, the techniques have been extensively developed and refined in these applications. There are many practical constraints to do with the various blends and multiple feedstocks, but the very high throughputs justify heat recovery projects – small percentage savings give a large absolute economic benefit, and a good rate of return on many capital investments. Complex heat exchanger network simulation and optimisation are justified, and network design software is widely used.

Oil refineries were studied in 3 of the original EEO-funded studies in the 1980s – at Gulf and Shell, plus a demonstration project at ICI implementing a novel heat exchanger design identified by the aromatics plant case study. Since then, numerous further studies have been made, and most significant companies have developed their own in-house capability. The processes involved are relatively similar throughout the industry. Obviously, heat exchanger networks have been a major source of energy savings, but process changes and utility system optimisation have also played a part.

8.7.2 Bulk chemicals – continuous

This is also a major traditional area for pinch analysis. As with oil refining, plants are large, continuous and largely confined to fluids (liquids and gases) – all highly beneficial for process integration. However, there is a greater variety of processes and plants than in oil refining. As mentioned in Section 1.2, the techniques were applied by ICI, Union Carbide and BASF in particular over a wide variety of their in-house chemical processes, leading to substantial savings from heat recovery, process change and utilities optimisation. EEO-funded studies in the inorganic sector at Staveley Chemicals and Tioxide, and in the organic sector at Coalite and William Blythe, all successfully identified savings from heat recovery, utilities improvement and process change.

Total site analysis with heat recovery via the steam system was a major potential saving at Coalite (see Section 9.5). Elsewhere, time dependent analysis could help with the integration of plants with different working hours.

8.7.3 Speciality and batch chemicals and pharmaceuticals

In contrast to the preceding sections, energy is generally a minor cost on these plants, especially when compared with the value of the products. Batch analysis techniques (Chapter 7) can be useful for debottlenecking (using Gantt charts) and optimising utility configurations (e.g. for heating and cooling vessels). EEO-funded studies were performed at Bush Boake Allen and Cray Valley Products in the specialty chemicals sector and Beechams and May & Baker in pharmaceuticals. In several cases, use of the time average model showed potential savings, but more rigorous methods such as the time slice model showed opportunities were less.

At Bush Boake Allen, feasible heat recovery projects were identified but on a 2–3-year payback. However, at Cray Valley Products, substantial savings were successfully achieved using the time event chart for debottlenecking.

8.7.4 Pulp and paper

The pulp and paper industry is a huge energy consumer; producing 1 kg of paper typically requires 2 kg of water to be evaporated. Even in a relatively modest producer like the UK, pulp and paper accounts single-handedly for energy use of over 100 PJ per annum (10^{17} J, 10^{8} GJ).

Heat recovery possibilities are limited, especially as there are many practical limitations in recovering heat from continuous sheets, but the industry is a major user of CHP based on steam turbines (Section 5.2), long pre-dating pinch analysis! The very high heat loads mean that steam turbines are well matched to the system, generally matching site power needs even with their modest power-to-heat ratio. Even here, however, gas turbines have made an impact.

One energy efficiency demonstration project, on preheating the felt rollers to paper machines with waste hot water, achieved the noteworthy feat of saving more than the theoretical maximum heat recovery! The felt holds the paper web against the main heating cylinders, and picks up moisture from the web, so has to be dewatered by squeezing followed by heating. The higher temperatures reduced surface tension and meant that more water was removed mechanically, thus reducing the evaporation load by more than the sensible heat added to warm the felt rollers. This is a classic example of process change.

8.7.5 Food and beverage

There have been a few studies in the food processing industry. Opportunities are restricted by product quality requirements and the prevalence of solids (making heat recovery difficult), but nevertheless energy saving projects were successfully identified. In some cases heat pumping (Section 5.3.1) has been a useful and economic technique, as processing is often carried out over long periods at moderate temperatures for product quality reasons. For example, heat pump dryers for fish are common in Scandinavia, where power costs are low because of the availability of hydro-electric power.

Many food processing operations are batch, which again inhibits economic heat recovery projects. Important exceptions are starch, sugar and edible oil processing, which are all continuous and operate on the medium to large scale. Savings have been found in all these areas, with EEO-funded studies at British Sugar and Van den Berghs & Jurgens. In the latter, heat recovery between processes was successfully achieved using the site hot water system, in a similar way to methods using the site steam system, as described in Section 5.4.

Modern dairies are already well integrated using plate exchangers with very low ΔT's, often down to 1–2°C. There are not many streams and these are well matched in flowrate and heat capacity, so the process can be virtually one huge pinch region. For example, in continuous pasteurisation the hot and cold milk exchange heat and only a small additional utility load is required. In separators, the milk is cooled down and separated into cream and buttermilk, and these are then reheated by the incoming milk as parallel streams in a single plate or plate-fin exchanger. Again, only a very small refrigeration load is needed, covering the final 1–2°C of cooling.

In brewing, the main opportunities are mainly on the hot water side, and are a rare example of successful economic heat recovery from a batch process.

Spirit distilling (whisky, gin and vodka) use substantial amounts of energy in the distillation columns, but the high product quality requirements impose severe constraints; even small shifts in column pressure or temperature will be rejected as they can affect trace components important for taste. Opportunities exist elsewhere in the process, notably with drying of the spent grain for which a hot air dryer heated by the exhaust from a gas turbine is very suitable (Section 5.6.1); a very successful project in the early 1980s was at Scottish Grain Distillers (Port Dundas, Glasgow). In some cases, heat pumping by mechanical vapour recompression has been used to recover heat from the exhaust vapour from the still, even in malt whisky production where product quality is paramount.

8.7.6 Consumer products and textiles

One bulk process in the consumer products area is detergent manufacture, which involves solution preparation and spray drying. The EEO sponsored a study at Procter and Gamble which identified some heat recovery opportunities (limited as the streams involved were viscous and fouling) and a beneficial process change on the spray dryer. The latter was implemented but the heat exchangers were not, partly because the economics were affected by a plan to install a CHP scheme burning waste derived fuel (WDF) which would have reduced the effective cost of low pressure steam to virtually zero. The CHP plant was not eventually installed for a mixture of practical and economic reasons, including fears about potential emissions and odour problems.

Other consumer products are manufactured in similar ways to food products or specialty chemicals, and similar constraints apply.

Textiles are included here as they do not fit comfortably into any other category. An EEO funded study on the Courtaulds man-made fibres plant showed that there were few opportunities for direct heat recovery (fibres being a far from ideal configuration for this) but did show opportunities in the utilities systems, particularly for replacing high-pressure steam by cheaper low-pressure steam.

8.7.7 Minerals and metals

Smelting is the process of producing metals from their ores, for example copper and zinc. The reactions are often highly exothermic and very high temperature exhaust streams (over 1,500°C) are produced. These sound highly promising heat sources but are very dirty and dust-laden, hence practical heat recovery is almost impossible, and quenching or scrubbing is used instead. Nevertheless, some of the heat can still be potentially recovered, if there are other suitable process duties on site (e.g. medium-temperature steam raising).

Iron and steel production also involves exceptionally high temperatures, the hottest streams being molten iron and steel. Heat recovery from hot exhaust gases from the blast furnace to heat the incoming air has been common for many years; as the blast furnace reactions are exothermic, the inlet blast can be considerably cooler than the exhaust gas, and the latter is also dust-laden and aggressive. Hence, brick-lined recuperators (Cowper stoves) are used; the hot exhaust heats the brick for several minutes and is then switched to another unit, while cold air is directed into the hot chamber. Studies carried out at British Steel identified further theoretical savings which, however, were not felt by the customer to be practicably achievable.

Cement manufacture involves the breakdown of calcium carbonate to quicklime (calcium oxide) and slaked lime (calcium hydroxide). This process is highly endothermic, so the exhaust gases from the kiln are considerably cooler than those going in, and theoretical heat recovery is low. Practical heat recovery from the hot, dust-laden exhaust stream is even more difficult. The requirement for high-temperature hot gases which need not be particularly clean make gas turbine CHP systems a possibility, although supplementary firing is needed to get the gas back up to the 1,200°C required in the kiln, and the power produced may have to be exported; a gas turbine with relatively low power production and high exhaust temperature might therefore be chosen.

China clay (kaolin) is another example of a high-volume mineral with substantial process heat demands. It is mined using high-pressure water jets and, even after the resulting slurry has been mechanically dewatered as far as possible, there is still a high heat load for evaporation and drying, with no corresponding waste heat source. Again, this is a classic application for a gas turbine CHP system.

8.7.8 Heat and power utilities

In general, pinch analysis reveals few additional opportunities in the stand-alone power generation industry. Steam turbine systems have been optimised over many years by trial and error, adding economisers, intermediate steam levels, boiler feed-water heating and other refinements as described in Section 5.2.3.1, so that the process eventually got to the bottom point of the "learning curve".

Gas turbines, with their better power-to-heat ratio, have become increasingly common in recent years, especially in combined cycle format which has the highest efficiency of all. Low gas prices during the 1990s and early 2000s helped, although this no longer applies at the time of writing (2006). A first look at the process diagram of a gas turbine suggests that there is a huge missed opportunity for heat integration. Cold air is fed into a compressor, heated to 1,000°C and passed through the gas turbine, emerging at 500°C to perform process or steam-raising duties. It seems obvious to preheat the air coming into the compressor to reduce the temperature lift. However, heating the air at atmospheric pressure reduces its density and changes the performance of the compressor. When the overall calculations are done, the extra power consumption of the compressor or the reduced power output from the gas turbine outweighs the heat recovered to the inlet gas. This is a classic example of needing to understand and model the complete process in context, rather than just look at heat duties. Heat can be recovered to the inlet gas, but only after the outlet from the compressor, where temperature driving forces and recoverable heat are much less.

Of course, major benefits are achieved if the waste heat from power generation can be used for CHP as described in Chapter 5, either for process heating duties or space heating. Scandinavia and other European countries have had district heating schemes for some years, a concept which has only recently been taken up by a few British cities, although Southampton's combined geothermal and CHP scheme (Section 5.6.2) is a noteworthy exception.

8.7.9 Buildings

In most buildings, heat demand for space heating and domestic hot water usually substantially exceeds heat rejected. Hence the pinch is usually at ambient temperature, and pinch analysis is not normally needed to establish this! As a consequence, any potential above-ambient heat source will be worth exploiting if heat recovery can be made economic. Driving forces are generally low; on the other hand, payback times of many years may be permitted. In summer, air conditioning may give a substantial cold utility demand, but this is below ambient so the pinch location is unchanged.

The pinch study on the Basingstoke Hospital site, described in Section 9.6, is a good general guide to what can be expected from analysing buildings. The pinch was at ambient temperature, and heat demands substantially exceeded heat sources. Time-dependent analysis was valuable in showing the effect of day/night variations and working hours of individual buildings on heat loads, and CHP opportunities were identified, particularly for distributed CHP using gas engines.

Another common application is of heat pumps to recover heat from warm damp air in swimming pools; the temperature lift is small and payback times of a few years have been achieved.

Exercises

Rather than performing another exercise, why not try the techniques out for yourself on a real plant? If you work in industry, take your process, plant or site and analyse it using the methodology from this book, following the outline in Section 8.2. Form a heat and mass balance, extract the stream data, obtain the targets, composite curves and pinch, consider process changes and design a heat recovery network. If you are in college or another situation where you do not have access to real plant data, consider any other process data which you can find, for example the IChemE's example design project on methyl ethyl ketone (Austin and Jeffreys 1979).

References

Austin, D. G. and Jeffreys, G. V. (1979). *The Manufacture of Methyl Ethyl Ketone from 2-Butanol; a Worked Solution to a Problem in Chemical Engineering Design*. IChemE, Rugby, UK.

Brown, K. J. (1988). Process Integration Initiative. *A Review of the Process Integration Initiatives Funded under the Energy Efficiency R&D Programme*. Energy Technology Support Unit (ETSU), Harwell, Oxon, UK.

UMIST (1996). *Course on Process Integration*. Lecture 21, Data Extraction.

9 Case studies

9.1 Introduction

Any theory, however ingenious and rigorous, is of little use unless it can be effectively applied in practice on real industrial equipment. Throughout the development of pinch analysis, research and application have gone hand-in-hand and this has ensured that the techniques are practical and usable. This final major section of the Guide illustrates the application of the integration techniques in "real-life" case studies. All of the case studies described are based on engineering designs performed in industry. For space and confidentiality reasons, many details are omitted. However, it is hoped that the material will help the reader to appreciate the use of integration techniques in context. Five studies have been selected to illustrate the breadth of applicability of the techniques, from small-scale batch processes to large complex plants and even a non-process application. Three of these appeared in the first edition of the User Guide (Linnhoff *et al.* 1982), although development in the techniques has given new insights on them; the other two are more recent. Hopefully, out of the studies presented, the reader will recognise one or more as familiar territory.

9.2 Crude preheat train

Our first case study has been widely publicised in the past, and deservedly so, because it has a place in history as the first pinch study performed on a major operating plant. The techniques were pioneered in industry by ICI plc, and this study took place at one of its associated companies.

The fractionation of crude oil into its major components such as, naphtha, gasoline, kerosene and fuel oil is a common process, being a major step in oil refining. In the fractionation process to be described here, a facility needed uprating by 25% to handle increased demand. Design studies carried out by a contractor suggested that it was not possible to increase the throughput of the plant without installing a new fired heater, and that this seemed to be the cheapest capital option. However, space in the plant was very restricted and the only location which could be found for the new heater (a hot oil circuit) was away from the main plant and on the opposite side of a busy site road, over which the hot oil would have had to be carried on a pipe bridge. Not surprisingly, this was unacceptable for operability and safety reasons! After conventional techniques to try to redesign the plant with reduced energy consumption had failed, the (then newly developed) pinch analysis techniques described in this Guide were tried, virtually as a last resort and under considerable time pressure. The initial targeting analysis (performed over a single weekend!) showed that it would be possible to achieve the 25% increase in throughput without increasing the size of the current heater at all, and network design techniques then showed a choice of practical layouts which would achieve this.

9.2.1 Process description

The flowsheet is shown in Figure 9.1, in simplified form. The crude oil feed stream is preheated in three sections by interchange with the hot fractions returning from the distillation columns. The first section runs from storage to a desalter unit, the second from the desalter to a preflash column which separates out some light naphtha, and the third from the bottom of the preflash to the crude tower. Process heating is provided by a fired heater, which preheats the crude into the crude tower and provides reboiling for the stripper. The new hot oil circuit was to be installed immediately before the fired heater.

Figure 9.2 shows the network grid diagram for the contractor's design, which was virtually identical in layout to the existing flowsheet. The design achieved more or less full utilisation of the existing heat interchange equipment without increasing the size of most exchangers, although the large air-cooled heat exchanger on the column overheads would have needed modification. The obvious first attempt was to try to reduce the heater load to its previous level by increasing heat recovery, installing larger heat exchangers and squeezing the temperature driving forces. However, this proved unexpectedly difficult. There is not enough heat in the fuel oil (hot stream 1) to satisfy the increased heat requirement for the crude tower feed (stream 10) and a new match to the gas oil (stream 2) is needed. Likewise streams 1–4 do not contain enough heat to supply stream 9 with the existing layout of heat exchangers (readers may try a few attempts at playing with the temperatures of the various matches to satisfy themselves of this!) and further heat must be recovered from stream 1 with another new match. The resulting network shown in Figure 9.3 now has four matches where the ΔT_{min} is as low as 7°C and not only requires two new exchangers but also requires a huge increase in total surface area – about 70% over the contractor's design! (see Table 9.2 for comparison). Since the gain in throughput was only 25%, this put the economics of the whole project in jeopardy. There was thus a very strong incentive to find any way to reduce the capital cost of the final network.

9.2.2 Data extraction and energy targeting

The network grid in Figure 9.2 shows the stream temperatures and match heat loads. Because of the project time pressures, it was not possible to do a detailed computer

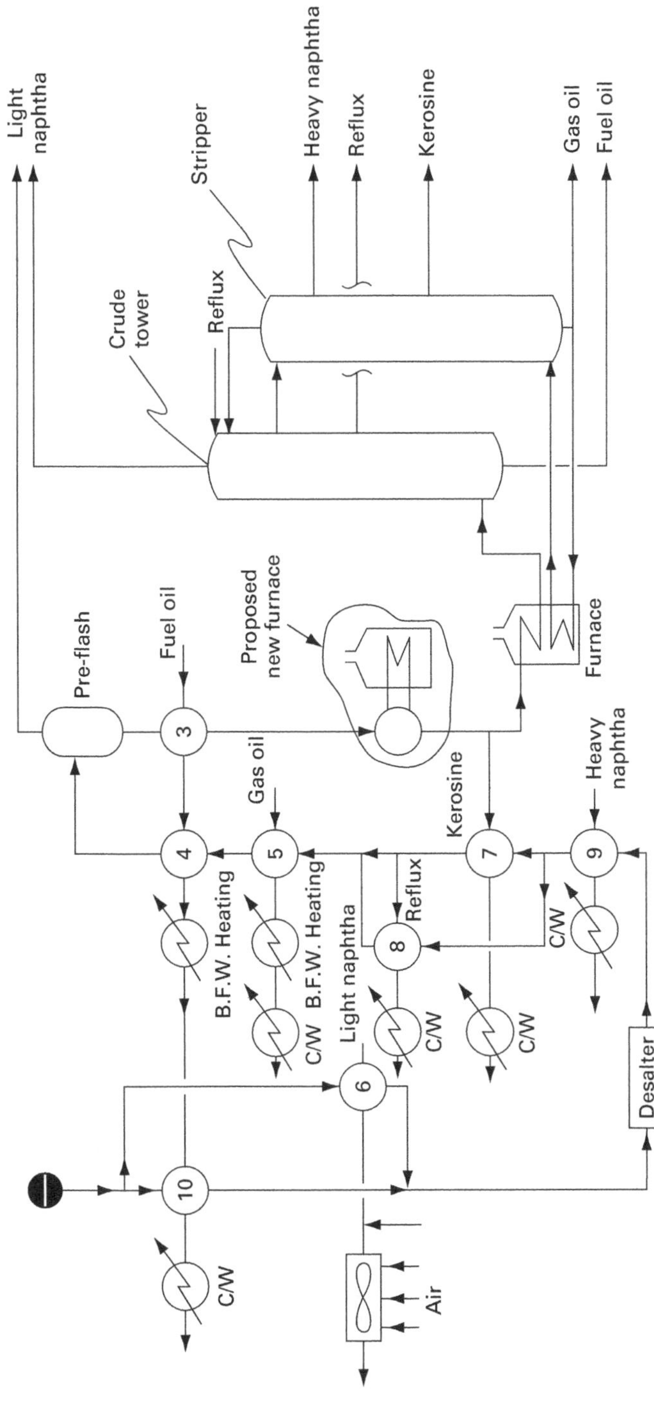

Figure 9.1 Flowsheet of crude oil fractionation preheat train

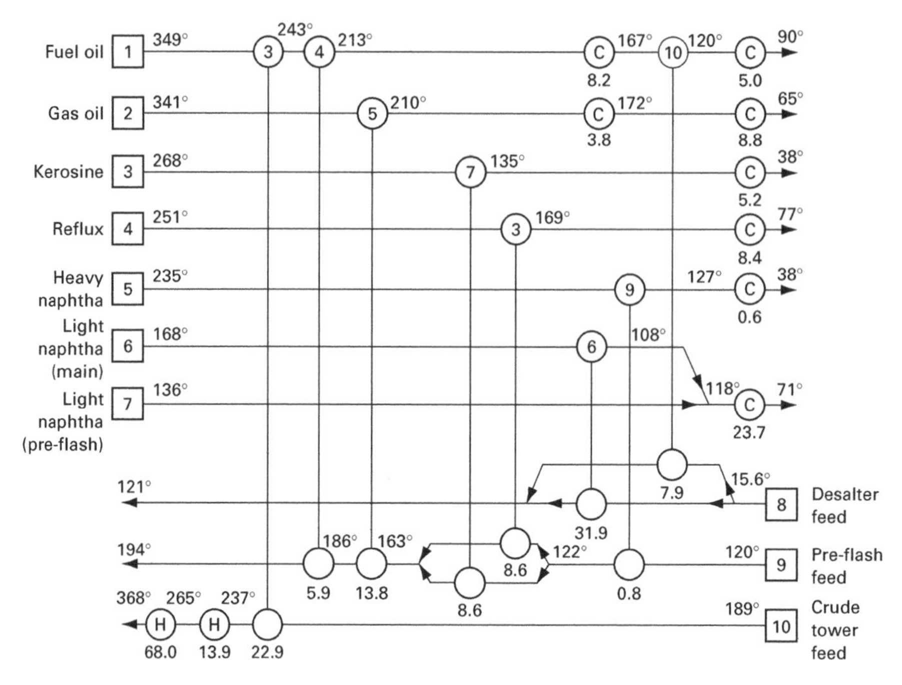

Figure 9.2 Network grid diagram for crude oil preheat train (contractor's design)

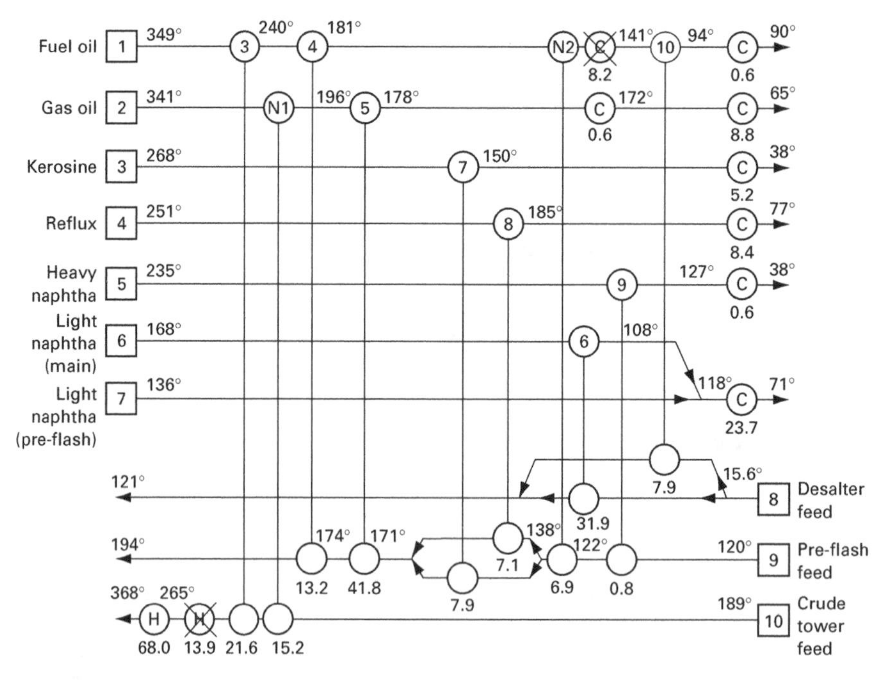

Figure 9.3 Grid diagram for network with increased heat exchange

simulation of the stream T/H profiles. Instead, data was extracted from the contractor's flowsheet in the manner described in Chapters 3 and 8, using the given design heat loads and temperatures. This gave the stream data listed in Table 9.1.

The choice of streams is mostly obvious, but we might wonder how many sections the crude feed should be split into. However, the desalter works at a closely constrained temperature and the preflash is operating as a separator (change in

Table 9.1 Stream data for crude oil preheat train

Stream	*Temperature* *T (°C)*	*Enthalpy* *H (MW)*	*Heat capacity flow rate* *CP = (ΔH/ΔT) (MW/K)*
1	349	49.8	0.215
	243	27.0	0.197
	213	21.1	0.178
	167	12.9	0.168
	90	0.0	
2	341	26.4	0.105
	210	12.6	0.100
	172	8.8	0.087
	111	3.5	0.076
	65	0.0	
3	268	13.9	0.065
	135	5.2	0.054
	38	0.0	
4	251	17.0	0.105
	169	8.4	0.091
	77	0.0	
5	235	1.4	0.008
	127	0.6	0.007
	38	0.0	
6	168	43.1	0.600
	136	23.9	0.478
	118	15.3	0.410
	108	11.2	0.303
	71	0.0	
7	136	12.6	0.256
	118	8.0	0.210
	108	5.9	0.159
	71	0.0	
8	15.6	0.0	0.379
	121	39.9	
9	120	0.0	0.400
	122	0.8	0.422
	163	18.1	0.600
	186	31.9	0.725
	194	37.7	
10	189	0.0	0.477
	237	22.9	0.496
	265	36.8	0.660
	368	104.8	

composition) so these should be break points between streams. There is no need to break up the three resulting streams further, even though the CP is varying considerably with temperature and partial vaporisation is occurring.

The smallest values of ΔT_{min} in the contractor's design are 7°C at the cold end of match 9, and 13°C at the cold end of match 7. However, ΔT_{min} values in other matches are much higher than this. Hence it was decided, for the "first look" at energy targeting to take a global ΔT_{min} value of 20°C. Calculating the Problem Table on this basis gave a hot utility requirement of 60.7 MW. The plant prior to uprating was consuming 68.0 MW, and the contractor's proposals required an extra 13.9 MW, that is, a total heat input of 81.9 MW! Hence the calculated target indicated a potential saving of about 35%.

From this point in the study, the prospects for finding a revamp which avoided using the extra fired heater appeared very good indeed and provided a tremendous stimulus to the operating company's and the contractor's engineers to find such a design.

The message is reinforced by looking at the variation of both energy use and heat exchanger area with ΔT_{min} (Figure 9.4 and Figure 9.5 respectively). Both the contractor's network and the increased-area variant are clearly a long way above the

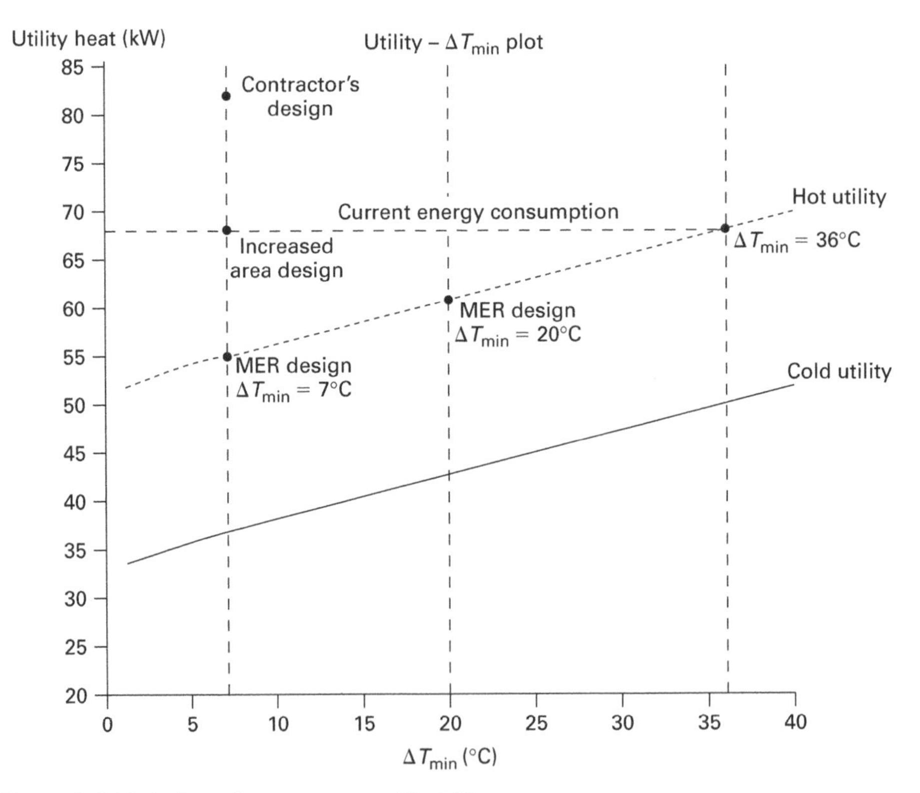

Figure 9.4 Variation of energy use with ΔT_{min}

ideal energy consumption for the plant at a ΔT_{min} of 7°C, and they are also significantly above the area targets, which suggests that the existing heat exchangers are not deployed in the most effective way. Again, this gives strong encouragement that a more cost-effective design can be found.

The question that may be puzzling us is, how did an experienced contractor apparently get it wrong? And why does the network layout of Figure 9.2 give such poor results? There seems nothing obviously wrong with it, and the streams have been matched roughly in descending order of temperature, which should logically give the best temperature driving forces and best heat recovery. However, when we plot the composite curves (Figure 9.6) and the grand composite curve (GCC) (Figure 9.7) we begin to get some clues. The pinch, at a shifted temperature of 173°C, is not sharp; there is a very long region of constrained temperature driving forces on either side, and this means that any non-optimal match in this region is very likely to transfer heat across the pinch or severely squeeze the driving forces in other neighbouring matches.

9.2.3 Pinch identification and network design

Calculating the Problem Table for $\Delta T_{min} = 20°C$, a heating requirement of 60.7 MW, a cooling requirement of 42.5 MW, and a pinch at 173°C are obtained. On inspection, it can be seen that this pinch is caused by the onset of vaporisation in the crude

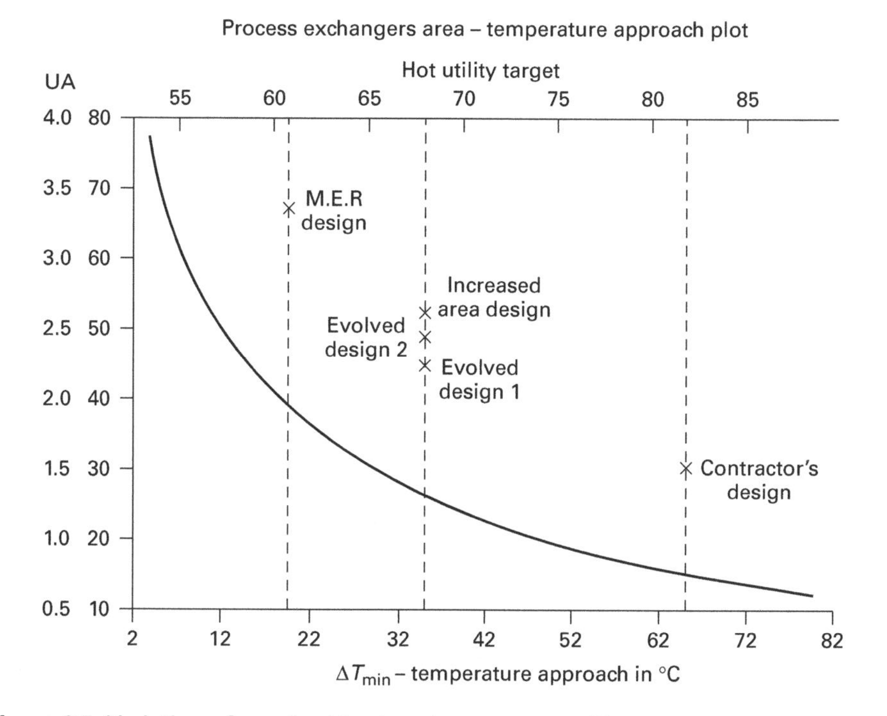

Figure 9.5 Variation of required heat exchanger area with energy use

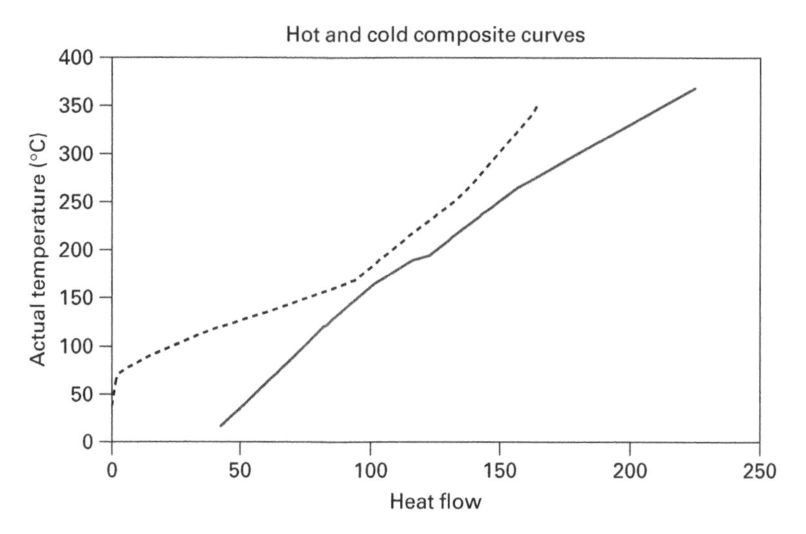

Figure 9.6 Composite curves for crude preheat train

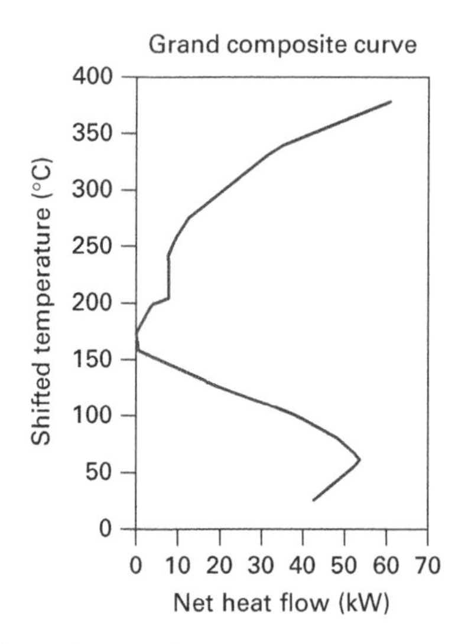

Figure 9.7 GCC for crude preheat train

feed stream, 9. We will now design an (maximum energy recovery) MER network using the "Pinch Design Method", keeping in mind that we want to maximise compatibility with the existing plant.

It is of interest initially to see how the contractor's network matches with the pinch principles. Figure 9.8 plots the pinch temperature locus on the network, and it can be seen that three exchangers partially cross the pinch and two coolers are partly above the pinch. These violations of our "Golden Rules" explain why the

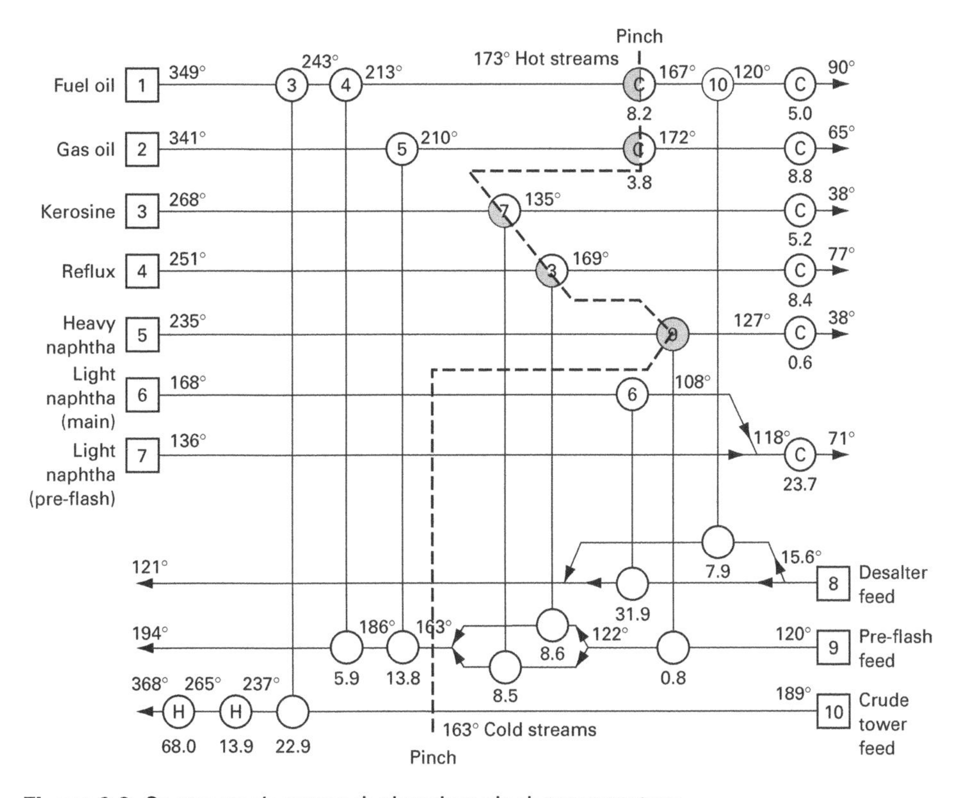

Figure 9.8 Contractor's network showing pinch temperature

contractor's design is well off the energy target, despite the use of low ΔT_{min} values on some matches.

Network design proceeds as follows.

9.2.3.1 *Above the pinch*

Figure 9.9 shows the stream set above the pinch. The first point to realise is that, because there are five hot streams and only one cold stream at the pinch, the cold stream must be split five ways (from the rule that all hot streams above the pinch must be matched). This would probably be impractical, so one or more of these splits must be evolved out. As a first simple evolution the load on stream 5 (the heavy naphtha) above the pinch, which is small in comparison with the net heating duty (less than 1%), can be ignored, reducing the required number of stream split branches on streams 9–4. These are the essential pinch matches.

The first design decisions are shown in the right-hand half of Figure 9.10, with stream 5 removed and the load on the process heater adjusted accordingly. Note that the four matches against the split stream 9 (the feed stream to the preflash tower) are basically present in the existing plant (although not against so many split branches). Having decided on these matches, match number 3 is added because it already exists (and is obviously sensible). We now assign loads to these matches. The loads on

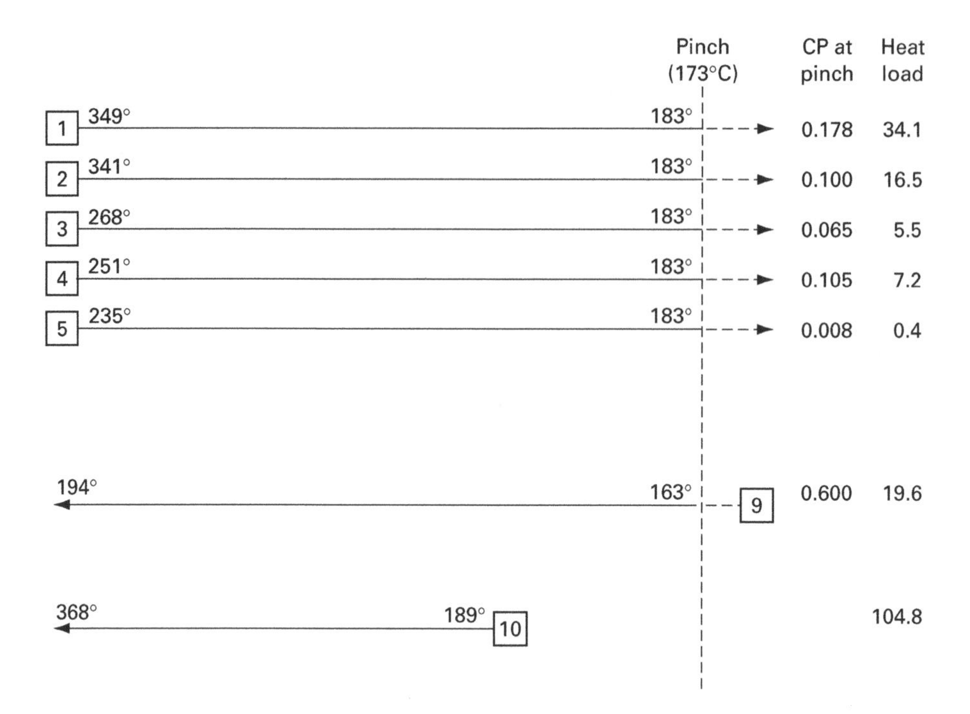

Figure 9.9 Stream grid above the pinch

matches 7 and 8 are maximised to "tick-off" streams 3 and 4. Match 3 should not cool stream 1 below 209°C (because stream 10 supply temperature is 189°C). This dictates the maximum load on match 3, which becomes a design decision. If stream 1 is cooled to 209°C in this match, the load on match 4 is fixed. This in turn fixes the load on match 5 by enthalpy balance on stream 9. Fixing the load on match 5 allows us to calculate the temperature of stream 2 on the hot side of the match. It comes out to be 206°C which means that a residual cooling duty of 14.18 MW is left on stream 2. Because we have obeyed the pinch design method, this is exactly the size of "hole" that we find is left to be filled on stream 10 (having placed the minimum load of 61.1 MW on the fired heater). Hence a match of this load is required between stream 2 and 10. However, if we try to perform a sequential match on stream 10 in either order, we get a temperature infeasibility. We could have deduced this from the GCC (Figure 9.7) – we are still in a region of very low heat flow, so we should treat this as a "near-pinch" and reapply the rules for matching at the pinch. Since there are two hot streams and one cold, we have to split stream 10, placing match 3 and the new match on parallel branches, as shown in Figure 9.10.

This is our completed above-the-pinch design. Note that it only requires one basic new match compared with the existing plant.

Finally, note that there is a small violation of ΔT_{min} at the cold end of the new match in Figure 9.10. However, remember that the ΔT_{min} of 20°C was chosen arbitrarily in the first place.

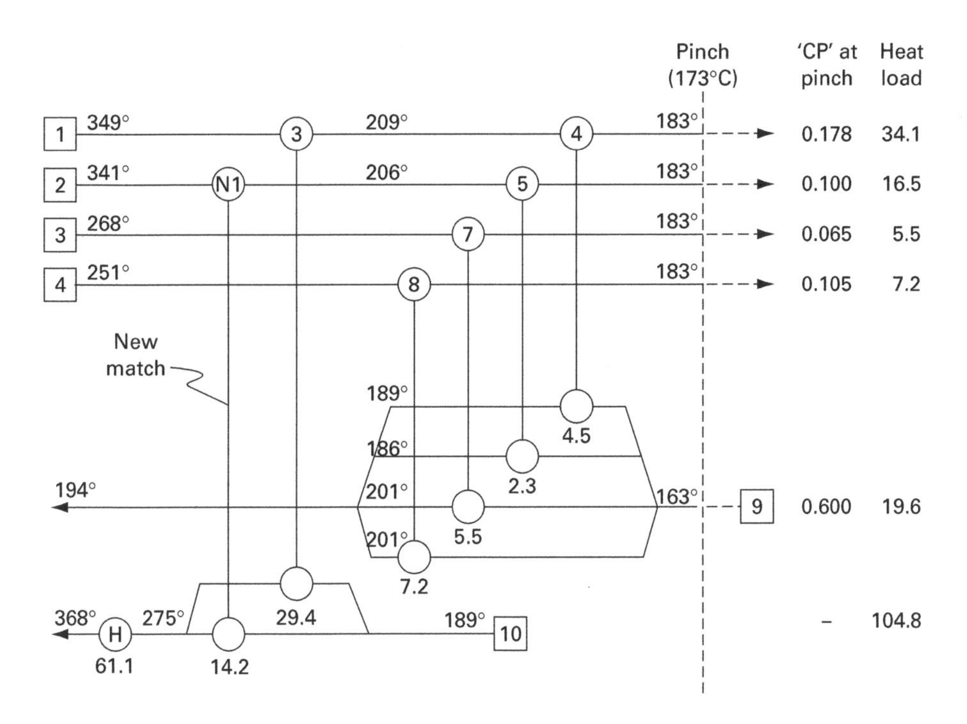

Figure 9.10 Completed MER network above the pinch

9.2.3.2 *Below the pinch*

Figure 9.11 shows the stream set below the pinch. Once again, the four-way split of stream 9 is required, this time because it is the only way to fulfil the requirements on CP for each match, and yielding the same topological set of four matches, 4, 5, 7, 8. The other existing matches 6, 9, 10 and all the coolers are also included as shown in Figure 9.12. If we assign the "base case" loads to matches 6, 9 and 10 (i.e. the matches away from the pinch), then the loads on the pinch matches 4, 5, 7 and 8 turn out as shown. Notice that because the sum of the CPs of streams 1 to 4 is almost exactly equal to the CP of stream 9 at the pinch, minimum driving force is maintained throughout the pinch matches. This means that there is no flexibility in choice of branch flow rate in the split stream 9.

9.2.3.3 *Complete MER design*

The completed MER design is obtained by merging the two systems "above" and "below", with the result shown in Figure 9.13. The split stream 9 branch flows calculated in the two halves are not compatible. However, because more flexibility exists in the above-the-pinch splits, the branch flows calculated for below the pinch are adopted for the combined design. This then means that the target temperatures on the individual branches are changed in the combined design (compare Figure 9.10). However, they remain feasible against the hot stream temperatures.

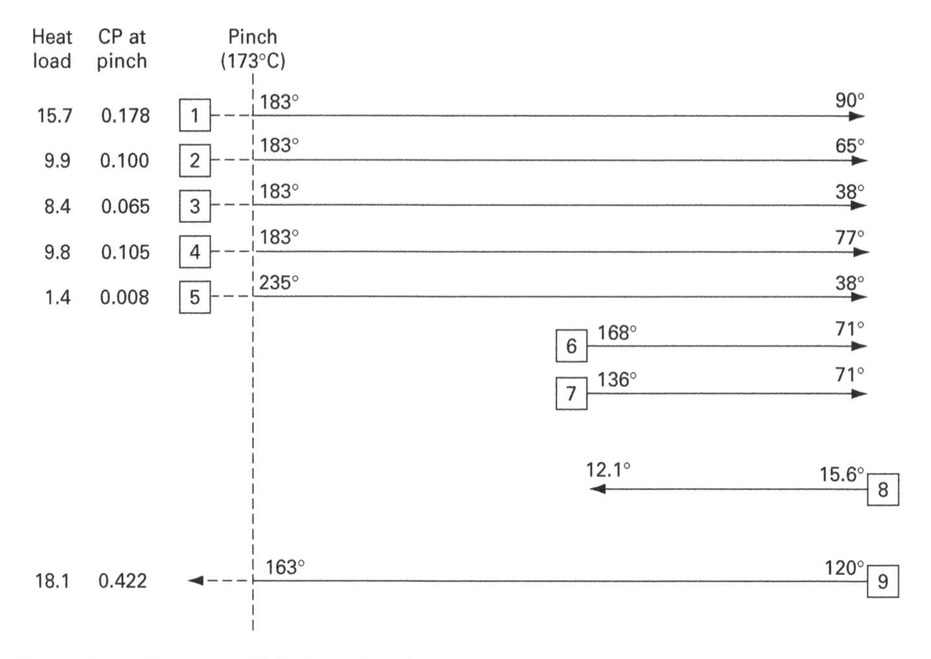

Figure 9.11 Stream grid below the pinch

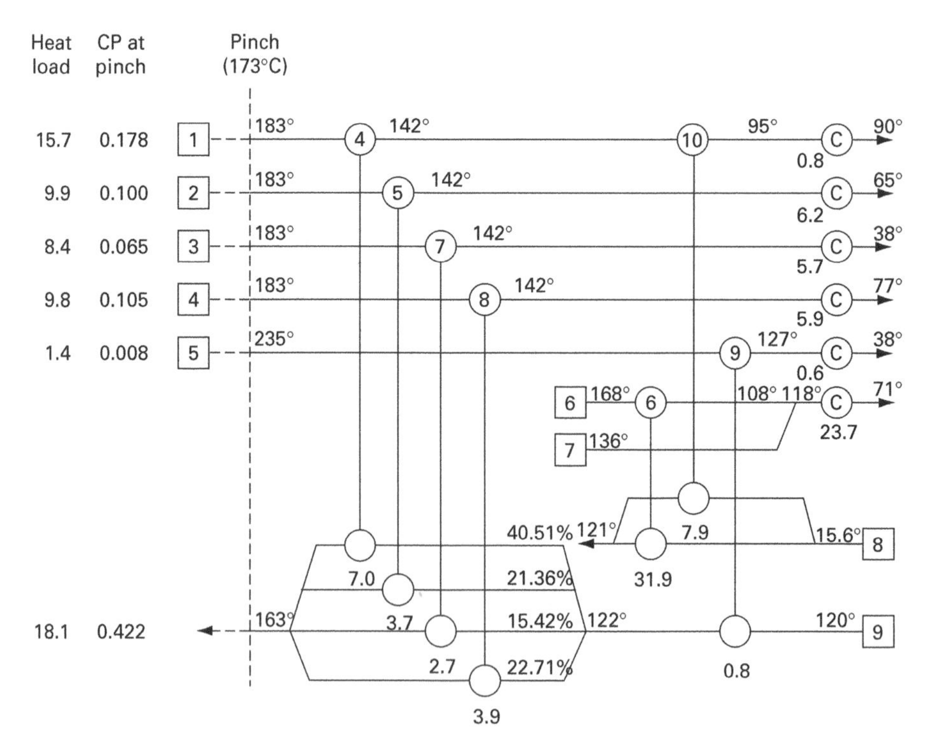

Figure 9.12 Completed MER network below the pinch

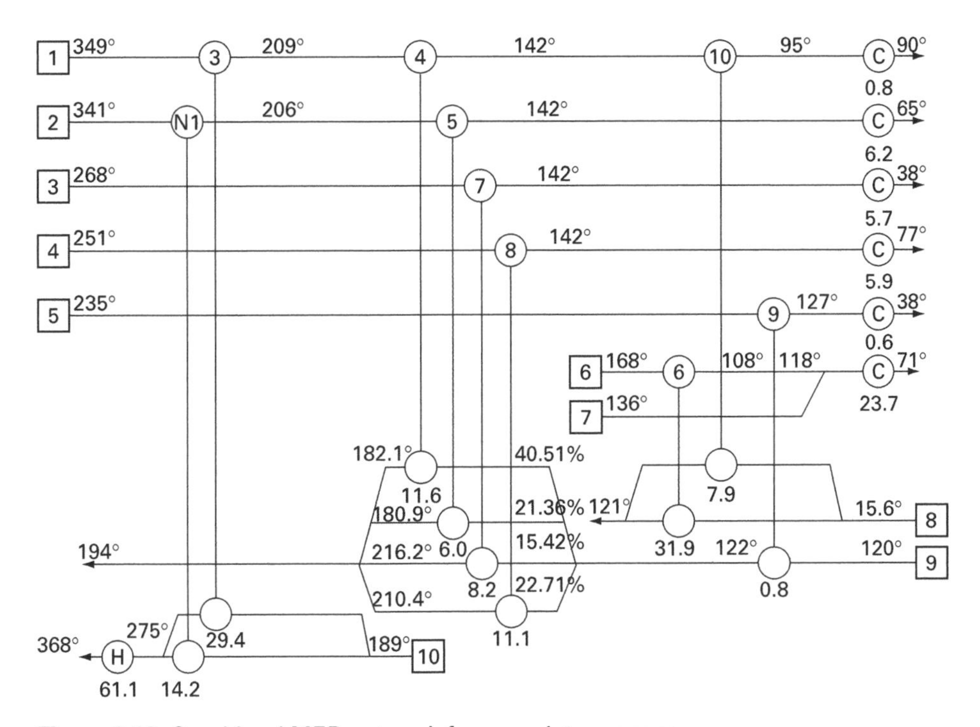

Figure 9.13 Combined MER network for complete process

9.2.4 Design evolution

The MER design shown in Figure 9.13 achieves a 10% energy saving over the existing plant and a 25% saving over the contractor design. Topologically, the only difference between the MER design in Figure 9.13 and the existing plant is one new match (labelled "N1"). The design therefore appears a promising starting point for the evolution of a good revamp scheme which avoids the need for supplementary heating.

The next step is to evaluate the areas of the heat exchangers. In this case, we do not have information on the heat transfer coefficients (although it could be back-calculated from the known exchanger areas) but it is sufficient to carry out a "UA analysis". Values for UA ($=Q/\Delta T$) for the contractor's "base case" design and for the synthesised MER design are shown in columns 2 and 4 of Table 9.2. Values are given for the interchangers and for the big cooler (the air-cooled heat exchanger on stream 6/7). It can be seen that the MER design pays a heavy penalty in terms of additional area and number of matches in need of modification. Indeed, it requires more additional area than the increased-area variant (column 3) which was judged unacceptable on cost grounds. An obvious strategy to adopt in evolving this design is to increase the heat input to the fired heater up to the maximum possible on the existing equipment, that is 68.0 MW. In other words, we will "relax" the design just to the point where supplementary heating becomes necessary.

Table 9.2 Comparison of "UA" values for different network designs

	Contractor's design	*Increased area design*	*MER design*	*Evolved design 1*	*Evolved design 2*
Energy use	81.9	68.0	61.1	68.0	68.0
Maximum stream splits	2	2	4	4	3
UA for matches					
N1	–	0.380 (new)	0.393 (new)	0.332 (new)	0.332 (5 mod)
N2	–	0.230 (new)	–	–	0.210 (new)
3	0.288	0.393 (mod)	0.714 (new)	0.337	0.337
4	0.159	0.512 (mod)	0.549 (mod)	0.412 (mod)	0.476 (mod)
5	0.152	0.150	0.286 (mod)	0.147	–
6	0.462	0.462	0.462	0.462	0.506
7	0.196	0.195	0.293 (mod)	0.198	0.193
8	0.132	0.115	0.454 (mod)	0.241 (mod)	0.234 (mod)
9	0.022	0.022	0.022	0.022	0.022
10	0.111	0.165 (mod)	0.180 (mod)	0.111	0.111
UA all exchangers	1.522	2.624	3.353	2.262	2.421
Air cooler	0.550 (mod)	0.550 (mod)	0.550 (mod)	0.550 (mod)	0.392
Total overall UA	2.072	3.174	3.903	2.812	2.813

By adopting this strategy, the design in Figure 9.14 is obtained. We have left the network structure unchanged, but have opened out the driving forces (i.e. increased the gap between the composite curves) to reduce heat exchanger area. The corresponding UA values are listed as "Evolved Design 1" in Table 9.2. The match loads and the stream temperatures are chosen for maximum compatibility with the existing plant. This allows matches 3, 5, 6, 7, 9 and 10 to remain un-modified. Match 4 has been split into two parts for easy piping. The existing part of match 4 is left on the hot end of the full stream 9. The new part is situated on a split branch. Not only is the extent of plant modification reduced in the design of Figure 9.14, but also the amount of additional area required is considerably less than for the MER network or the increased-area variant.

Two potential problems with this design remain, however. These are the four-way split of stream 9 and the need for an expensive modification of the big air cooler on streams 6/7. This last difficulty can be overcome by adding area to matches 6 and 4, allowing load to be shifted round the loops from the air cooler to any of the water coolers on streams 2, 3 and 4. The effect of eliminating one of the stream split branches is shown in Figure 9.15. The branch chosen is that carrying match 5, with match 5 therefore being completely eliminated. Match 5 is chosen for elimination because it carries the smallest load amongst the branch matches (and hence its removal causes least upset of driving forces amongst the remaining

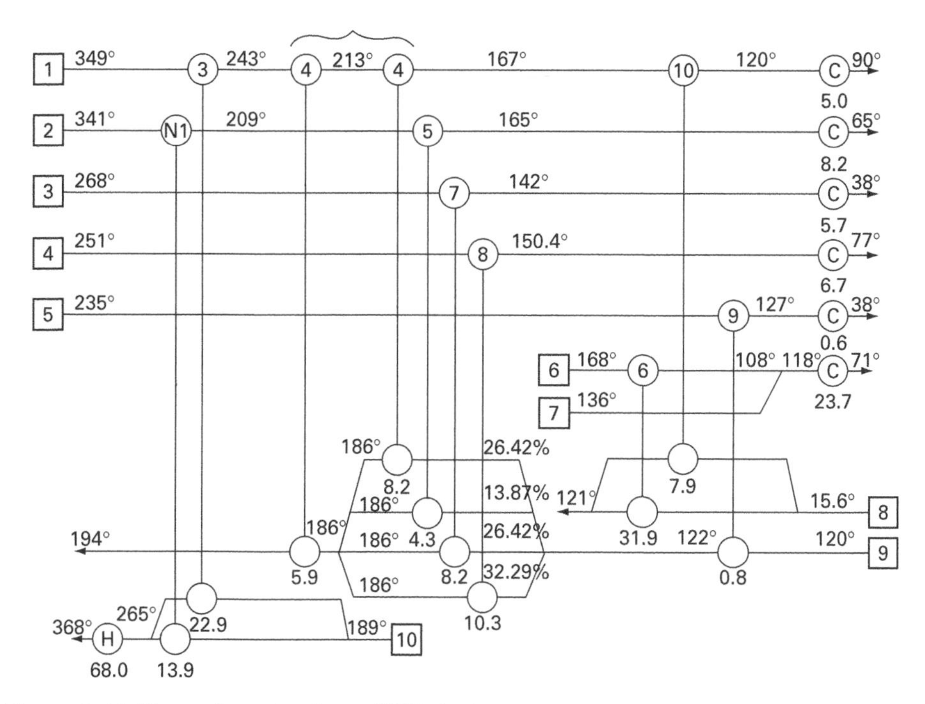

Figure 9.14 First relaxation from MER design

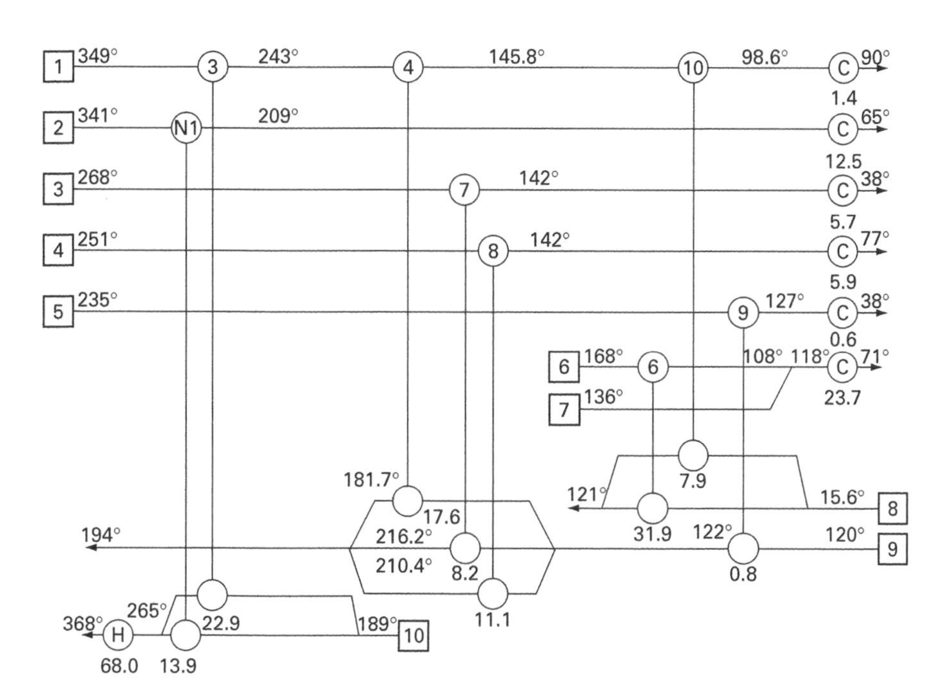

Figure 9.15 Elimination of four-way stream split

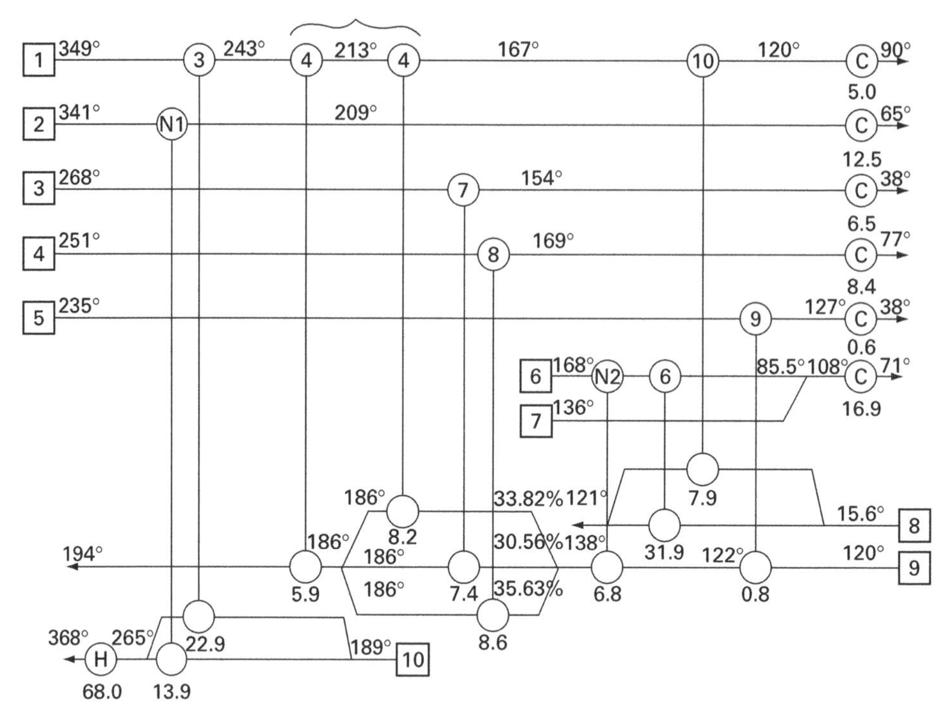

Figure 9.16 Second evolution of network design

matches) and because the existing shell of match 5 can be re-piped relatively easily for utilisation in the new match N1. However, it can be seen in Figure 9.15 that this decision has a considerable "squeezing" effect on driving forces, returning the design to the situation where matches 7 and 10 require modification. In order to avoid this and ease the driving force squeeze in the matches to stream 9, an extra source of heating for stream 9 must be found. The only candidates available (on temperature grounds) are streams 2 and 6. Although the former has a higher-supply temperature, its low CP means that it will struggle to meet the demand from stream 9. Hence a new match N2 is introduced between streams 6 and 9, leading to the design in Figure 9.16, with matches 7 and 10 restored to their un-modified state, and the large air cooler not now needing modification.

We still have problems, however, as N2 reduces the temperature of stream 6 from 168°C to 157°C so that match 6 has to be increased in size. This problem can be overcome by splitting stream 6 and placing the new match N2 on one branch, and match 6 on the other. The effect is to maximise driving force in match 6 sufficiently for the existing unit to cope (only 10% increase in UA). Stream 6 is an overhead vapour stream, and large diameter vapour lines to exchanger N2 would be needed, at high cost. Instead, heat can be taken from the top of the crude tower in a liquid "pump-around", forming the split branch for the new match N2. The overhead vapour stream is reduced in mass flow due to the pumparound, and internal vapour condensation occurs within the column where the liquid is reintroduced, reducing vapour flows in the topmost section. This is a process change, and column performance must be recalculated to ensure that the required splits and compositions can still be obtained.

There is an interesting conceptual point here. In Section 6.4.4 it was pointed out that pumparounds should go above the pinch and above the GCC to save energy, whereas this pumparound is just below the pinch. However, it is being installed not to reduce the targets further, but to assist with equipment design by opening out the driving forces. A pumparound above the pinch would have to be mounted further down the column where temperatures are higher, but this would disturb column operation more severely and, because of the very constrained nature of the GCC, very little additional heat would be recovered.

Finally, it should be noted that the opening out of driving forces due to the relaxation means that a stream split is no longer necessary on stream 10, as matches 3 and N1 could now be in series. However, the driving forces on match 3 would be reduced (with ΔT at the cold end falling from 54°C to 26°C), giving a much larger exchanger. After further consideration, it was decided to retain the stream split.

The UA values for the Figure 9.16 design (with the stream 6 "pumparound" modification) are given in the final column of Table 9.2. Let us compare this design with the "four-way split" design of Figure 9.14. It requires the same number of exchanger modifications and the same amount of area. It requires one more match but one less stream split. At this stage of the study it is not possible to say which design is better. However, it seems clear that the two candidate designs have major advantages over the contractor's design and the increased-area variant. The reason is that the pinch design method has ensured that the design is correct where it matters most – in the region near the pinch – and has identified the vital multiple stream split which makes best use of the very limited temperature driving forces in this region. Without this, it would be impossible to reach the targets without extensive cyclic matching requiring many small heat exchangers.

9.2.5 Design evaluation

Next in any study, a more detailed check of equipment performance and system operability is required, followed by engineering design specification and costing. However, a rough estimate of heat exchanger costs could be gained at this stage by using the UA values and the results are shown in Table 9.3.

Table 9.3 Additional area and costs for required matches (1982 prices)

Match	*$\Delta(UA)$ ($=Q/\Delta T_{LM}$) (W/K)*	*C (£/(W/K))*	*C_T (£)*
N1	1.80×10^5	0.104	18,700
N2	2.10×10^5	0.102	21,400
4	3.17×10^5	0.098	31,100
8	1.02×10^5	0.110	11,200
			82,400

After detailed evaluation, the design chosen for construction was the network of Figure 9.16. It is shown in flowsheet form (including the new pump-around) in Figure 9.17. The techniques described in this Guide saved energy to a value of about £1 million per year (at 1982 prices) compared with the contractor's design. This gave a payback of a few months on the new and enlarged exchangers, even allowing for installation costs. In addition the design was safer (due to the elimination of a fired heater) and, perhaps surprisingly, more operable. The presence of the three-way split on stream 9 meant that if the crude feedstock was changed, yielding a different balance between light and heavy fractions, the branch flows could be adjusted to compensate.

The design in Figure 9.16 was installed and was fully satisfactory in operation, achieving the expected rates and savings.

9.2.6 Conclusions

The main points highlighted by this study are:

(i) The Pinch Design Method generated a network which was substantially better than that obtained by any previous methods of heat exchanger network design.
(ii) The targeting stage gives a rapid initial assessment of the scope for change and the likely difficulties which will be encountered in obtaining a solution.
(iii) The network design method can be used systematically to produce good "revamp" designs, even where the existing heat exchanger network is complex. It allows a productive interaction with the engineer's experience (a good example is the use of the pump-around in the preferred solution).
(iv) Designs produced by proper use of the method are elegant, sometimes yielding both energy and capital savings.
(v) A higher degree of process integration does not necessarily cause control problems. If the integration is well balanced the controllability can be enhanced.
(vi) Parallel stream splitting is a practical tool for improving energy recovery and operability.
(vii) The Pinch Design Method can be employed to give good designs in rapid time and with minimum data.

9.3 Aromatics plant

9.3.1 Introduction

The plant concerned in this study was part of one of the largest aromatics complexes in Europe. It was commissioned in 1969 and used state-of-the-art conventional technology. The original study was again performed by ICI and reported in the first edition of the User Guide. Since that time, it has also been subject to a large amount of analysis by leading international researchers, and it has become something of a standard test case for network design. Intriguingly, not only the network but the

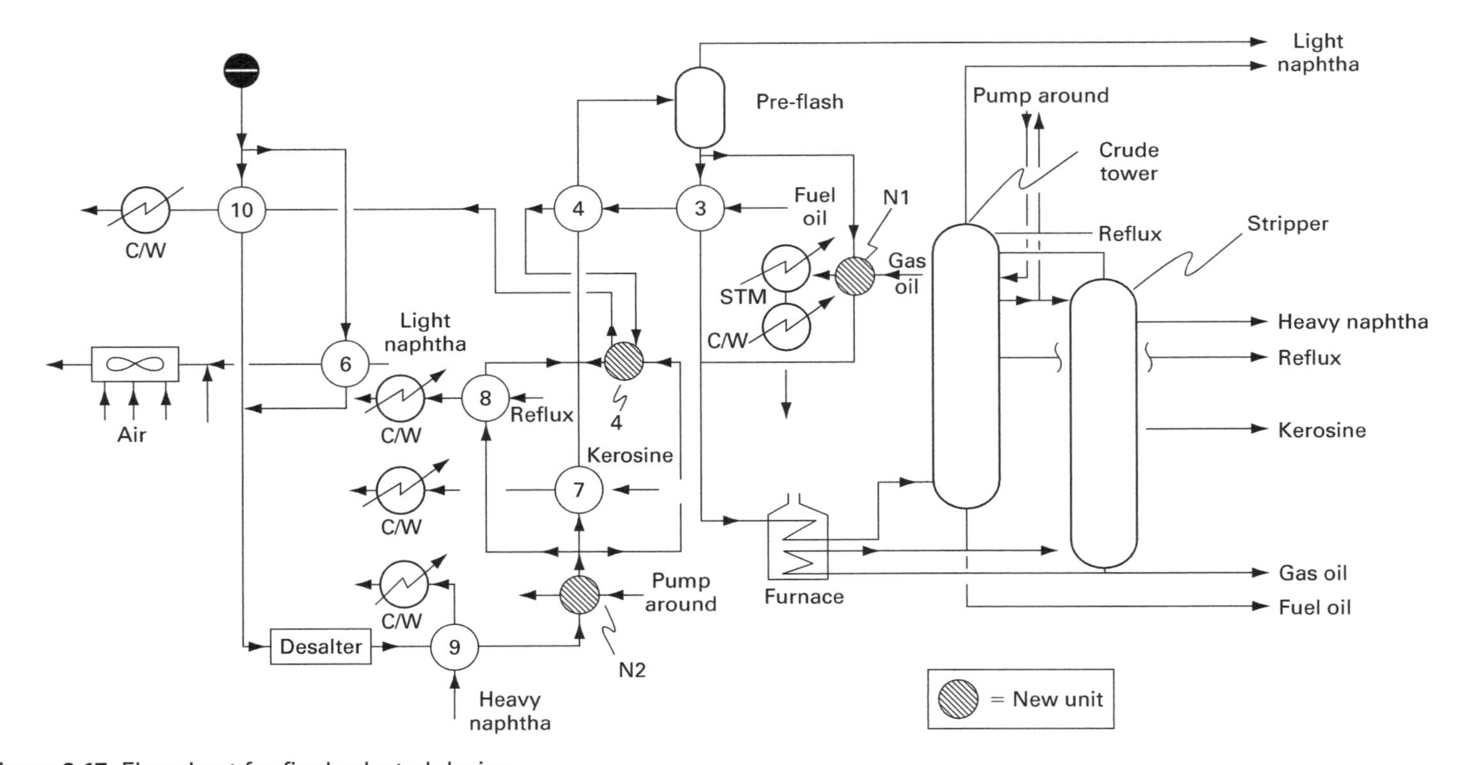

Figure 9.17 Flowsheet for final selected design

stream data seems to have undergone evolution in the successive reporting! The data quoted below returns to that given in the IChemE Guide, which most closely resembles the details of the actual plant.

9.3.2 Process description

A schematic diagram of the process flowsheet is shown in Figure 9.18. The feedstock is a central fraction of naphtha containing chiefly paraffins and cycloparaffins which are reformed into product containing paraffins and aromatic compounds. The process can be described as follows, indicating the streams which will be extracted:

Stream 1 (cold): The feed is vaporised (H1) and passed through a desulphurisation reactor (R1).

Stream 2 (hot): Heat is recovered from the reactor effluent in two interchangers (A, B) prior to condensation (C1) and gas separation (F1).

Stream 3 (cold): The liquid from the separation stage is re-heated by reactor effluent (B) and fed to a stripping column (D1) in which the light ends and sulphur-containing compounds are removed.

Mixing: The desulphurised naphtha stream from the column is mixed with recycle gas.

Stream 4 (cold): The two phase mixture is preheated in a series of process interchangers (D, C). The mixture is finally raised to the reaction temperature of 500°C by a radiant furnace (H2) fired by a mixture of gas and fuel oil.

Reactions: The reactions take place in a pair of reformers (R2, R3).

Stream 11 (cold): Between the reformers, the mixture is re-heated to reaction temperature by a fired heater (H3).

Stream 5 (hot): The reformer effluent, at 490°C, is cooled in interchanger C and then passed to exchanger X, which heats other cold streams (actually the reboilers of columns D1 and D2).

Stream 6 (hot): The mixture emerging from X is cooled further in three exchangers which preheat the feed (D, E) and provide the heat source for other process requirements (F, heating cold Stream 10). Final cooling and gas separation takes place in C2 and F2.

Stream 7 (cold): The gas recycle is compressed (P1) and preheated (E) prior to mixing with the liquid reformer feed.

Stream 8 (cold): The liquid from the flash drum is passed to a column for stabilisation (D2) and a conventional feed/tails interchanger (G) is installed to reduce the reboil requirement by adding feed preheat.

Stream 9 (hot): The reformate stream passes through exchanger G and is finally cooled in C3 prior to storage.

9.3.3 Stream data extraction

It is not difficult to extract temperature and heat load data for the eleven streams given above from the flowsheet. The "base case" heat exchanger network (shown in

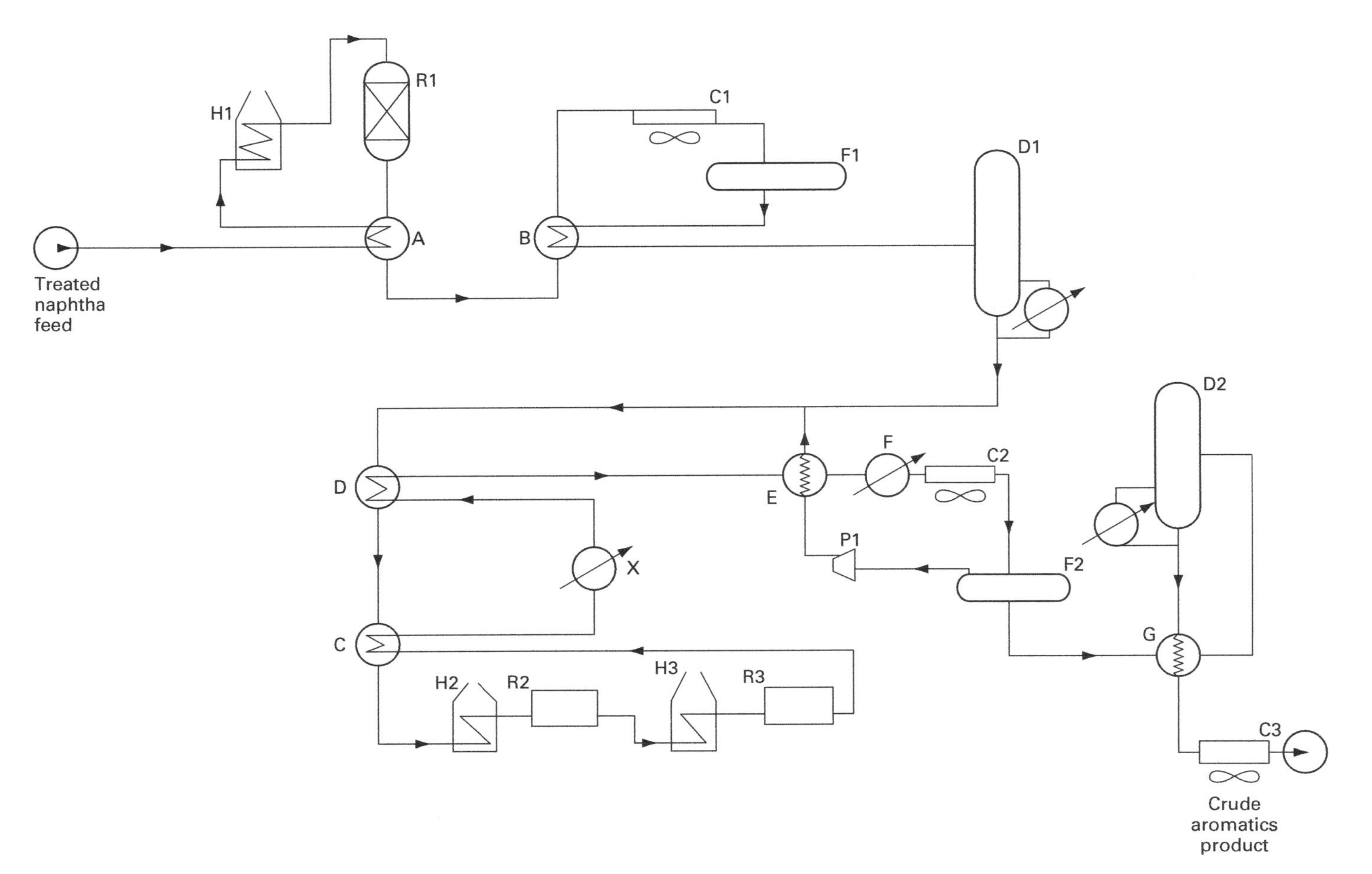

Figure 9.18 Schematic flowsheet of naphtha reformer (aromatics plant)

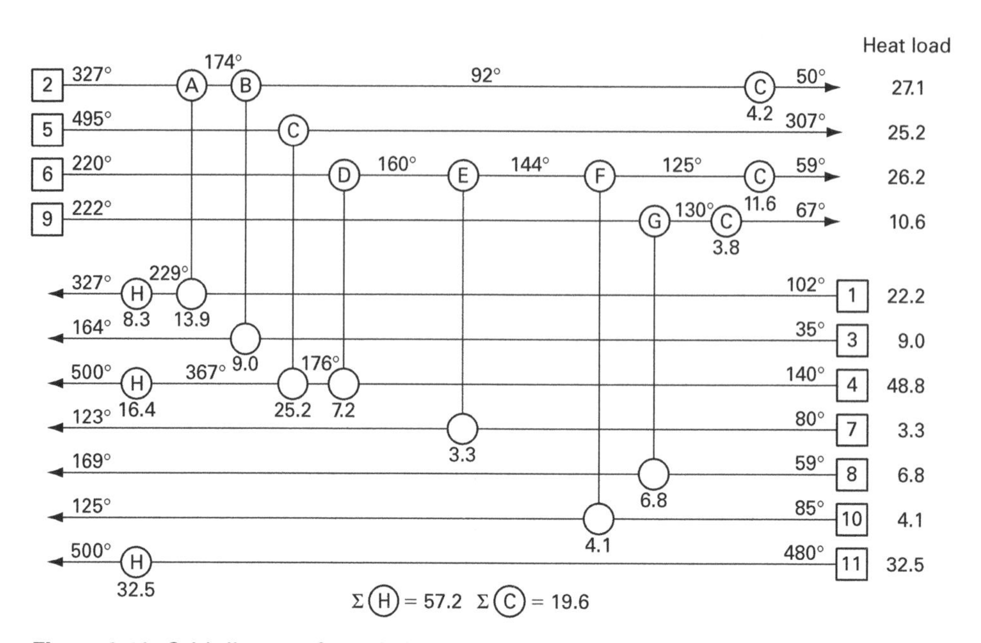

Figure 9.19 Grid diagram for existing heat exchanger network

flowsheet form in Figure 9.18) is represented in grid form in Figure 9.19. The grid shows all the flowsheet heat exchangers, heaters and coolers with their heat loads and corresponding stream temperatures (in °C). Heat loads were given in thousands of tonne-calories per hour in the original data (ttc/h) and have been left in these units; any set of units can be used for a pinch analysis as long as they are self-consistent. (1 ttc is the heat required to warm 1 tonne of water by 1°C, which equates to 4,187 kJ. Hence 1 tc/h = 4.187 GJ/h = 1.163 MW = 1,163 kW. Note that confusion often arises in practice as to whether "calories" are based on 1 kg or 1 g of water.)

Many of the streams in the process have pronounced non-linear temperature/enthalpy profiles (due to partial condensation, for example). For accurate calculations, this non-linearity was represented by temperature/enthalpy data points generated by modelling the streams using a physical properties computer program. For this simplified description, the only data points used have been those at the end of exchangers, using known flowsheet heat loads and temperatures. This linearised data is shown in Table 9.4. Heat capacity flowrates have been back-calculated. Rather unusually, they fall with increasing temperature for some streams (usually they would rise slightly) because of the partial vaporisation/condensation of the mixture.

This was the stream data as quoted in the IChemE Guide 1st Edition, but is it correct? Look at the flowsheet and see if you can spot any anomalies or pitfalls. For a reminder, take a look at Sections 3.1 and 8.4 on stream data extraction.

Firstly, have all the streams been considered? The reboilers on column D1 and D2 will have a significant heat load but have not been included in the analysis.

Table 9.4 Stream data for aromatics plant

Stream no	Supply temperature (°C)	Target temperature (°C)	Temperature/enthalpy data		
			T (°C)	*H (ttc/h)*	*CP (ttc/h/K)*
1	102	327	102	0.0	0.1094
			229	13.9	0.0847
			327	22.2	
2	327	50	327	27.1	0.0908
			174	13.2	0.1098
			92	4.2	0.100
			50	0.0	
3	35	164	35	0.0	0.0698
			164	9.0	
4	140	500	140	0.0	0.200
			176	7.2	0.1319
			367	32.4	0.1233
			500	48.8	
5	495	307	495	25.2	0.134
			307	0.0	
6	220	59	220	26.2	0.120
			160	19.0	0.2062
			144	15.7	0.2158
			125	11.6	0.1758
			59	0.0	
7	80	123	80	0.0	0.0767
			123	3.3	
8	59	169	59	0.0	0.0618
			169	6.8	
9	220	67	220	10.6	0.0756
			130	3.8	0.0603
			67	0.0	
10	85	125	85	0.0	0.1025
			125	4.1	
11	480	500	480	0.0	1.625
			500	32.5	

(There are no condensers on these columns because they are strippers and the off-gas is removed without condensing). They form the heat duty X against which the hot stream from the reformer (streams 5 and 6) is matched. To be rigorous, the reboiler loads on D1 and D2 and the whole of the reformer effluent stream (including the section between 307°C and 220°C which is matched against X) should be included in the stream data. However, if it is decided that these existing matches must be retained at their existing size and heat duty at all costs, they can be omitted from the stream data. This is "targeting with constraints" or "remaining problem analysis" (Section 4.5), and it reduces the possibilities for heat exchange. For the time being, we will work with the original stream selection from the 1st Edition, but we will note later what would happen if these streams were included.

Secondly, there is a mixing point between the bottoms from D1 and stream 7, producing stream 4, and from the rules in Section 8.4.1 we should model this as isothermal mixing. In fact stream 7 is entering at 123°C and the mixed stream 4 leaves at 140°C. The rigorous way of modelling this would be to raise stream 7 to the temperature of the D1 bottoms and start stream 4 at that temperature. Again, we have not bothered to change the original data, but we should at least note that the target temperature of stream 7 is "soft" data and that we can easily allow it to change if we see benefits from the network design point of view. There would be a change in energy target between the isothermal and non-isothermal mixing cases if these streams are close to the pinch or in a region of low net heat flow – we will find out in the next section if this is the case.

Finally, there are "zonal" effects which will be discussed in more detail in Section 9.3.8.

9.3.4 Energy targeting

The smallest observed approach temperature in the heat exchangers is 10°C, at the hot end of exchanger B. Thus, 10°C was used as the initial estimate for ΔT_{min}. However, it should be noted that the ΔT_{min}'s on other matches range from 19°C (match F) to 128°C (match C), so that using a ΔT_{min} of 10°C can be expected to give us lower energy but a much higher area requirement than the existing network. As this was a high tonnage process with high energy usage, though, there is a considerable financial incentive to achieve all or part of any savings identified.

Targeting over a range of ΔT_{min} gives the results shown in Table 9.5.

At ΔT_{min} = 10°C the hot utility target is 46.5 ttc/h. The current hot utility usage is 57.2, representing a 23% excess energy usage above the minimum. Conversely, 57.2 ttc/h corresponds to a relatively high ΔT_{min} of 42°C, suggesting that the existing area is not distributed in the most appropriate way. Targets increase by 3–4 ttc/h for every 10°C increase in ΔT_{min}.

However, the composite curves (Figure 9.20) show us that there is a very extensive region where the curves are close together and almost parallel over the

Table 9.5 Variation of energy targets and pinch temperatures with ΔT_{min}

ΔT_{min} (°C)	Pinch (shifted/actual temperatures) (°C)	Hot utility (ttc/h)	Cold utility (ttc/h)
0	220 (220–220)	42.66	5.06
10	145 (150–140)	46.52	8.92
20	150 (160–140)	50.44	12.84
30	155 (170–140)	53.49	15.89
40	122 (142–102)	56.50	18.90
50	127 (152–102)	60.43	22.83
60	130 (160–100)	64.19	26.59

80–220°C temperature range – as in the crude distillation unit, we have a "pinch region" rather than a sharp pinch, and great care will be needed with network design over this entire range. The GCC (Figure 9.21) confirms this, highlighting the large region of low net heat flow. As well as the true pinch at 145°C shifted temperature, there are near-pinches at 215°C and 107°C. There is a big high-temperature heat load (stream 11, overcoming the reactor endotherm) which cannot be matched against any hot stream, but there is plenty of opportunity to recover heat to other cold streams at lower temperatures.

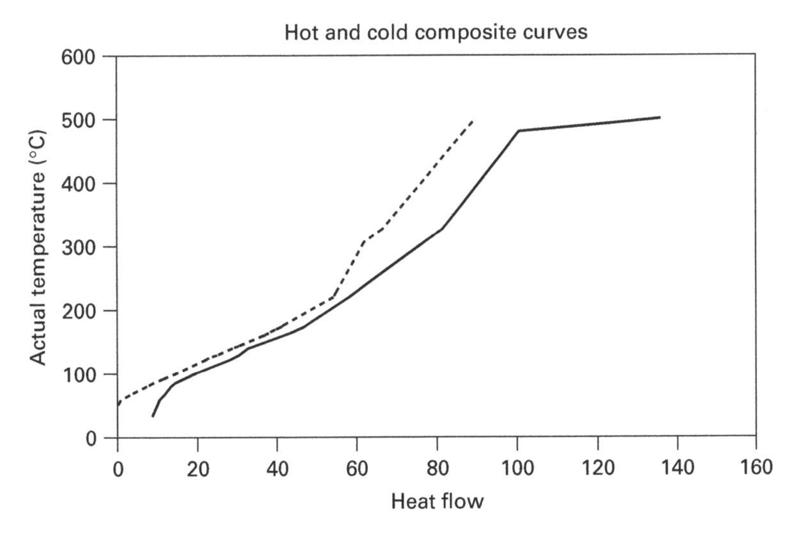

Figure 9.20 Composite curves for aromatics plant

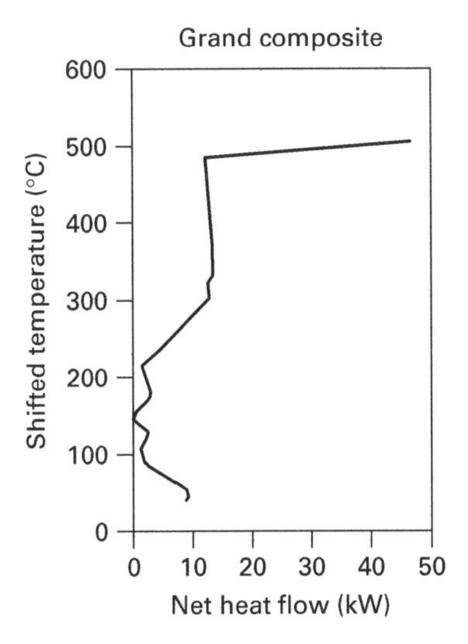

Figure 9.21 GCC for aromatics plant

9.3.5 Design of an MER network

Knowing the pinch temperature of 145°C (shifted temperature, corresponding to 150°C for hot streams and 140°C for cold streams), it is a simple matter to identify the stream sets for "above the pinch" and "below the pinch" (Figures 9.22 and 9.23). The pinch is caused by stream 4 starting at 140°C with a relatively high-heat capacity flowrate, that is 0.2 (ttc/h)/°C.

We can now produce an MER design, but should bear in mind that, as this is a retrofit situation, we should re-use as many of the existing exchangers as possible, even if this means introducing otherwise unnecessary stream splits and/or ΔT_{min} violations!

9.3.5.1 *Above the pinch*

From Table 9.6 we see that hot streams 2, 6 and 9 each require to be cooled to 150°C with one of the four cold streams at the pinch (1, 3, 4 or 8). For temperature feasibility each one of these matches must have $CP_{HOT} < CP_{COLD}$, and we can see that this is very difficult to achieve. Ideally stream 2 should be matched with stream 4, and stream 9 should be matched with stream 1. However, stream 6 requires a split and would have to be matched not only with streams 3 and 8 but also with a split of stream 4. This is not ideal.

An alternative is to say that the CP values of streams 2 and 1 are almost identical, while that for 6 is not too far above 4, and that for 9 is not too far above 3.

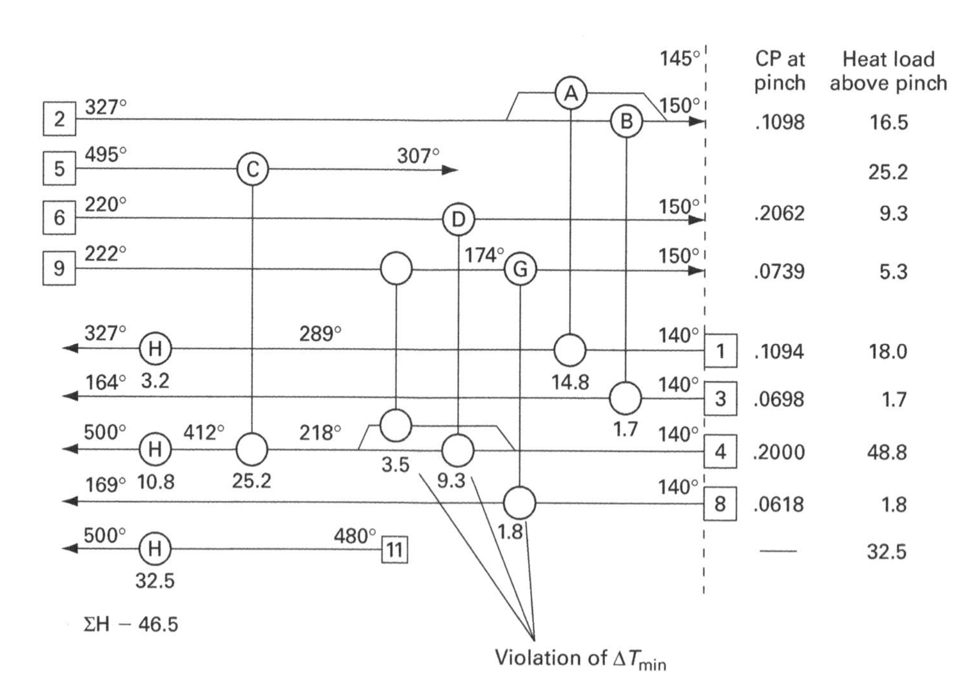

Figure 9.22 MER design above the pinch

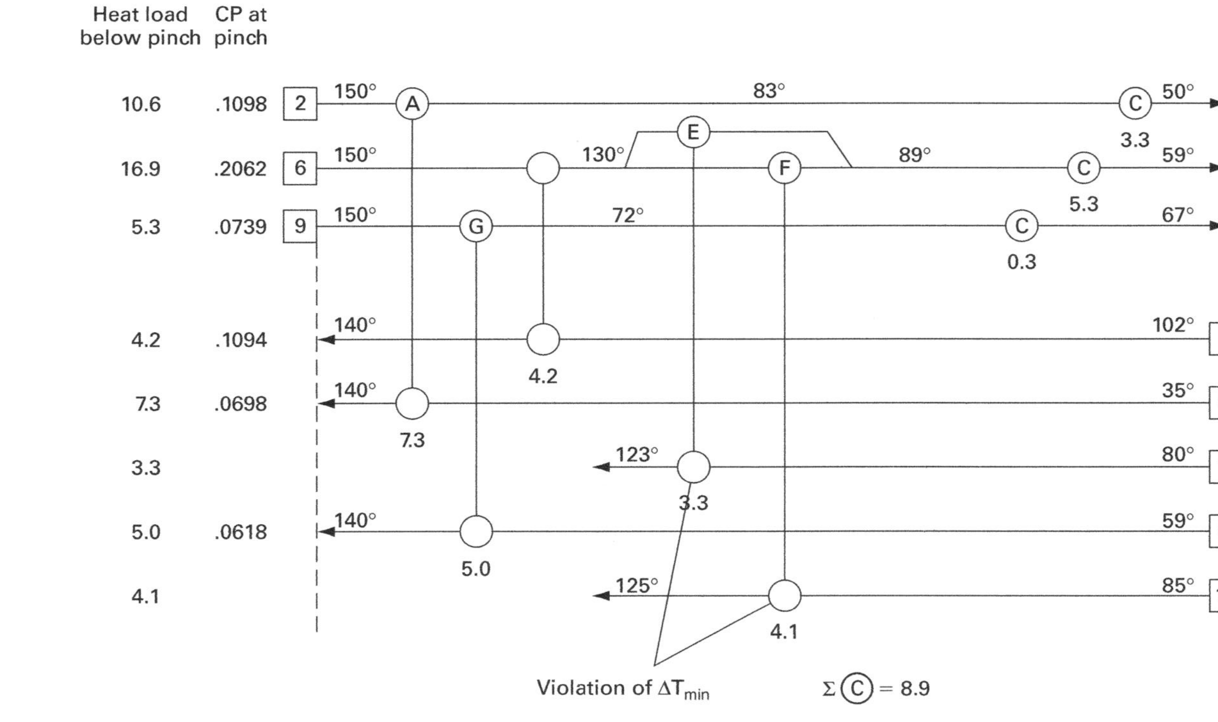

Figure 9.23 MER design below the pinch

Table 9.6 CP-values for streams at the pinch

Hot stream number	*CP value*	*Cold stream number*	*CP-value*
Above the pinch – all hot streams must be matched, $CP_{HOT} < CP_{COLD}$			
6	0.2062		
2	0.1098	4	0.2000
9	0.0756	1	0.1094
		3	0.0698
		8	0.0618
Below the pinch – all cold streams must be matched, $CP_{HOT} > CP_{COLD}$			
6	0.2062	1	0.1094
2	0.1098	3	0.0698
9	0.0756	8	0.0618

Therefore, if these streams were matched, any violation of ΔT_{min} should not be too great. Either 6 or 9 (or both) could also be split and matched against 3. The preference would be to split the stream which has the biggest temperature range above the pinch (to minimise the ΔT_{min} violation) or the largest CP (to put the smallest heat load on the exchanger with low driving forces). On the second criterion, stream 6 should be split in preference to stream 9.

We might ask whether these awkward stream splits are essential; could a different choice of ΔT_{min} help? However, we know the pinch is caused by streams beginning, ending or changing CP, and there are no other hot streams beginning or ending in this temperature range. Stream 6's CP does fall to 0.120 above 160°C, so for a ΔT_{min} of 20°C or more one would at least lose the split on stream 4 (as streams 3 and 8 are now sufficient to balance 6). But this is no great gain.

For a new design one of these alternatives would give the essential pinch matches. For a retrofit study we must think again, since this design would render the existing matches A, B and G redundant. Therefore, we want to match stream 2 with streams 1 and 3, stream 9 with stream 8 and stream 6 with stream 4.

The heat in stream 2 is recovered through matches A and B, with the requirement for a new stream split. The heat in stream 6 is recovered through match D (even though the match will now inevitably violate ΔT_{min}, as $CP_{HOT} > CP_{COLD}$) and the heat in stream 9 is partially recovered by stream 8 (again with an inevitable ΔT_{min} violation) via match G. Following these design decisions, and "ticking off" of streams, the design problem is now reduced to that of recovering the heat in stream 5 and the residual heat in stream 9 (between 222°C and 174°C).

The available heat sinks are streams 1, 4 and 11. Examination of these sinks quickly shows that the only candidate sink for the heat in stream 9 is stream 4, and only then after a stream split to bypass exchanger D! This new match violates ΔT_{min}. Because of its small driving forces and the requirement for a stream split, it is hardly surprising that the existing design misses this match! It could also be identified by designing away from the near-pinch at 220°C.

The remaining design problem, that is, heat recovery from stream 5, is easily solved using the existing match C. The residual heating requirements (total load = 46.5) are supplied by utility heaters.

9.3.5.2 *Below the pinch*

Streams 1, 3 and 8 in Figure 9.23 must be heated to 140°C by process interchange. Table 9.6 shows that no stream splits are required and indeed there is some flexibility in choosing stream matches. Existing matches (also used above the pinch) B (2 vs. 3) and G (9 vs. 8) are obvious, feasible design choices, leaving the requirement for a second new match between streams 6 and 1. These three matches satisfy the heating requirements of streams 1, 3 and 8. Streams 7 and 10 are now supplied with heat via existing matches E and F with stream 6. Again, stream splitting is essential for feasibility, but even so matches E and F now violate ΔT_{min}. This is because we are still in the "pinch region" of low net heat flow, with a near-pinch at 107°C.

An alternative design to that is shown in Figure 9.23 would be to match streams 2 and 1 (equivalent to exchanger A) and match 6 and 3. However, stream 3 has a higher load than stream 1, so if it is "ticked-off", matches E and F become not merely violations, but infeasible (negative ΔT_{min} at the hot end). This can be overcome by supplying the low-temperature heating needs of stream 3 from stream 1, but this network (Figure 9.24) then requires an extra exchanger compared with Figure 9.23.

There are alternative networks which would not violate ΔT_{min}. However, these designs would be less compatible with the base case.

9.3.5.3 *Completion of design and energy relaxation*

At this stage in the study, having generated separate above and below the pinch designs, detailed evaluations of the feasibility of stream splitting and of increasing the surface area of existing units was undertaken. It was decided that stream splitting would not be attractive and the design was relaxed by importing more energy. Apart from eliminating stream splits, this relaxation aimed to increase ΔT_{min} (and decrease surface area requirements in general) and to merge the duplicate matches B and G.

The resulting design, labelled Phase III, is shown in Figure 9.25. The hot utility usage of 52.1 corresponds to 12% above minimum. Further energy relaxation to eliminate the relatively small new match between streams 6 and 1 (opening out driving forces on matches E and F), and also to further minimise the additional surface area required in existing matches, leads to the design labelled Phase II in Figure 9.26. (Phase I, completed before the pinch study, had been the installation of F to recover heat to an adjacent plant.)

Phase III represents an "add-on" energy saving project which is fully consistent with Phase II, but was not installed immediately as it had a longer payback time. Thus, the pinch design method, followed by energy relaxation, established a strategy for the phased improvement of the aromatics process heat recovery – a valuable

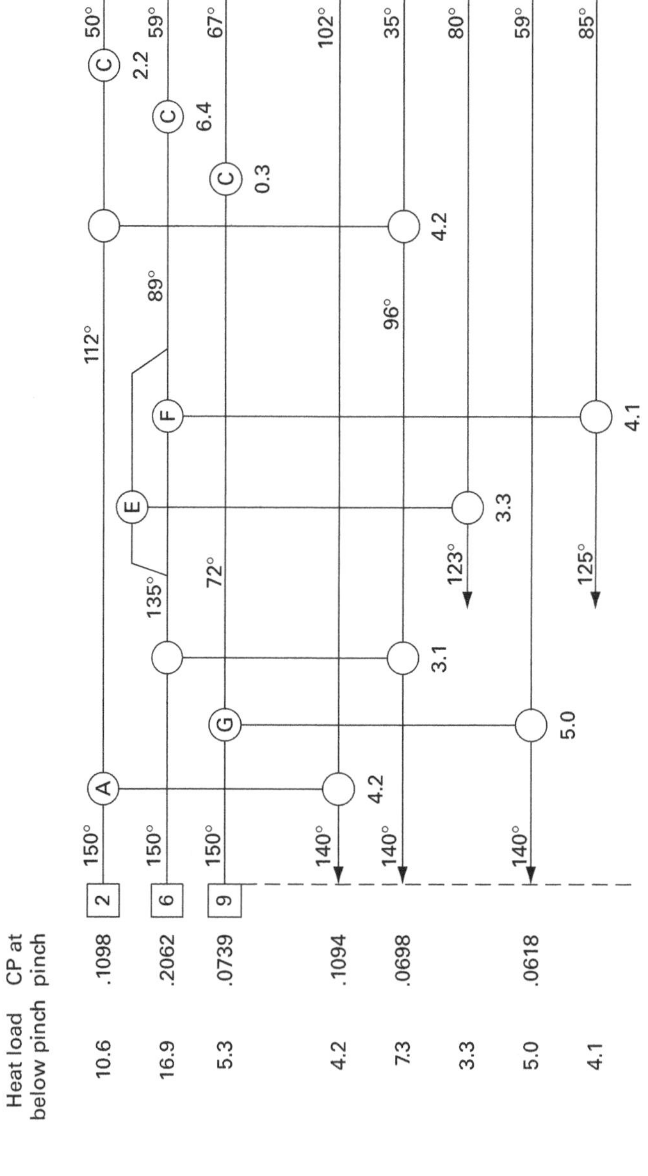

Figure 9.24 Alternative below-pinch MER design

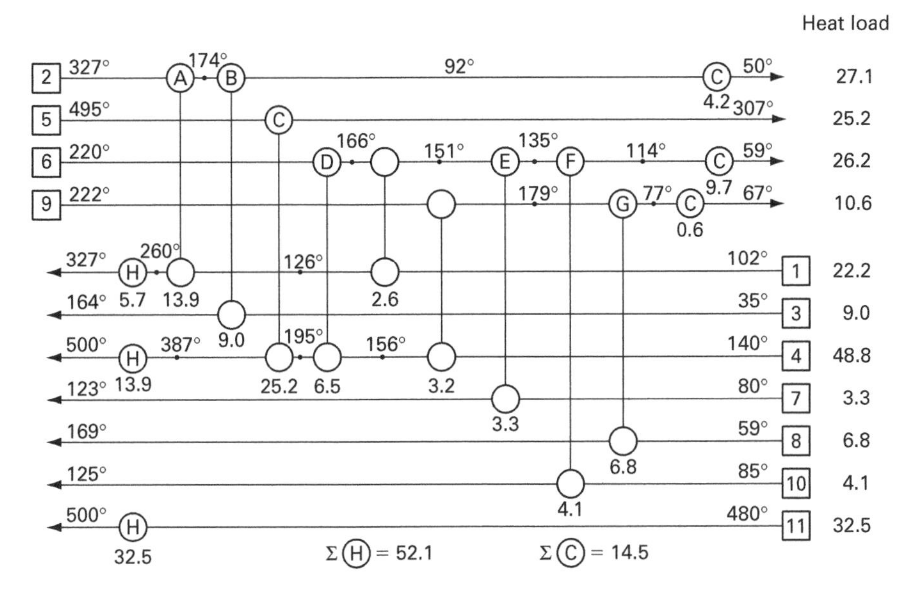

Figure 9.25 First network relaxation (Phase III proposal)

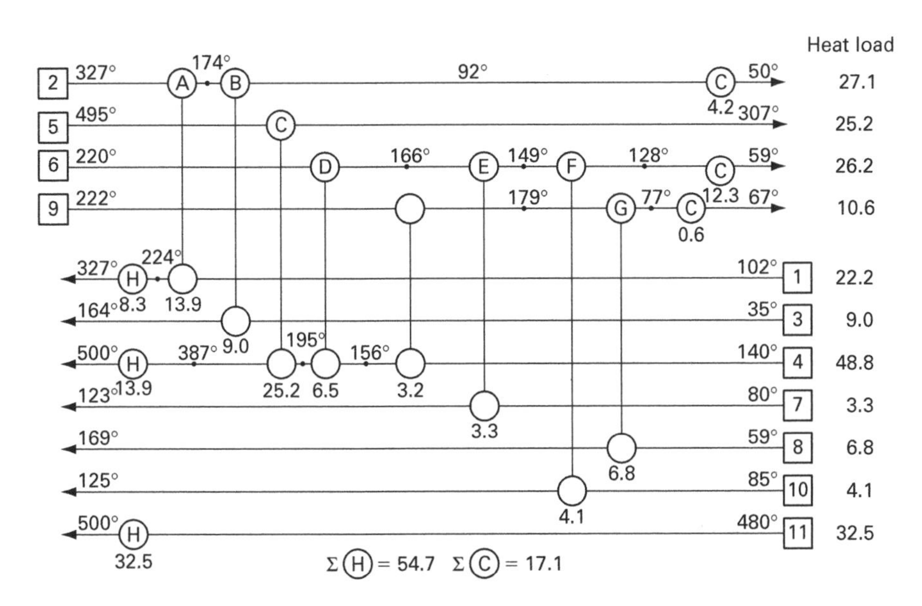

Figure 9.26 Second network relaxation (Phase II proposal)

technique, particularly in an economic climate where capital for projects is restricted. Phases II and III and the minimum energy design are compared with the base case in Table 9.7. In the event Phase III was superseded by a process change which achieved similar energy savings more simply – see Section 9.3.7.

Table 9.7 Summary of schemes proposed for aromatics plant

Scheme description	*Energy use ttc/h*	*Target ttc/h (ΔT_{min} = 10°C)*	*% above target*
Base case	57.2	46.5	23%
Phase II initial networks	54.7	46.5	18%
Phase II installed	55.2	46.5	19%
Phase II and III	52.1	46.5	12%
With process change old	52.2	43.8	19% new, 12%

Phase II involves just one new match – between streams 9 and 4 – and it was decided to install this. However, we must remember that we have designed the network using a ΔT_{min} of only 10°C, which now occurs at two further locations – the cold end of exchanger D and the hot end of exchanger G – in addition to the hot end of exchanger B which was in the base case. So temperature driving forces have been squeezed and it is essential to check whether additional area will be required on other heat exchangers. A "UA" analysis, summing the values of $Q/\Delta I$ for each match, gives a value of 1.22 for the base case and 2.01 for Phase II – an increase of no less than 65%. Part of this is accounted for by the new exchanger, but the majority comes from an increased area requirement on matches D and (above all) G, where the ΔT has been much reduced although the duty is the same.

It was decided that the amount of additional work needed to enlarge existing exchangers was unacceptable. The network was therefore relaxed further; the load on the new match was reduced, and hence the energy savings fell from 2.5 to 2.0 ttc/h. There are complicated knock-on effects throughout the network. The driving force on G has been reduced without additional area, so the heat load falls. This means that the temperature of stream 8 entering D2 will be less, so the reboiler duty will be higher. This in turn will come out in an increased load on exchanger X and a lower hot stream exit temperature, which means that the supply temperature of stream 6 will fall. For tracking these complex trade-offs, it is very useful to have a process simulator, network simulator or spreadsheet. However, the net result was that the predicted saving of 2.0 ttc/h was indeed achieved, as shown by ICI monitoring and targeting studies based on recorded fuel oil and steam use, before and after the modification.

9.3.6 Network design based on existing layout

For a situation such as this where there is an existing heat exchanger network, an alternative design route is possible. This is to examine the existing network, identify the pinch violators and examine methods to remove them as simply and cheaply as possible. This has particular attractions in the present situation, where the ban on stream splitting means that any practical network will be substantially different to the MER network.

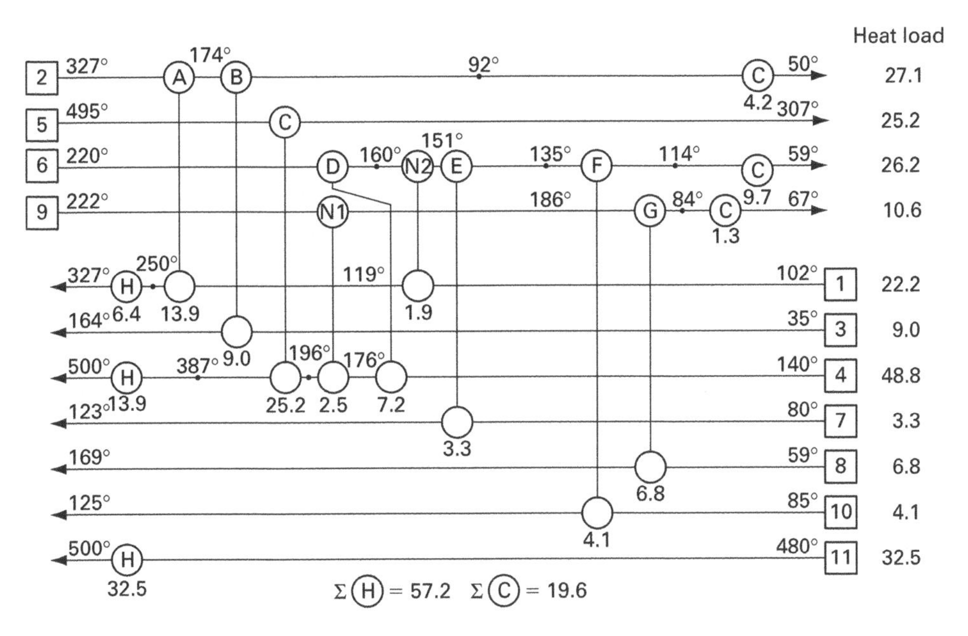

Figure 9.27 Network developed from identification of pinch violators

Examining the existing network with the knowledge of the pinch location shows us that there are four pinch violations, all due to heat exchangers transferring heat across the pinch. The culprits are:

Exchanger A – 4.0 ttc/h
Exchanger B – 0.5 ttc/h
Exchanger E – 1.7 ttc/h
Exchanger G – 4.5 ttc/h

It can be seen that the total, 10.7 ttc/h, is exactly equal to the amount by which the process was calculated to be off target in Section 9.3.3.

Knowing the pinch violators, we can now consider ways to eliminate the worst culprits, which are clearly A and G. A is heating stream 1 below the pinch (102°C to 140°C) so if we can find an alternative heat source for this stream, the spare heat from A can be absorbed above the pinch and reduce the load on heater H1. Conversely G is taking above-pinch heat from stream 9 to heat stream 8 below the pinch, so we need to find a cold stream above the pinch to use up this heat. The obvious choice is stream 4, as there is plenty of driving force in match C and some extra preheat can easily be fitted in below it. Returning to stream 1, we find the only remaining hot stream with some obvious spare heat below the pinch is stream 6 above exchanger F, although only a moderate-temperature drop is possible without causing a ΔT_{min} violation on F. Nevertheless, it is very simple to produce a network incorporating the two new matches as shown in Figure 9.27.

Comparing this network with that of Figure 9.25, we find that the two new matches predicted by both methods are between the same pairs of streams. The only significant difference between the networks is the order of the matches on streams 6 and 4. There is little to choose between them in energy terms and the decision can be made on other grounds, such as the physical location of the plants relative to each other, the change in exchanger duties and the amount of repiping needed. The network with both exchangers N1 and N2 recovers 0.7 ttc/h less than the Phase III network for no obvious gain in return. However, if only exchanger N1 is installed, the resulting network recovers the same energy as Phase II and has the major advantage that the heat duties and temperatures for exchangers D, E and F are completely unchanged. The balance of driving forces through the network is also better and as a result, the total UA value is 1.80 compared with 2.01 for the Phase II network recovering the same amount of energy. This means that, when the network is relaxed to avoid enlarging exchanger G, less additional heat will be required. This network would therefore have given slightly better energy savings than the actual Phase II installation, even though the only difference is the order of exchangers D and N1 on stream 4. However, other process reasons militated against it – the required pipe runs were longer, and since the new exchanger N1 was of a novel design (see next section), the lower temperature driving forces in it compared to Phase II were felt to be undesirable.

9.3.7 Practical process design considerations

Pinch analysis can provide the stimulus to explore new solutions or novel equipment. Two examples arose on this particular plant.

Firstly, the Phase II heat exchanger is of an unusual design which needed considerable design effort to ensure that it would perform as predicted. This duty consisted of the partial vaporisation of a two phase stream. The design adopted was a vertical rod baffle exchanger with vapour belt and shellside liquid injection nozzles. The constraints of plant layout forced the use of a vertical exchanger. Thereafter, it was necessary to consider pressure drop, vibration and liquid distribution, particularly slugging and their effect on the process design. The complexity of the resulting design reflects the lengths to which the designers had to go to in order to have confidence that the duty would be satisfactorily achieved.

In many circumstances, the design team, when faced with such a problem, might have chosen to overcome the problem by altering the heat exchanger match without a quantitative knowledge of how this would affect the energy recovery network. Targeting and network design using pinch techniques not only indicates the appropriate matches to achieve minimum energy but also quantify the energy penalty involved in pursuing other network configurations, giving the process designer a definite incentive to achieve the optimum network design.

The annual savings generated by installing the Phase II exchanger were of the order of £0.5 million p.a. and the payback achieved was around 6 months.

Secondly, we note that we require additional area on several exchangers, including B. The cold stream here is undergoing partial condensation and is then re-heated prior to becoming the vapour feed for the next column. The ICI engineers therefore asked – do we need to condense the entire stream? Obviously it is necessary to separate off some liquid, but any vapour present could be passed directly to the following column D1. To achieve this it would be necessary to separate the vapour from the liquid in stream 2. A special "knock-out pot" was designed for this, which was sufficiently novel to be patented. The result was that the flows through exchanger B and the heat load on it were greatly reduced, so that the existing area could handle the duty easily and indeed the ΔT_{min} on the match fell to practically zero. Technically, this was a process change as it altered the stream data. Its implementation further reduced the energy requirement to around 12% above the original target (though in fact it further reduced the targets). This change came to light as a direct result of the systematic and rigorous procedures necessitated by the network analysis. In fact, it gave considerably better economics than installing the Phase III heat exchanger, and was therefore implemented instead. The novelty of the application was such that it was supported financially under the UK Government's Energy Efficiency Demonstration Scheme and a monitoring report was published as EEDS Expanded Project Profile 204 (1988). The energy saving was 3 ttc/h, and a cost saving of £578,000 p.a. was achieved (1982 prices) at a payback of 11 months. The main capital cost, apart from installing the knock-out pot, was for enlarging exchanger A to handle a higher heat load at a lower temperature difference.

There were two additional benefits from this process change. Firstly, the start-up time for the plant was reduced by about 1 h, because of the lower flows needing to be heated. Secondly, the load on the air-cooled condenser on stream 2 was much reduced; not only did this save energy, but it also overcame maintenance problems due to the original condenser having to run on absolute maximum load in the summer.

9.3.8 Further considerations

The aromatics plant has been subject to a great deal of study in recent years and some different possibilities for networks have been advocated (though none take account of the process change described above). Ahmad and Linnhoff (1989) performed an economic evaluation on the plant and concluded that the optimal ΔT_{min} was close to the original estimate of 10°C, because of the very high mass flows and absolute energy costs. They then proceeded to design a network based on this ΔT_{min} and retaining the prohibition on stream splitting, and ended up with two new exchangers plus two new shells on existing exchangers (Figure 9.28). For example, a second shell was added to exchanger G to enable it to maintain its heat load at lower ΔT. A variant design, closer to MER, had four new exchangers. Although exchanger capital costs were high, energy cost savings were substantial.

Amidpour and Polley (1997) pointed out that a drawback with this design was that it involved a number of heat exchangers between streams which were a long way apart geographically. In fairness, this was also the case on the existing plant – notably for match X – but the problem was accentuated by the Ahmad and Linnhoff design. Moreover, it would have involved longer pipe runs, higher heat losses and higher pressure drops, and the additional capital cost of pipework and pumps would have been significant. The plant is thus a good candidate for a "zoning" approach as

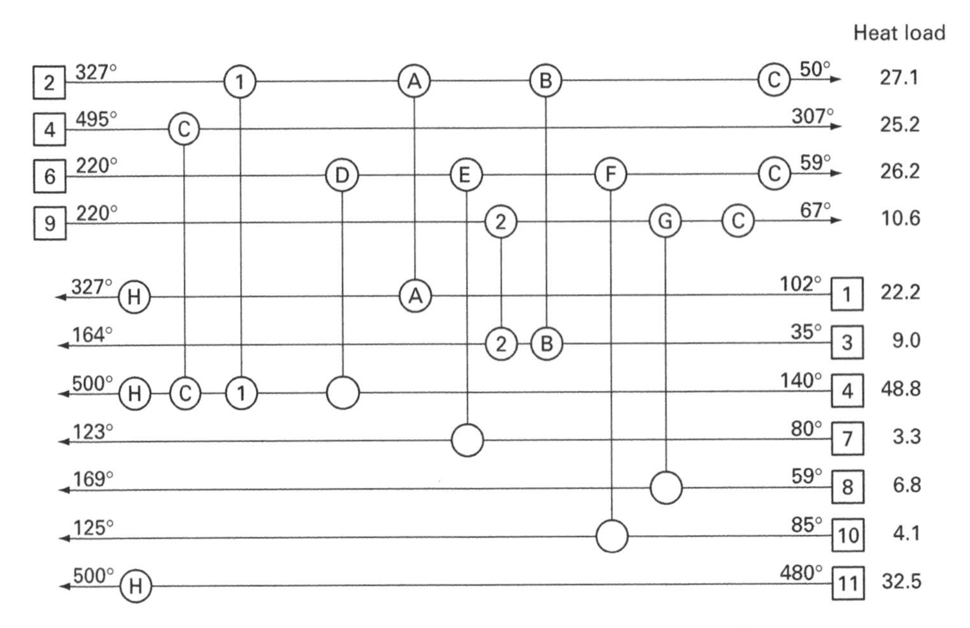

Figure 9.28 Alternative network design by Ahmad and Linnhoff (Temperatures and heat loads omitted for clarity)

described in Section 3.5.1. One obvious possibility is to separate the streams in the top half of the flowsheet, before D1 (streams 1–3) from those in the bottom half (streams 4–9). Targeting these areas separately gives hot utility targets of 5.8 (pinch 107°C) and 44.4 (pinch 215°C), a total of 50.2 ttc/h as against the combined value of 46.5. Therefore, combining the two sections together gives a target saving of only 3.7 ttc/h, which is less than the energy penalty we have already accepted by banning stream splits and relaxing the network. So there is no real incentive to transfer heat between the two separate zones. Looking back at the networks, the only exchanger which crossed between zones was the new unit between streams 6 and 3 which appeared in the MER design and Phase III but disappeared in Phase II. So the final proposed designs in Figure 9.26 and Figure 9.27 are consistent with zonal integrity. In contrast, the two new matches in the Ahmad and Linnhoff design both transferred heat between zones; 2 vs. 4 and 9 vs. 3.

Although grid diagrams do not give any information on the location of streams, the MER design could take account of physical location. In this plant, the streams are numbered in order and it may be expected that the stream 1 pipe run will be close to stream 2 but could be considerably further away from stream 9. This can affect the choice of matches, that is prefer to match streams with similar numbers. It can be seen that this rule holds with virtually all the streams which were matched in the original design, and indeed in many cases the feed and product for the same vessel are matched.

A final interesting implication is that, if zonal integrity was demanded from the beginning, the networks for each zone would be designed separately starting from

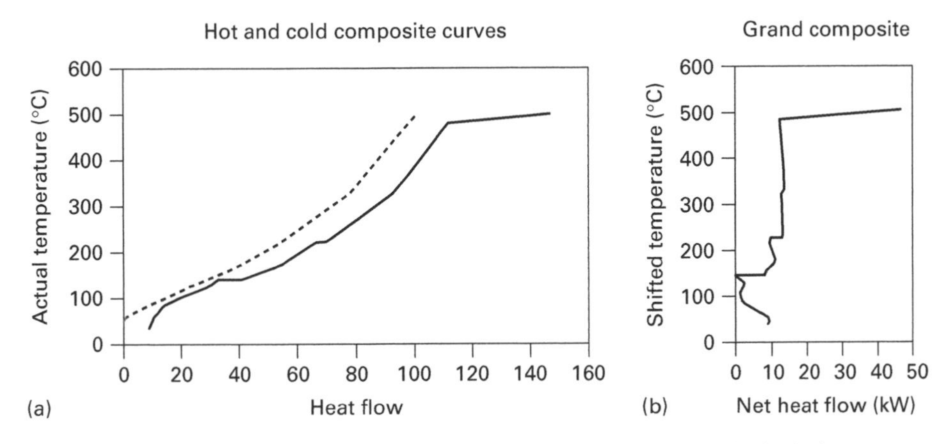

Figure 9.29 Composite curve and GCCs with additional streams included

the pinch temperature of the zone rather than the overall pinch. This would change the MER network structure, especially for zone 2 (streams 4–11) (see exercises, Section 9.8).

9.3.9 Targeting and design with alternative stream data

It was mentioned earlier that a different choice of streams during data extraction would have given different targets and networks. We will now see what the effect would have been. The heat loads and temperatures quoted below are best estimates, as full information is not available for this section of the plant.

The "missing" portion of streams 5 and 6 between 307°C and 220°C had a heat load of 11.2 ttc/h (corresponding to a CP of 0.1287). This was matched with two reboiling streams, D1 at 140°C (heat load 8.0) and D2 at 222°C (heat load 3.2). If we add these streams into the analysis and retarget, the energy target and pinch temperature are unchanged (all these streams are above the pinch). However, the shape of the composite curve and GCCs change as the driving forces are opened out above the pinch; see Figure 9.29. Alternative above-pinch network designs can be generated such as the one shown in Figure 9.30. The extra flexibility has allowed a significantly simpler network design to be generated with no $\Delta T_{\min}$ violations, while maintaining zonal integrity. There is no requirement for a split on stream 4; the main change is that half of the X match has moved down to below exchanger D, as has the new match between 9 and 4. The overall network resulting after relaxation (Figure 9.31) is subtly different in structure from the Phase II design. Driving forces are significantly better, and the UA value has now fallen to 1.53 (although this does not allow for the fact that exchanger X1 will also have to be enlarged as the driving force across it has dropped). In particular, the driving force across exchanger D is much higher and it could recover even more heat than shown. In a new plant design this option would be worth serious consideration, although the final choice between the two forms of the X match would also be

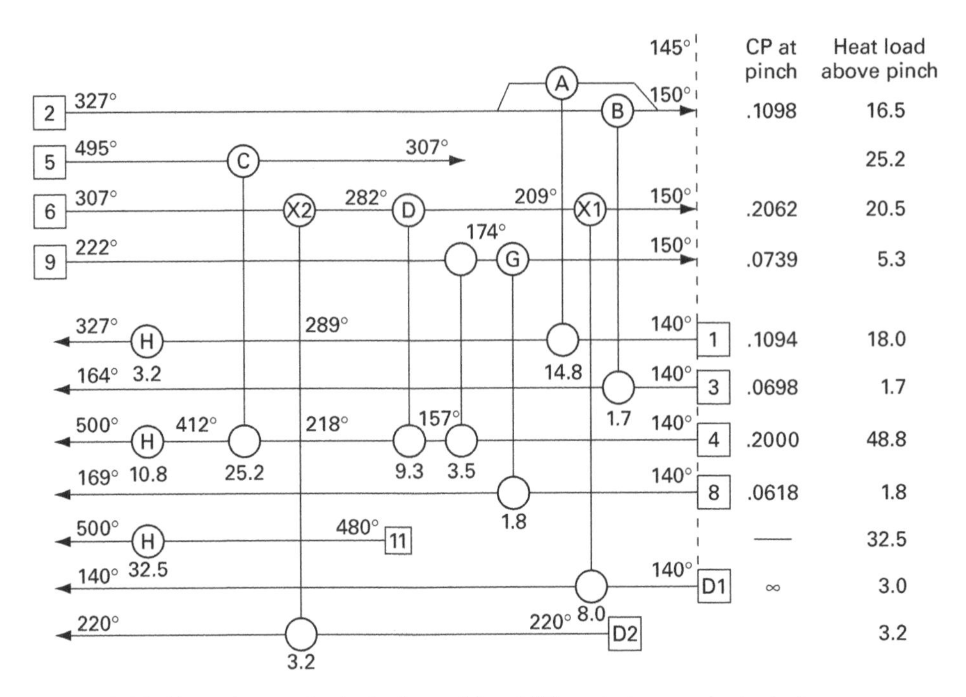

Figure 9.30 New above-pinch design with additional streams included

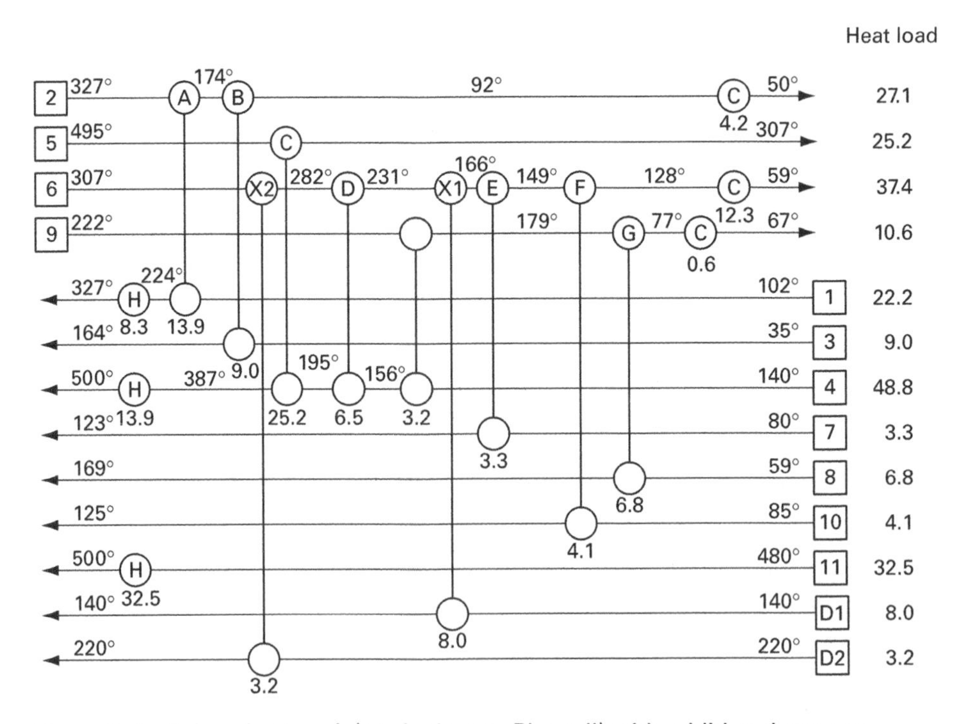

Figure 9.31 Relaxed network (equivalent to Phase II) with additional streams

made on the basis of vessel location, piping costs, etc. For the retrofit system, however, X will be left as it is, because there is not sufficient incentive to change it.

If the non-isothermal mixing is removed, stream 7's target temperature rises to 140°C and its heat load by 1.3 ttc/h to 4.6; the heat load on stream 4 falls correspondingly to 47.5 and the CP of the first portion falls to 0.1639. This does affect the targets, as a cold stream which was above the pinch has been moved below the pinch, which will give a gain in heat recovery and a fall in utility targets (remember the plus–minus principle!). The new hot utility use is 45.3 ttc/h, a saving of 1.2. However, the pinch region has become even tighter; the pinch has moved to 107°C and there are virtually two other pinches at 145° and 215°, with heat flows less than 0.2. Hence, getting this energy recovery in practice is very difficult; four cold streams now exist below the 145°C near-pinch instead of three, so that a stream split is definitely required to recover the extra energy. So it is not surprising, when we look at the evolved networks without stream splits in Figures 9.26 and 9.27, that changing the mixing temperature gives no energy saving and no change in basic network design. The only effect is to shift some of the heat load from exchanger D to exchanger E, which may slightly reduce the total heat exchanger area required by giving better temperature driving forces.

9.3.10 Conclusions

Overall, the phased energy saving plan, with the "relaxed" Phase II and the process change installed, saved over £1 million p.a. at a payback of well under a year.

Later developments in techniques have shown that some of the data extraction and targeting could be refined. This can significantly change the apparently best network design for a new plant. Even simple actions like changing the order of two exchangers on a stream can give a significant gain in energy recovery. However, it is a tribute to the original ICI engineers that over 20 years of intense academic research on this case study have not revealed a retrofit design significantly better in practice than the one which they eventually installed.

9.4 Evaporator/dryer plant

This study relates to a survey carried out in a food processing factory. After first stage processing, the product was fed in dilute aqueous solution to an evaporation and drying section. The latter section contained modern, continuous operation equipment, but the first-stage processing comprised old and inefficient batch equipment. It had been decided to rebuild completely the "front end" of the process, and so the opportunity was taken to perform an energy utilisation study to see whether efficiency improvements could be made in the total system, with any hardware changes to be made whilst the plant was shutdown.

As in the two preceding cases, this study was originally described in the first edition of this User Guide. However, for this process, newer methods for both the stream data extraction and the analysis of process change mean that the stages of analysis are quite different, although the conclusions are broadly the same.

9.4.1 Process description

The continuous operation evaporator/dryer section is shown in Figure 9.32. Crude feedstock is extracted by adding hot water. The extract solution (4% total solids) is fed into the continuously-operated evaporator/dryer plant. The evaporator uses flash tanks rather than the more common calandria in a vessel; this has some potential advantages in reducing fouling on heating surfaces, especially in the later stages when solids content of the slurry is high. It has three stages, the first two operating on the "multiple effect" principle. Flash vapour from the first stage drives the second stage. Also, some vapour from the first stage is recompressed by a steam ejector for re-use in driving the first stage (heat pumping principle). The steam ejector and the third stage are driven by utility steam (9.29 bar, saturated, 177°C). The flash vapour from the second and third stages is condensed against cooling water in a vacuum condenser. The evaporator produces liquor of 30% total solids which is fed to the dryer. The dryer operates with a continuous belt, carrying the product through seven temperature zones ranging from 30°C to 60°C. Internal drying tunnel air circulation is maintained by four fans, and heat input is via six steam radiators (using 6.53 bar utility steam, saturated at 162°C). Wet air from the dryer is vented at the midpoint (41°C) and from the end of the drying tunnel (60°C), but depending on ambient conditions, a controlled amount of the hottest wet air is recycled. This, together with fresh air make-up is dehumidified by contact with lithium chloride solution down to a humidity of 0.005 kg/kg at a controlled temperature of 30°C. The lithium chloride solution is regenerated by spraying over steam heating coils in a "return air" (55°C) stream.

9.4.2 Stream data extraction

The hot and cold stream data extracted from the flowsheet in Figure 9.32 are given in Table 9.8, divided into the two broad areas of the evaporator and the dryer.

The bottom three streams are optional, as explained below. Discounting their heat load of 478 kW, the total heat load on cold streams is 2,771 kW and on hot streams is 2,302 kW.

In the evaporator area, only those vapour loads which arise directly because of process duties are included. This means that at this (energy targeting) stage, we ignore the steam ejector (which is a heat pump), and stick to analysis of the basic process. The effect of heat pumping is only assessed after the process GCC has been established (see Section 5.3.1).

The various latent heat loads are handled as suggested in Section 3.1.3, by taking the initial temperature and setting the target temperature 0.1°C different, thus giving a very large CP.

Difficult questions arise in several areas:

(a) What should we do about the direct steam injection? It would be possible to simply ignore this and treat it as an unavoidable process stream, as was done in the original analysis. However, these flows do provide heat to the plant and this

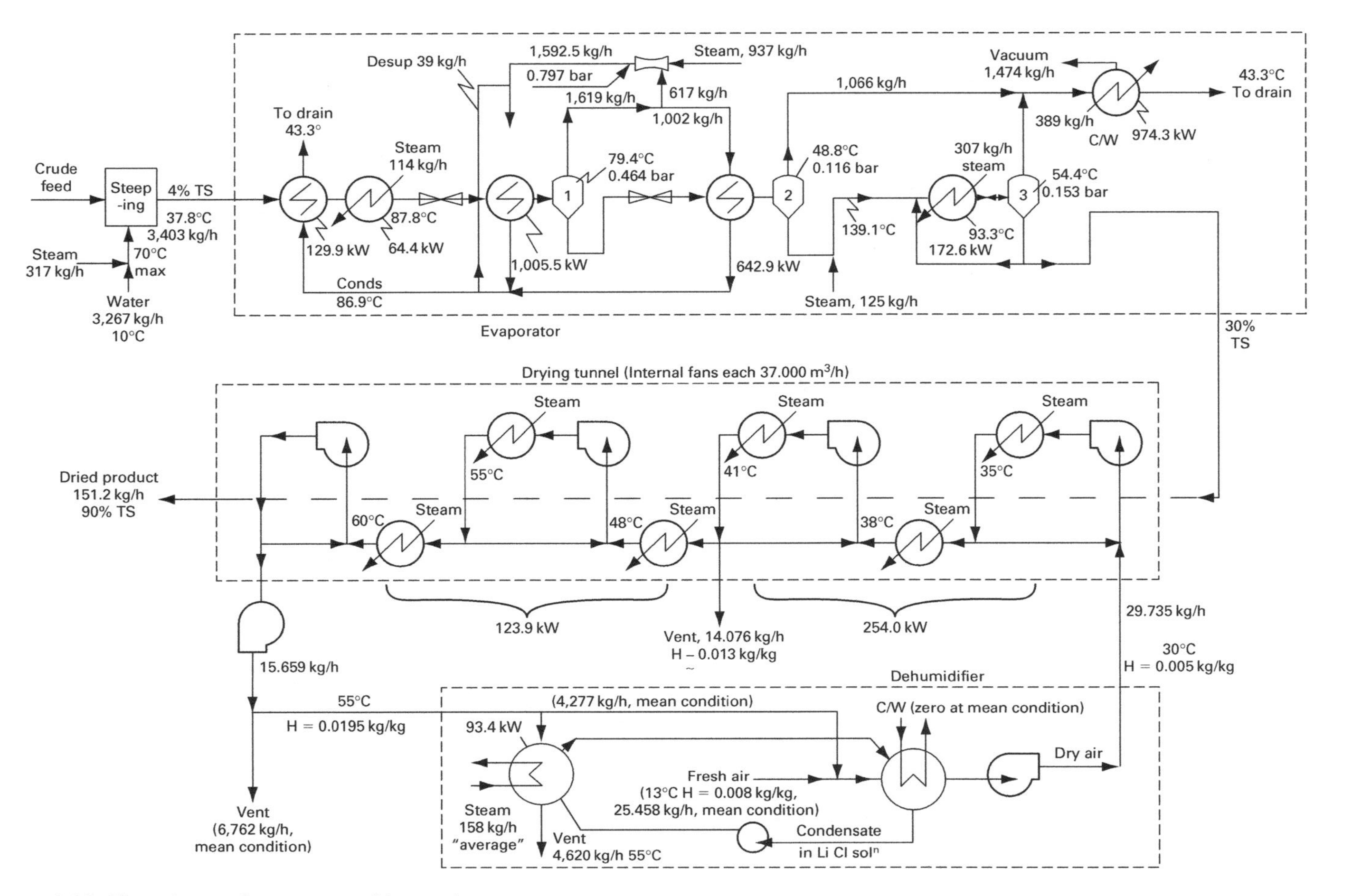

Figure 9.32 Flowsheet of evaporator/dryer plant

Table 9.8 Stream data for evaporator/dryer plant

Stream name	*Type*	*T-in (°C)*	*T-out (°C)*	*H (kW)*	*CP (kW/K)*
Evaporator area					
Process water	C	10	70	−183	−3.05
Feed	C	37.8	87.8	−198	−3.98
Flash 1 heating	C	79.4	79.5	−1005.5	−10,055
Flash 1 vapour	H	79.4	79.3	1039	10,390
Condensate 1	H	86.9	10	232	3.017
Flash 2 heating	C	48.8	48.9	−643	−6,430
Flash 2 vapour	H	43.3	43.2	714	7,140
Flash 3 heating (i)	C	48.8	(54.4)	−6.5	−1.169
Flash 3 heating (ii)	C	(54.4)	93.3	−263.5	−6.772
Flash 3 vapour	H	43.3	43.2	260	2,600
Condensate 2/3	H	43.3	10	57	1.714
Dryer area					
Regenerator	C	55	55.1	−93.5	−935
Zone 1 dryer heating	C	41	41.1	−254	−2,540
Zone 2 dryer heating	C	60	60.1	−124	−1,240
Dryer vent 1	(H)	41	13	149	5.307
Dryer vent 2	(H)	60	13	140	2.977
Regenerator vent	(H)	55	13	189	4.500

heating could be inappropriately located, so if possible they should be included in the analysis. This would be true even if it were essential for the process mass balance for the additional water to be added. In fact, since the steam is being injected into an evaporator system whose whole purpose is to remove excess water from the slurry, it would seem appropriate to avoid steam injection as far as possible!

(b) How do we handle the flashes? The feed to flash tank 1 is heated to 87.8°C and then passed through a throttle valve. The following steam heater is evaporating liquid at 79.4°C, which becomes a cold stream at constant temperature, and tank 1 simply acts as a separator. The same is true of flash 2. However, in flash 3 the throttle valve comes after the heater, so the cold stream is a sensible heat stream over a range of temperatures, and vapour flashes off from the hot liquid after it passes through the throttle valve, giving a corresponding drop in temperature. Whether these flashes are appropriately positioned with respect to the pinch can be considered later by process change analysis.

(c) How do we deal with mixing streams? We need to avoid placing unnecessary constraints on the system. Consider the bottom product from flash tank 2. This is heated by direct steam injection from 48.8°C to 139.1°C, then mixed with a much larger recycle flow of bottoms liquid from Flash 3 at 54.4°C, giving a mixed temperature of 68°C; finally this mixture is heated to 93.3°C by steam. For a start, the non-isothermal mixing could be avoided and the liquid at 139.1°C could be treated as a hot stream being run down to 93.3°C. But in fact the constraints on the system are far less than this; we simply need to provide a given flow of liquid as feed to Flash 3 at 93.3°C, of which some comes from Flash 2 bottoms at 48.8°C and most from Flash 3 bottoms at 54.4°C. In other words, a single cold stream can be used in the analysis, with a fairly low CP between 48.8°C and 54.4°C and a much higher one between 54.4°C and 93.3°C. The total heat load on the stream is equal to the heat supplied in the existing flowsheet by the steam injection (125 kg/h, 97 kW at 2,600 kJ/kg) and the indirect steam heater (173 kW), giving a total of 270 kW. If the steam injection can be partly or totally abolished, the water content of the streams will drop and this should reduce the evaporation load either on Flash 3 or the following dryer section, but this can be covered by iteration in a second-stage analysis.

(d) What do we do about the condensate which is run off to drain? This water is at 43.3°C and still contains heat which could be potentially recovered. For the purposes of this analysis, we will treat condensate as additional hot streams, with target temperatures of 10°C. Bear in mind that, if these streams are not needed for heat recovery, it is not necessary to put a cooler on them; so the cold utility load will be lower than shown. The choice of whether or not to include these streams is fairly arbitrary; the alternative is to leave them out at this stage and include them as a process change later.

(e) Likewise, the warm wet air from the dryer is vented to atmosphere. There is no environmental problem with this, but we should note that heat could potentially be recovered from these streams. Since they are at relatively low temperatures and heat recovery from them would be difficult, we will leave them out of the analysis for the time being although they have been noted in the stream data. If the pinch temperature turns out to be sufficiently low that heat recovery from one or more of these streams could be worthwhile, we can introduce them later as a process change.

(f) The condensate from flash tanks 2 and 3 emerges at 43.3°C, but the steam was generated from these flash tanks at 48.8°C and 54.4°C respectively. It would appear that some throttling has taken place in the vapour lines. For the present we will assume that the steam condenses at the lower temperature of 43.3°C, which will give us a conservative estimate for our targets, but we can bear in mind that improvements in the vapour lines might allow us to recover some of the latent heat from this steam up to 11°C hotter.

It may be noted that the analysis in the original IChemE Guide ignored the direct steam injection streams and treated the feed to Flash 3 as a single heat load at 93.3°C, which is higher than necessary and could potentially conceal possibilities for heat recovery.

In the dryer area, the six drying loads have been combined into two to simplify the analysis. We see later that this has no effect on the result of the analysis. Also, the heating duties in the two zones are assumed to take place at the highest-air temperature leaving each zone. This is because the small air temperature rise which occurs across each heater is unlikely to be exploitable by use of a counter-current heat exchanger. The load on the regenerator is the yearly average load. When it comes to the revamp design, we must remember that the regenerator load can rise to as much as 317 kW on hot and humid days.

9.4.3 Energy targeting

The ΔT_{min} for the existing process is 5.5°C on the heat exchanger between the condensate and the feed stream. However, this ΔT_{min} may be over-optimistic for some matches. In particular, any match involving the gas and vapour streams in the dryer area will probably require a much larger ΔT_{min} because of the low heat transfer coefficients on the gas side. A larger ΔT_{min} contribution could be assigned to these streams, say 15°C.

Adding up all the current steam duties on the plant excluding heat pumping in the ejector, we find 280 kW from direct steam injection, 1887.5 kW from indirect steam heaters and 471.5 kW from the dryer zone heaters, giving a total of 2,639 kW. Of this, 1,039 kW can be attributed to the steam from flash tank 1, so net hot utility is 1,600 kW. Likewise, cooling in condensers accounts for 974 kW and rundown of condensate to drain is another 157 kW, so total cold utility is 1,131 kW. Comparing this with the total hot and cold stream loads deduced from Table 9.8, we find that heat recovery is currently 1,171 kW (comparing results from both hot and cold streams gives us a useful cross-check on our data).

The composite curves are shown in Figure 9.33. It is instantly obvious that we have a very different situation to that which pertained on the crude preheat train and aromatics plant. The pinches and near-pinches are very localised and temperature driving forces elsewhere are high.

Calculating the Problem Table (with a ΔT_{min} of 5.5°C) and plotting the results as a GCC yields the graph shown in Figure 9.34. The net hot utility requirement is 1,517 kW, *without* any heat pumping. This compares with a net heat requirement for the existing design of 1,600 kW (remembering that heat in the utility steam condensate and the latent heat of Flash 1 vapour is utilised). Inspection of the network in the light of pinch principles shows that the difference of 83 kW comes from heating the process water below the pinch by utility steam. The sharp pinch suggests that heat pumping could give substantial further energy savings. The pinch is caused jointly by the entry of the cold feed stream at 37.8°C and the vapour from flashes 2 and 3 at 43.3°C.

9.4.4 Heat pumping strategy

The first point to note about the GCC shown in Figure 9.34 is that it is dominated by the large latent and "pseudo"-latent heat changes in the system. The pinch is very sharp and this suggests that heat pumping may be possible, for example, by mechanical or thermal vapour recompression across each evaporator effect. (Section 5.3.1 describes the different types of heat pump). For heat recovery, the only directly usable latent heat source is Flash Vapour 1. Flash Vapour 2 occurs below the pinch. Hence, to save more energy, heat from Flash Vapour 2 must be "lifted" above the pinch.

However, there is a difficulty. Only 347 kW can be pumped across the pinch at minimum temperature lift. Further saving can only result from heat pumped to a level hot enough to supply Flash 1. This represents a temperature rise of at least 40°C

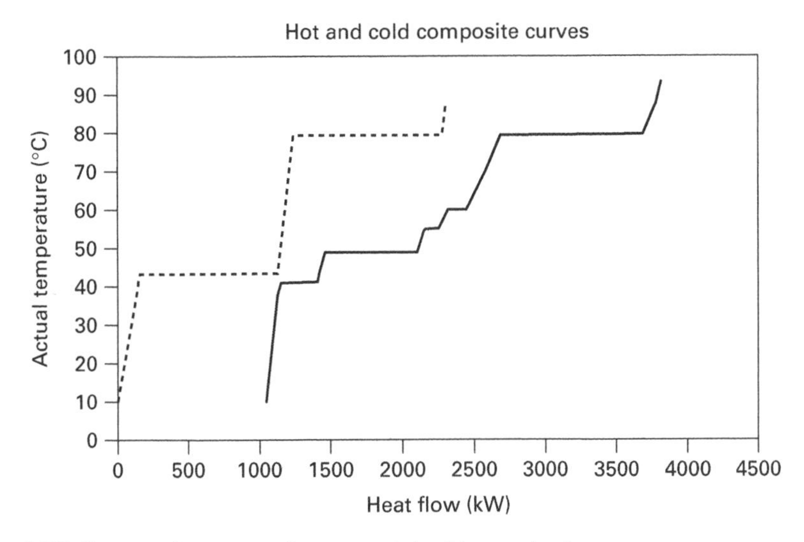

Figure 9.33 Composite curves for evaporator/dryer plant

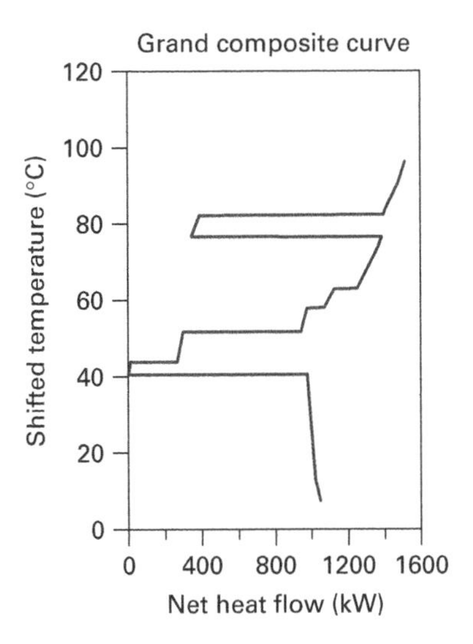

Figure 9.34 GCC for evaporator/dryer

(probably higher, allowing for the ΔT on the evaporator and condenser) which is rather high for a heat pump. The situation can be improved by making use of the pocket and adopting a *two-stage* heat pumping strategy. Heat from the Flash Vapour 1 source is pumped to supply Flash 1, leaving a "hole" for hot utility supply at a lower level. Thus if a quantity of heat X is pumped out of Flash Vapour 1, then the scope for heat pumping from Flash Vapour 2 at minimum temperature lift (i.e. into Zone 1 and Flash 2) becomes $(347 + X)$ kW. The limit on X will come when one of the source or destination streams is used up; as can be seen from Figure 9.34, the required temperature lift increases sharply when Zone 1 and Flash 2 (and small associated sensible heat streams) have been fully satisfied. Heat pumping potential is roughly 900 kW of upgraded heat, and $X \sim 550$ kW.

In fact, heat pumping from Flash Vapour 1 up to supply Flash 1 is already carried out in the existing design by means of a steam ejector. As can be seen from Figure 9.34, it is not placed across the pinch and therefore does not actually save any energy at present that could not be saved by conventional heat recovery methods! The benefits are that it pulls the vacuum on Flash 1 and upgrades some heat, giving it a wider range of possible applications. Thus some of Flash Vapour 1, after upgrading, can perform the Flash Heating 1 duty; otherwise, it would have had to be transferred to the dryer section and used there. In the existing plant, the steam ejector only takes 38% of the heat available in Flash Vapour 1 (390 kW out of 1,039 kW) because thermocompressors require a large amount of "driver" high pressure steam to upgrade the low-pressure vapour; that is $X = 390$. We can therefore upgrade 737 kW from Flash Vapour 2 at minimum temperature lift. This heat would be used for Dryer Zone 1 and Flash Heating 2. Since we have to use the full amount of driver steam in the thermocompressor, the heat loads within the pocket must be met by heat exchange to achieve the targets. Hence, the regenerator, Zone 2 and the remainder of Flash Heating 2 must be heated by Flash Vapour 1 instead of site steam as at present.

If a mechanical compressor were adopted instead of the ejector, then the potential for energy saving becomes greater because more heat can be pumped from Flash Vapour 1 (over 1,000 kW potentially) and a correspondingly larger amount can be pumped from Flash Vapour 2 at minimum temperature lift. In this case, the limit will be the heat available from Flash Vapours 2 and 3 which is 974 kW. The additional saving looks attractive, but we should remember that in addition to the capital cost, we are now using expensive power to perform the heat pumping rather than cheap driver steam.

Both these schemes give worthwhile savings and could reduce the total energy consumption of the plant by up to 60%! Before we decide on a heat pumping strategy, however, we should remember that there are other ways of changing the temperature of usable heat sources, as described in Chapter 6. So we should look at process change analysis and see if this gives us some alternatives.

9.4.5 Process change analysis

The obvious question in an evaporator system dominated by latent heat loads is whether the effects are well placed in relation to the background process. Therefore, we split the GCC and take out the three evaporator effects. Effects 1 and 2 plot as rectangular boxes whereas effect 3, where sensible heat is taken in and the vapour is generated by flashing, plots as a slightly distorted triangle. There is also a small unbalanced heat load for both effects 1 and 2, which is left in the background process.

The split GCC is shown in Figure 9.35. Obviously effects 2 and 3 are working across the pinch. However, we see that there is a large pocket immediately above

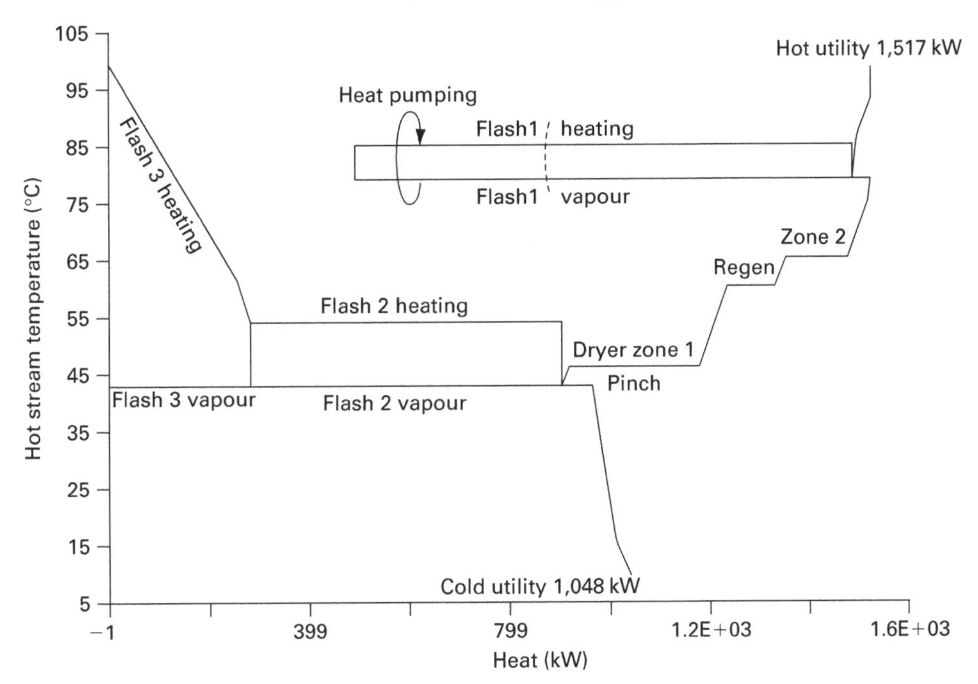

Figure 9.35 Split GCC – existing process

the pinch, and a relatively small increase in temperature would allow the effects to fit partly above the GCC. So there is a definite incentive to explore process change options.

If Flash 3 vapour could be condensed at a slightly higher temperature, it could supply the lowest of the heat demands, Dryer Zone 1. Remembering that Flash Tank 3 was at 54.4°C, if we can remove the pressure drops in pipework so that the vapour actually condenses at this temperature, it can fulfil the whole of this duty. This would save 260 kW of heat, as shown in Figure 9.36.

However, the practical implementation of this raises a problem. Currently, the Zone 1 duty at 41°C is fulfilled by steam heating at 162°C, giving a ΔT of 121°C. This would fall to at most 13°C if the heat came from Flash 3 vapour, so that nearly a ten-fold increase in heat exchanger area would be required! Again we have to remember that these are air heating streams and the heat transfer coefficients will not be as good as for liquid–liquid exchangers, so the additional capital cost could be alarmingly high. (We could have allowed for this in targeting by giving these streams a higher ΔT_{min} contribution.)

Can we go further? We could try to increase the temperature of effect 2 to fit it into the pocket. To do this, we would need to increase its pressure – often expensive. But in this case the entire evaporator train is under vacuum, so all we need to do is reduce the vacuum pulled. Indeed, since we want to maximise driving forces,

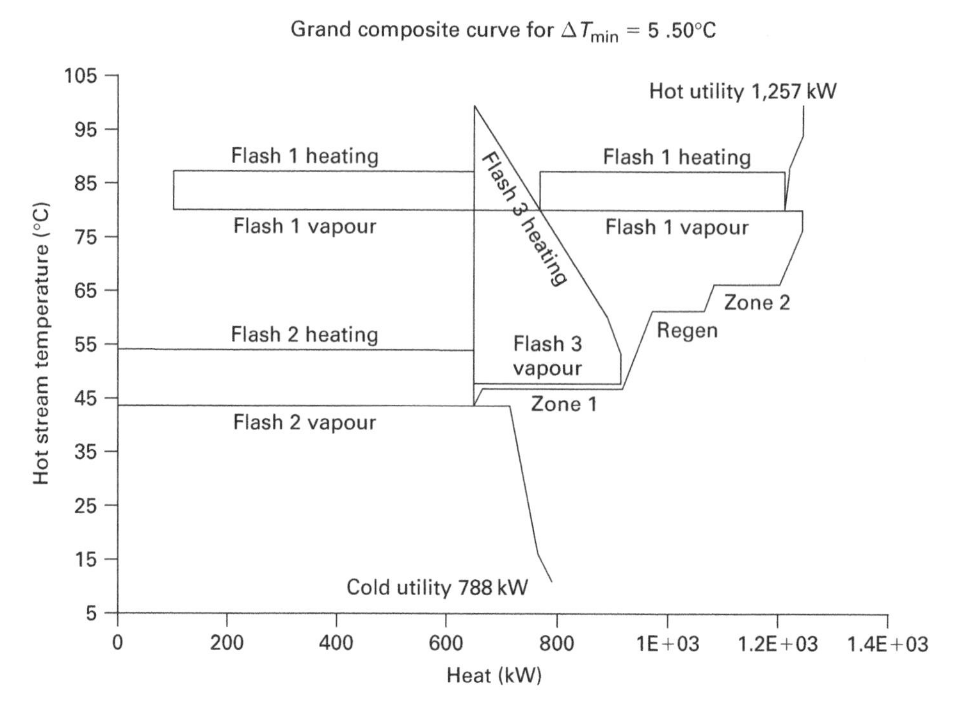

Figure 9.36 Split GCC with Flash 3 repositioned

we can also increase the temperature of effect 1, and can indeed reduce the vacuum to zero so that this tank is at 100°C. This has the further advantage that effect 1 is now hot enough to heat effect 3 – the driving force is below 7°C but this does not matter so much on this match with a condensing stream on one side, a liquid stream on the other and relatively high heat transfer coefficients. In fact, driving forces could be improved by increasing the recirculation flow around Stage 3 (the recirculation is present for outlet concentration control purposes). This is because the CP of the recirculating stream is increased, so less temperature drop is needed to obtain the same heat load and the same amount of flashed steam.

The resulting GCC is shown in Figure 9.37. We have now saved most of the Effect 2 heat load – about 500 kW. Moreover, the pinch has moved so the existing heat pump is now in the right place and is giving an additional saving of 120 kW! (To keep driving forces high and reduce the additional heat exchanger area, a ΔT_{min} of 14°C was selected.)

In this simplified analysis, the unbalanced heat loads from the original effects have been left at their original temperatures; moreover, the effect temperatures are different so the condensate temperatures and sensible heating/cooling loads have also changed. Therefore, it is essential to correct the stream data and recalculate the targets for the final analysis stage. The advantage of the split GCC is that it has allowed us to rapidly screen a number of promising alternatives without having to do tedious recalculations at each step.

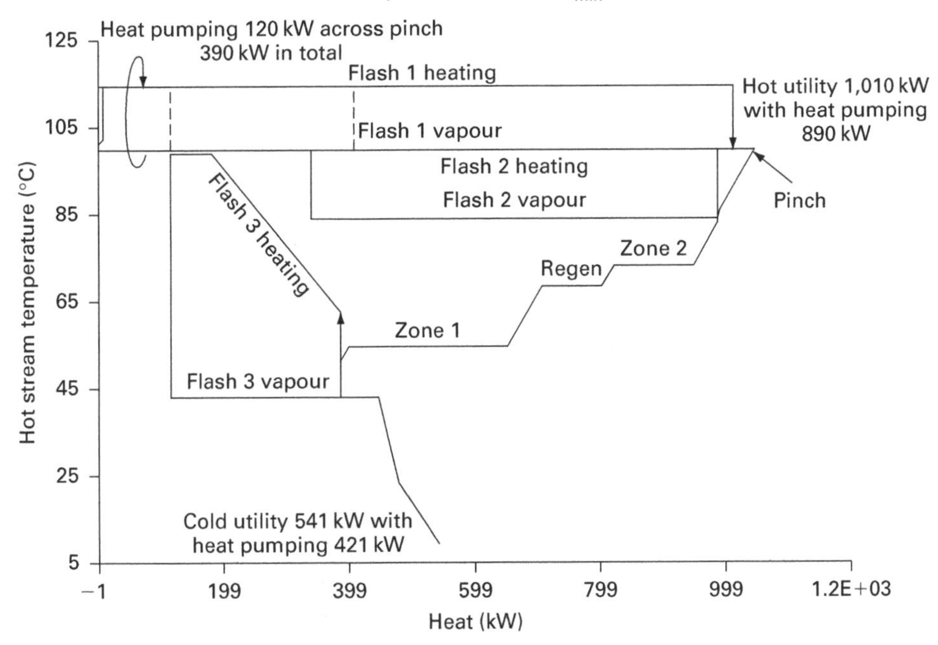

Figure 9.37 Split GCC with effects 1 and 2 repositioned

The other way of identifying this opportunity would have been to look at the composite curves (Figure 9.33) and to note that if the hot composite could be "lifted" a little, it could move to the right and the horizontal sections could sit more on top of the cold composite. The user can choose whichever of the composite curves, basic GCC and split GCC is most helpful to help him visualise the situation.

9.4.6 Selection of final scheme layout

We now have at least four schemes to choose from:

(a) Install heat pumping from Flash Vapours 2 and 3 to give low-pressure steam suitable for Flash Heating 2 and some dryer duties; use heat from Flash Vapour 1 for other dryer duties. Saving: up to 737 kW.
(b) As above, and replace the thermocompressor by a mechanical vapour recompressor to upgrade more heat. Saving: 974 kW.
(c) Increase the temperature of Flash 3 vapour slightly and use it to heat Dryer Zone 1, increasing the heater surface area substantially. Saving: 260 kW.
(d) Increase the pressure of the whole evaporator train and use heat from Flash Vapour 2 to heat the dryer. Saving: approximately 630 kW. (Could also be done in conjunction with scheme (c)).

A rough economic evaluation showed that scheme (d) was considerably best. The slightly lower savings compared to (a) and (b) were completely outweighed by the avoidance of expensive heat pumping equipment. The only heat pump required was the one already installed – the ejector on Flash 1. To keep driving forces high and reduce the additional heat exchanger area, a ΔT_{min} of 14°C was selected.

The final scheme adopted is shown in Figure 9.38. The changes compared with the existing layout are as follows:

(a) Flash steam 2 is piped to the dryer and used to heat Zones 1 and 2 and the regenerator, replacing utility steam.
(b) Flash 2 steam and its condensate is used to heat the process water, replacing utility steam (including the below-pinch heating).
(c) Part of flash steam 1 is now used to heat flash 3. The recycle rate has been increased so that the temperature before the flash is now only 86°C. The direct steam injection has been reduced, but not eliminated as it is still useful for sterilisation.

The stream data requires recalculation and the final energy use comes out at approximately 850 kW. This is 667 kW less than the original target and 750 kW below the current energy use of 1,600 kW, so the overall saving is no less than 47%.

The total evaporation across Stages 1 and 2 was held at 2,685 kg/h as before, so as not to change the solution concentration in Stage 3. This was felt desirable because of the temperature sensitivity of the product in concentrated solutions (another argument against increasing the temperature of Flash 3). The quantity of 0.6 bar steam available from Stage 2 for Dryer and Regenerator duty is 945 kg/h, which is 205 kg/h in excess of requirement (equivalent to 130.6 kW of "spare" heat at 86°C). However, this is only at the *average* yearly condition. On the hottest, most humid summer days, this surplus becomes a 186.4 kW deficit, which would then have to be made up from utility steam supply let down to 0.6 bar.

The scheme in Figure 9.38 only takes care of the latent (and "pseudo" latent) heat changes, based on the heat pumping analysis described previously. We must now check that the sensible heat sources and sinks are optimally interchanged. The stream data for this "residual" stream set are given in the diagram in Figure 9.39, along with the MER design that achieves the target predicted for this set. Notice that we have included the 130.6 kW of "spare" 0.6 bar steam from Stage 2, and the stream of process condensate ("Condensate 3") which is available at 86°C (previously, this was available as part of "Condensate 1" at 43.4°C). In the MER design, we have deliberately avoided matches to the hot Dryer Vent streams since such matches would be expensive and difficult to implement. The consequence of this decision is a small ΔT_{min} violation at the cold end of match 4.

Match 2 is insignificantly small, and so a sensible evolved version of the MER design is shown in Figure 9.40, with match 2 eliminated and match 4 merged into match 1. This causes the ΔT_{min} violation to become worse. However a simple and acceptable structure has resulted. Match 1 already exists in the plant (see Figure 9.32), and is simply required to carry a higher load (211 kW rather than 130 kW). Since all exchangers in the existing system are of the plate-frame type, the increase in size can be implemented cheaply and easily by adding further plates. Since the

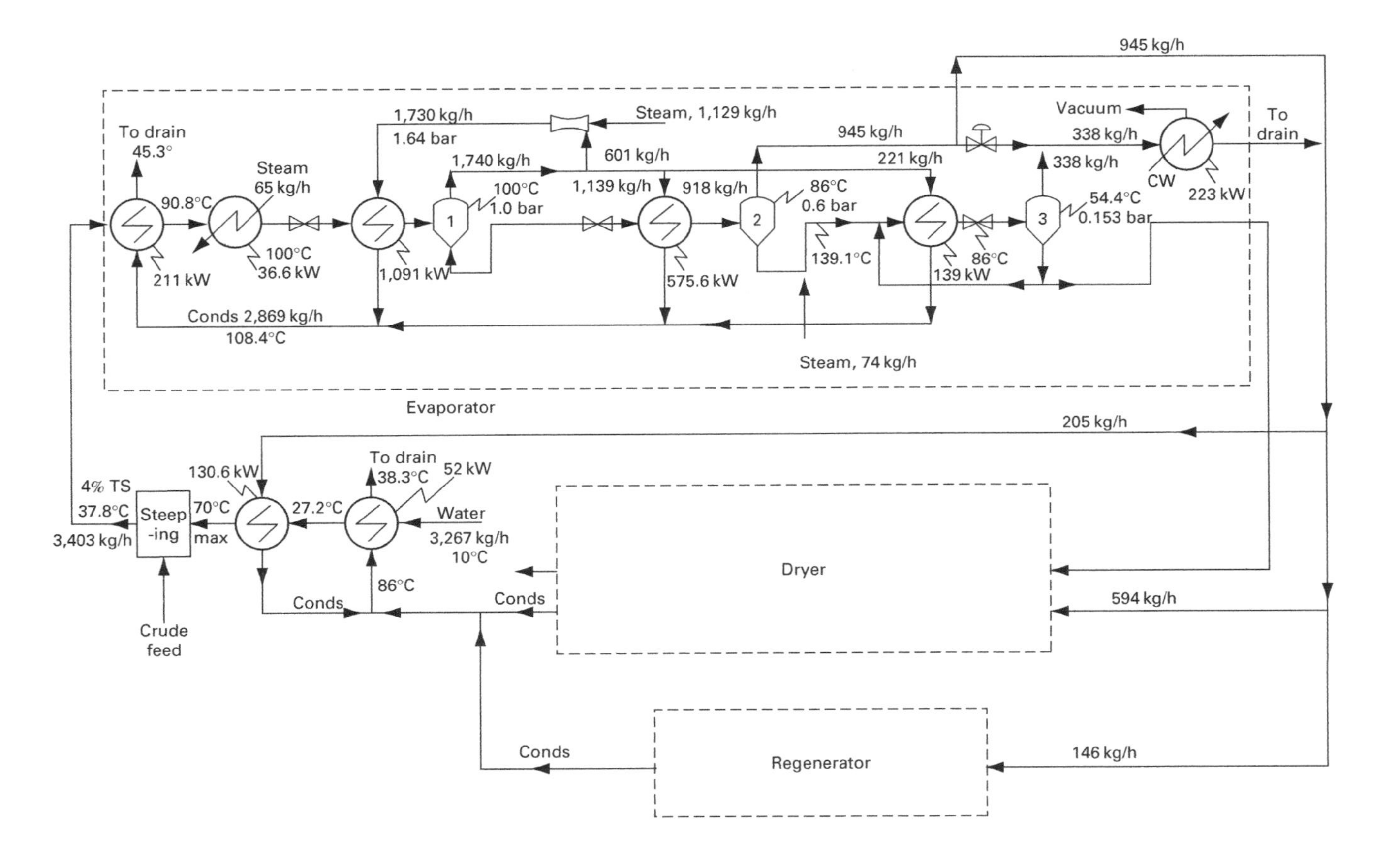

Figure 9.38 Final layout of modified plant

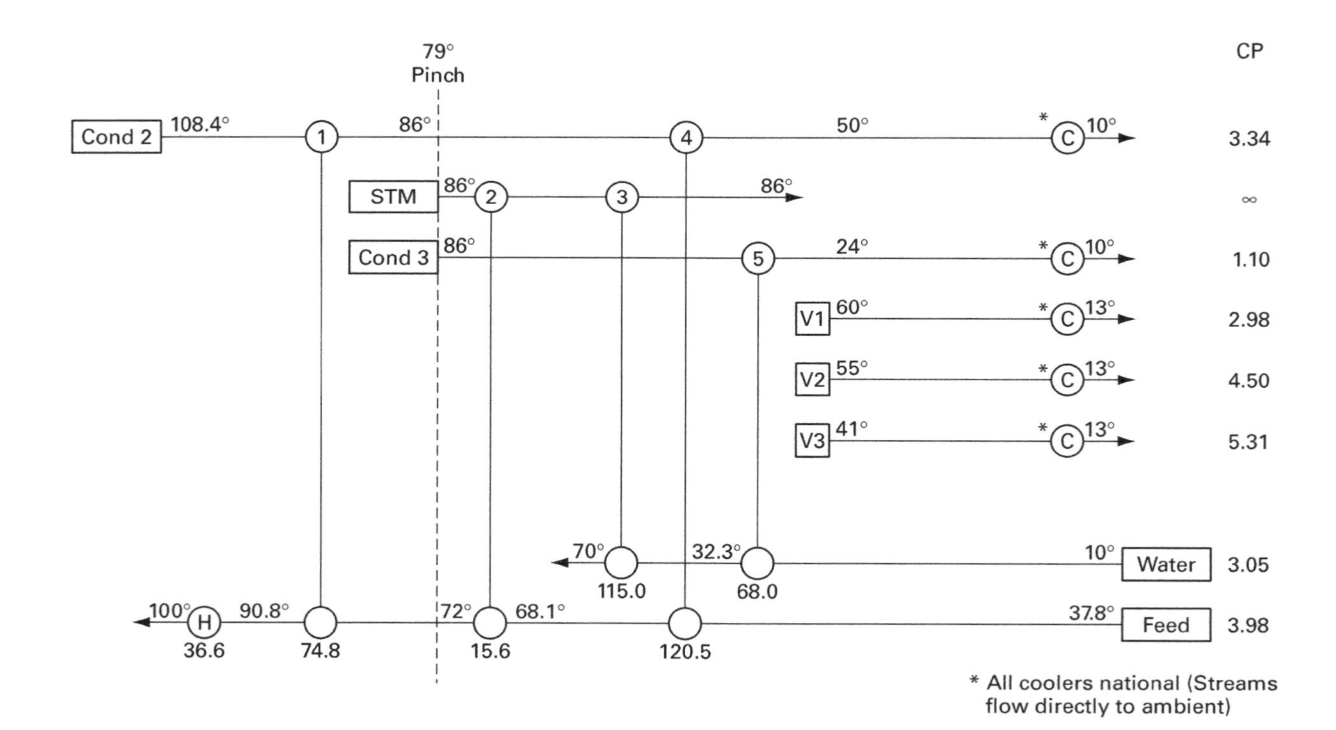

Figure 9.39 MER network design for sensible heat and residual streams

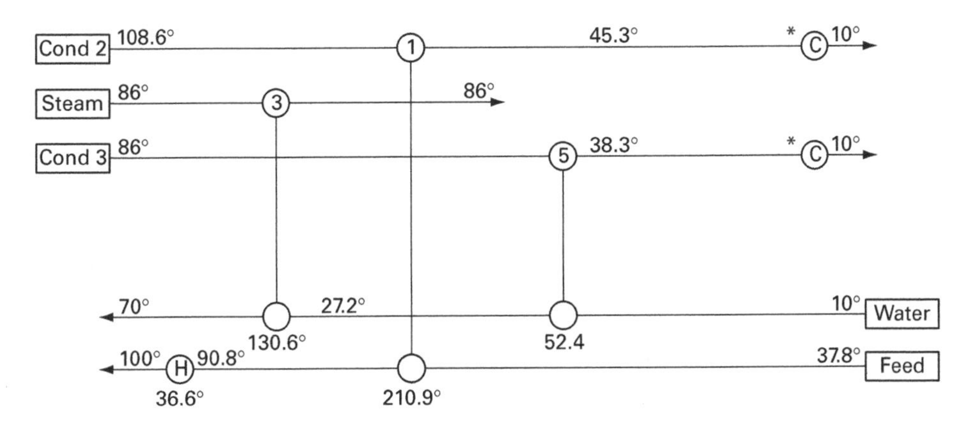

Figure 9.40 Network design for revised system

other two matches are both to the cold water stream, they can be implemented in a single new frame. The residual utility steam heating duty on the feed stream could be incorporated into the Stage 1 Flash Heater, thereby saving about 13 kW of utility steam (due to the heat pump). This is a small saving, and it is probably better to leave some load on the Feed Heater for operability reasons. The full revamp scheme is shown in Figure 9.38 with the corresponding network in Figure 9.40.

The improvement was worth £85,000 per year and the total installed cost for all the modifications was about £100,000 (all at 1982 prices). Around half of this cost is necessary for modifications to the dryer. Under present operation, the midpoint ΔT between steam and dryer air is 121°C, which becomes squeezed to 45°C in the revamp. Hence, a large amount of extra surface area is required, which in turn necessitates modifications to the drying tunnel and fans. An outline list of other modifications required is as follows:

(a) Change steam ejector.
(b) Control valve/pressure controller needed on the 0.6 bar steam system.
(c) Extra heat exchange area needed on Stage 3 Flash Heater and Feed Heater.
(d) Capacity of vacuum system needs increasing (due to increase in size of piping system under vacuum).
(e) Install 0.6 bar condensate system.
(f) Install new heat exchangers for cold water heating.
(g) Install let-down facility from utility steam system to 0.6 bar system.

With a Capital Grant available, the scheme achieved a payback of about 12 months.

9.4.7 Conclusions

The main conclusions to be drawn from this study are:

- The concepts and techniques of pinch analysis are not just for big plants in the bulk chemicals industry. They can yield surprising results on applications of much smaller scale.

- Process changes can yield even greater savings than development of heat exchanger networks (especially on small simple plants). The analysis method leads the user into a thorough understanding of his problem by the systematic application of thermodynamic principles.
- Heat pumping can be valuable, but both the type and placement of the heat pump should be selected based on the process heat loads as shown by the GCC.

9.5 Organic chemicals manufacturing site

9.5.1 Process description and targeting

This is a fairly simple study included as an example of the benefits of heat integration between separate processes. It was reported by Clayton (1988). Formation of the heat and mass balance and methods of stream data extraction were not discussed. All heat loads were reported in arbitrary heat units. The site concerned had three separate major plants, all with different pinch temperatures. There were relatively few opportunities for heat recovery within each individual plant; targets were at best 10% below current energy use. For example, the so-called TAF plant contained

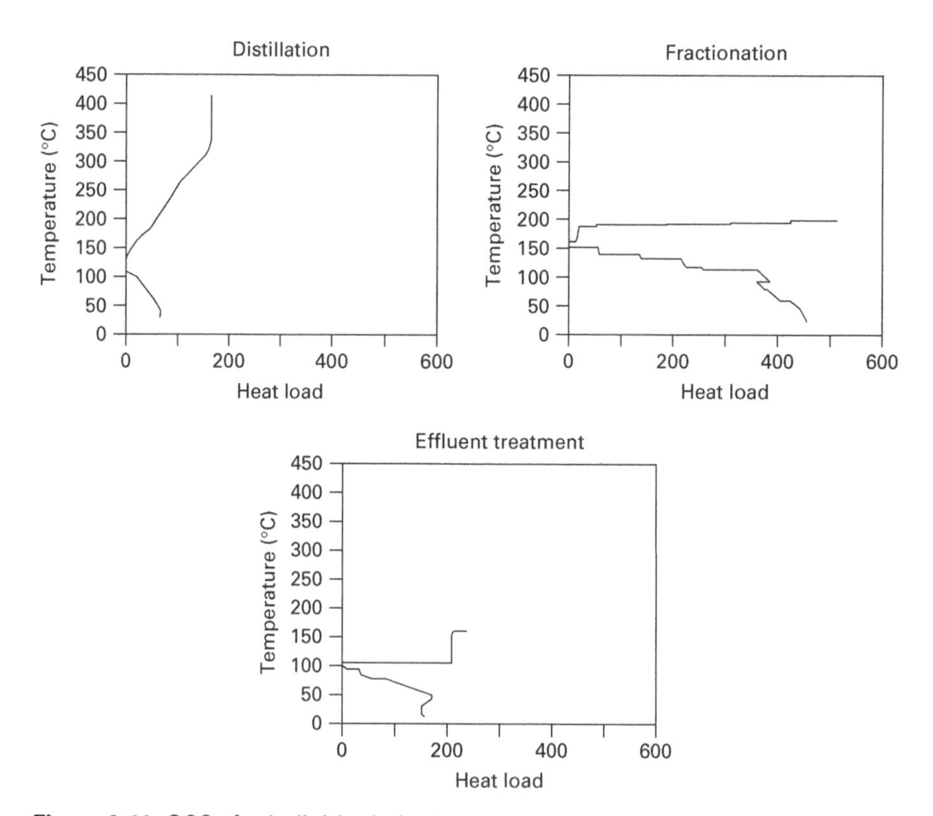

Figure 9.41 GCCs for individual plants

numerous distillation columns, all working across the pinch, but because of the heat sensitivity and high viscosity of the products handled there was no possibility of shifting the pressure and temperature of any of these columns sufficiently to allow one to heat another.

However, when the stream data set for the entire site was combined and the targeting process was repeated, a much larger saving was obtained – approximately 30%. This must be due to heat released from below the pinch in one plant being usable above the pinch of another. Inspection of the individual plant GCC's (Figure 9.41) shows that the TAF plant releases large quantities of heat in the 110–150°C range, and the ETP has a major heat demand just above 100°C. This can be checked by repeating the targeting for these two plants combined; the absolute saving is nearly as great as that for the overall site (268 heat units out of 295) and the percentage saving is even higher, 33%. The combined GCC for the two plants is plotted in Figure 9.42, while the GCC's are plotted back-to-back in Figure 9.43 (in effect, a split GCC). It is very clear from these plots how the heat recovery is achieved.

The current and target energy uses of the individual plants and the total site are given in Table 9.9.

9.5.2 Practical implementation

The potential saving is clear, but how can it be achieved? The TAP and ETP plants were a quarter of a mile apart, and piping streams from one location to the other to achieve direct heat exchange would have been completely uneconomic. Moreover, it would be undesirable to force the two plants to operate simultaneously at all times.

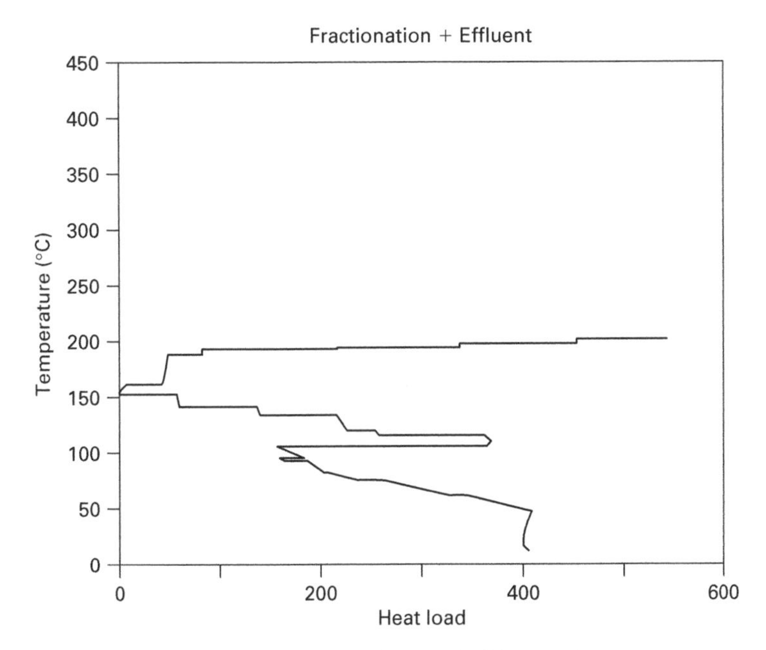

Figure 9.42 Combined GCC for fractionation and effluent plants

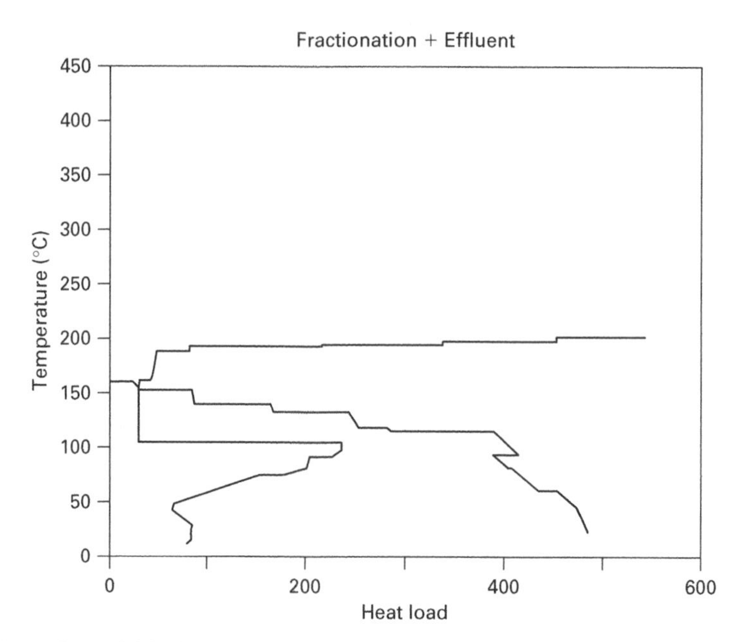

Figure 9.43 Split GCCs for fractionation and effluent plants

Table 9.9 Current and target energy use for organics site

Plant or process	*Current HU*	*Current CU*	*Target HU*	*Target CU*	*Pinch temperature (°C)*	*Heat saved*	
						Units	*(%)*
COD (Crude Oil Distillation)	187	87	167	67	110	20	10.7
TAF (Tar Acid Fractionation)	575	519	514	458	152	61	10.6
ETP (Effluent Treatment Plant)	238	158	238	158	98	0	0.0
Total site, processes not linked	1000	764	919	683	–	81	8.1
Total site, processes combined	1000	764	705	469	152	295	29.5
TAF and ETP plants linked	813	677	545	409	152	268	33.0

However, the heat demand of the ETP plant is due to very low pressure steam used for direct steam injection. Steam at this level could be raised from the condensers of the TAF plant and piped across the site using the existing low-pressure steam mains. In other words, the distillation columns could act as a supplementary boilerhouse! If the TAF plant was not working, the ETP plant could be supplied by site steam as before; if the ETP plant was shut down, the steam raised on the TAF plant could be used for other low-pressure needs (e.g. space heating) elsewhere on site. This is a classic example of total site analysis, using the site steam system to transfer heat (Section 5.4.3).

The main practical objection was whether the condensers of the distillation columns would operate properly. The driving force across them would fall from 120–150°C (heat exchange with cooling water) to 20–50°C (raising steam at 100°C or more). This corresponds to a very large increase in heat transfer surface area and a substantial capital investment. Nevertheless a payback period of 1.5 years was attainable.

9.5.3 Conclusions

- Zonal targeting is a valuable technique to explore the potential energy savings from linking two or more separate plants together – true process integration!
- We can use site steam mains to recover heat between plants in different locations, as an alternative to direct heat exchange with long pipe runs.

9.6 Hospital site

There are two features of particular interest in this case study (originally reported by Kemp 1990). Firstly, it shows pinch analysis being applied to a non-process plant – a group of buildings. Secondly, it illustrates the time-dependent analysis (as used for batch processes) in action.

9.6.1 Site description and stream data extraction

The main energy demands are for space heating (water-filled central heating systems), air systems (heated in winter, air-conditioned in summer), general hot water supplies and other miscellaneous uses, including the hospital laundry and incinerator. Heat loads change with occupancy and external conditions over a 24-h cycle.

The main block of the hospital is fully air-conditioned; ambient air is heated in winter or cooled in summer to give the desired temperature. Central heating is carried out by a hot water circulation system; the water is supplied at 80°C, returns at 71°C and is then re-heated in a gas-fired calorifier; the latter is therefore a cold stream. The central heating system is shut down during the summer. Domestic hot water must also be provided; daytime use is about five times that during the night. There are other areas of the complex, notably the kitchens, where daytime use of heat is much higher than night use.

Heat is supplied by site steam from a central boilerhouse about 1 km away; this boilerhouse also serves other buildings on the estate. Adjacent to the boilerhouse are the site laundry and an incinerator burning hospital waste, both of which operate during the day only. Figure 9.44 is a simple schematic plan of the site.

It can be seen that the stream data will vary considerably between night and day situations, and also from month to month. The two extreme cases of summer and winter conditions – January and July data – were studied separately. Day and night base cases were also used; but to supplement these, the day was divided up into time intervals like a batch process and the Time Slice Model (as described in Chapter 7) was used. Table 9.10 gives the stream data for the winter situation, when heat demand is highest.

Table 9.10 shows that there are considerably more cold streams than hot streams. Moreover, the heat available from the incinerator is not currently used. Current daytime energy use is 2,534 kW. The 110 kW of heat being recovered comes from a heat

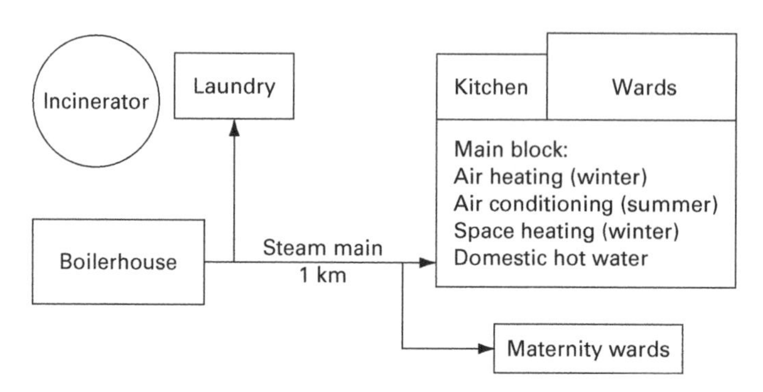

Figure 9.44 Sketch plan of hospital site

Table 9.10 Heat loads and times of operation for hospital site (winter)

Stream details and locations	*Times of operation*	*No. of streams*	*Hot streams total*		*Cold streams total*	
			Temperatures (°C)	*Heat load (kW)*	*Temperatures (°C)*	*Heat load (kW)*
Main air supply	24 h: Day	35	20–15	43	5–37	780
	Night	33	20–15	27	2–37	681
Space heating	24 h: Day	2			71–80	709
	Night	2			71–80	665
Domestic hot water	24 h: Day	1			5–60	244
	Night	1			5–60	48.5
Autoclaves	24 h	1			162	13
Kitchen	0530–0200	1			5–30	101
Laundry	0700–1630	3	37–30	67	18–26, 162	797
Incinerator	0900–1630	1	600–217	650		(a)
Boilers	24 h	1	235–60	(b)		
Summary						
Maximum load	0900–1630			760		2,644
Minimum load	0200–0530			27		1,407

Notes:
(a) The incinerators burn waste materials supplied at zero cost. This is a combustion reaction and therefore no cold stream requiring external heating exists.
(b) The heat recovery from boiler flue gas depends on the use of the boilers themselves, which in turn depends on the hot utility use. Therefore, the boiler flue gas is best treated as a utility stream rather than a process stream. Day period = (0600–1800), Night period = (1800–0600).

exchanger on the laundry drainwater and from heat recovery coils on a few of the air intake/extract pump units.

In some cases, notably the kitchen, the heat flows were large, random and highly intermittent. This gives a moderate average heat load. Heat recovery involving such streams would be very difficult because they would be present only occasionally and heat storage would be required within the cycle. Streams of this sort should be noted during data collection and heat recovery from them should be avoided if possible. It may be useful to repeat the targeting analysis leaving these streams out because it is unlikely that any cost-effective network will feature heat recovery from or to them.

In summer, the space heating loads disappear and the air heating loads fall to a small fraction of their winter values. Correspondingly, some air cooling is required by the air conditioning system, and this is provided through refrigeration. Currently 1,030 kW hot utility and 207 kW cold utility are required on a summer day. The dominant heat load is the steam required by the laundry.

9.6.2 Targeting using time intervals

The day is now divided into time intervals. The boundary times are 0200, 0530, 0600, 0700, 0900, 1630 and 1800. Targets for each time interval can then be calculated separately using the time slice model. Figures 9.45 and 9.46 show how the target

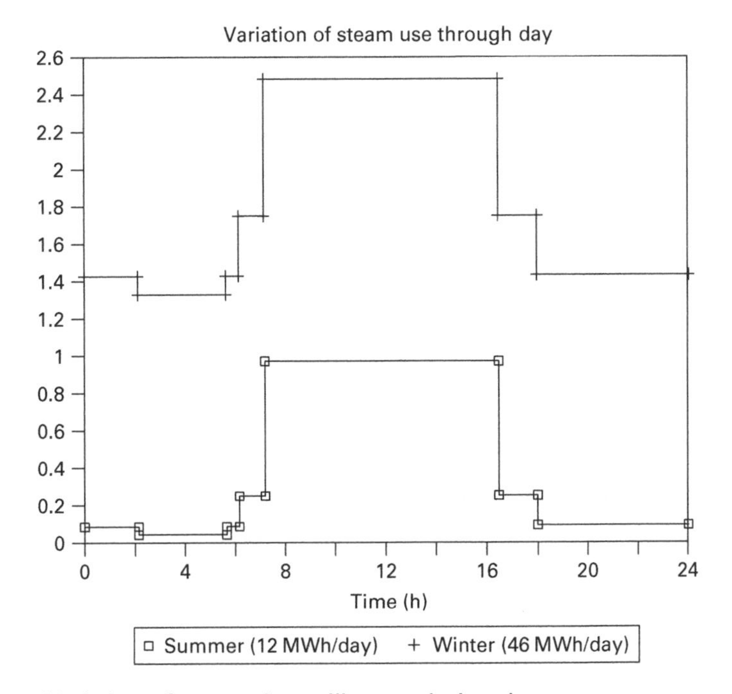

Figure 9.45 Variation of current hot utility use during day

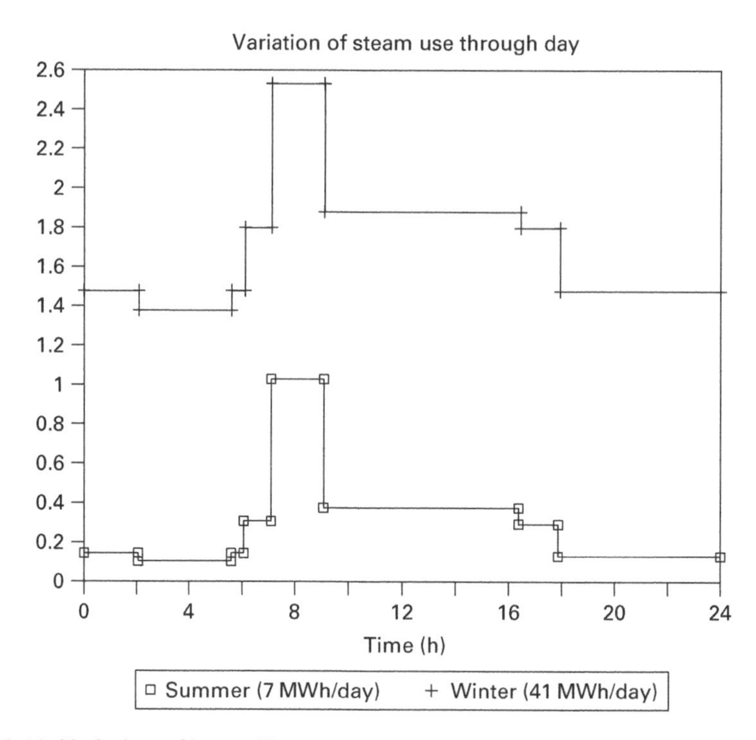

Figure 9.46 Variation of hot utility target over day

compares with the current use for both winter and summer cases. The values are identical except for the period when the incinerator is in operation, when 650 kW could be recovered from its exhaust.

The heat recovery targets are slightly higher during the day than the night, as would be expected. However, there is a sharp peak between 0700 and 0900 when the hot utility use reaches the current value (2,534 kW in winter). The presence of this transient high demand reduces the value of heat recovery, as an additional boiler would have to be started up specially to handle it.

Figure 9.47 shows the GCC at this peak demand time. The pinch comes at ambient temperature and the hot utility is much greater than the cold utility (which is in fact zero). Such a profile is typical of buildings.

9.6.3 Rescheduling possibilities

Rescheduling can be used not only to replace heat storage, but also to change the periods of high utility demand. There are two ways in which the transient could be removed:

(i) Move a cold stream currently in the 0700–0900 period to operate at a different time.

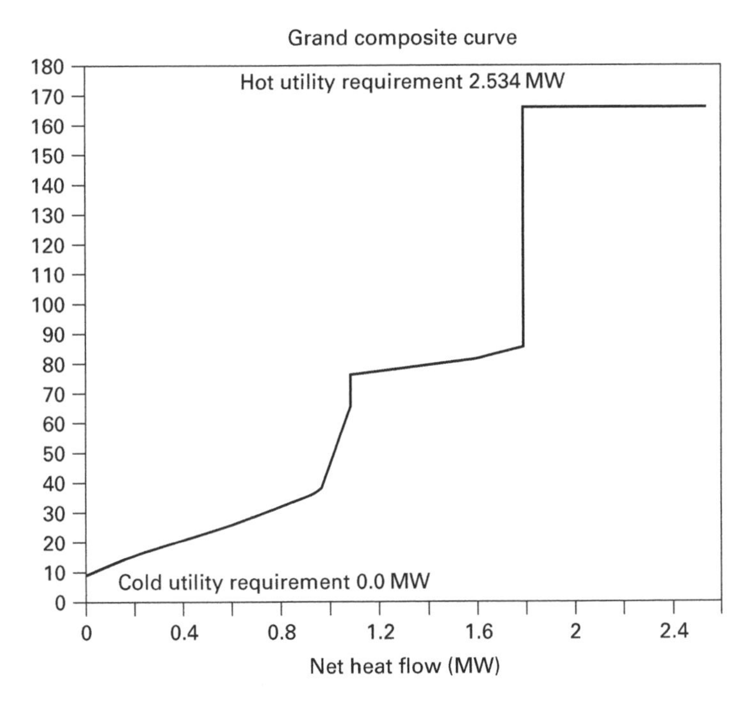

Figure 9.47 GCC for winter day (0700–0900 period)

(ii) Move a hot stream not currently in the 0700–0900 period to operate during that period.

Under (i), it is clearly impossible to shut down the central heating or air conditioning systems for 2 h, but the laundry operations could be scheduled to start 2 h later but still finish at 1630 (otherwise a corresponding transient appears after that time). However, the laundry is running at maximum capacity, so new machines would have to be installed at high capital cost.

Option (ii) could be achieved by starting the incinerator at 0700 and running it for 2 h longer – a Type 3 scheduling change. If the same amount of waste was burnt as before, the exhaust flowrate will fall and so will the heat recovery rate in kW. However, the total heat recovered over the day will be the same and the transient will be removed. In practical terms this means that one less boiler is needed. The steam demand for this case is shown in Figure 9.48.

The local authority also have the option of importing additional municipal waste and running the incinerator at full power for the whole of the period 0700–1630. This will increase the heat recovery summed over the day.

Since the existing incinerator was at the end of its useful life, the most cost-effective project was to purchase a new incinerator with an integral waste heat boiler. The marginal payback for installing the waste heat boiler with rescheduling was less than a year, even including the additional labour costs for the longer period of operation. The steam from the waste heat boiler could then be used for the laundry which was immediately adjacent.

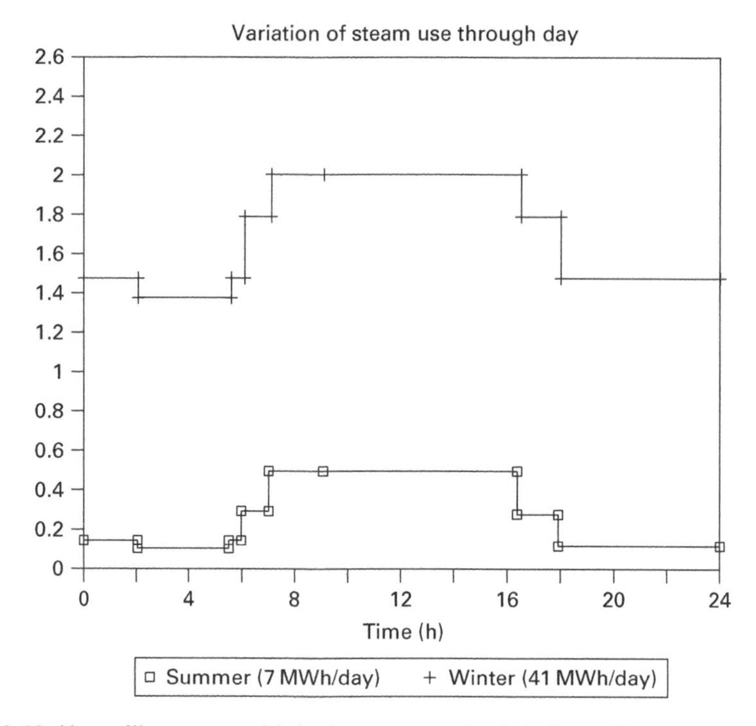

Figure 9.48 Hot utility target with incinerator rescheduled – variation during day

9.6.4 Process change possibilities

Figure 9.47 shows the GCC for the 0700–0900 time interval when the hot utility requirement is at its maximum, 2,534 kW. At this stage no rescheduling has been assumed.

The site has a threshold pinch at ambient temperature. This is usual for buildings as the requirements for space heating and other heat demands are rarely balanced by any significant hot streams. Therefore, if any additional hot streams can be introduced above the pinch, they will save energy. Two possibilities exist:

(i) Recover heat from more of the air exhausts.
(ii) Recover heat from drain water and warm effluent streams.

The first could be achieved by installing more heat recovery coils similar to those already present on a few air outlets. Each individual air inlet and outlet would require a separate exchanger, and there are 30 separate streams. The temperature driving forces are low. It was calculated that heat recovery coils were uneconomic to retrofit to the air system.

The drain water flow was intermittent and at variable temperatures which rarely rose above 20°C. Again there was no way to justify heat recovery from these streams in economic terms.

9.6.5 Opportunities for combined heat and power

The GCC, Figure 9.47, shows that heat is required over four different temperature ranges; 160–170°C for steam raising for the laundry and autoclaves, 80–85°C for space heating water circuits, 10–70°C for domestic hot water and 10–40°C for heating air supplies. Of the three main types of combined heat and power (CHP) system, diesel (or gas) reciprocating engines have several advantages where the pinch is near ambient temperature, as for most buildings including this hospital:

1. Because the pinch is at ambient temperature, the heat from the jacket cooling water can be recovered as well as that from the exhaust gases, more than doubling the recoverable heat.
2. The jacket water is at 85°C, ideal for supplying the space heating system.
3. The heat-to-power ratio of buildings is generally less than for process plants. The diesel engine has a lower heat-to-power ratio than a gas turbine or steam turbine and thus matches the site requirements better.
4. Diesel (or gas) engines can be obtained in smaller sizes than gas or steam turbines and thus give more flexibility where the heat and power loads are relatively low.

A further advantage on this particular site was that four standby generators of various sizes were already installed in the main block to provide 740 kW(e) emergency power. Conversion of these to provide heat as well as power was considerably cheaper than installing a new diesel engine, with a payback of under 2 years, and satisfied about half the site heat demand. It was important to ensure that they were in sufficiently good condition to work reliably and that maintenance would be sufficient. In fact, regular use improves the likelihood that they will successfully generate power in an emergency; it particularly benefits the alternators and electrical system.

Installing an additional 800 kW(e) diesel engine would provide all the residual site heat and power requirements. The marginal payback on this unit is 4.3 years; the overall payback would then rise to just over 3 years but the main block would be self-sufficient in heat, and the long steam mains with their large associated heat losses would not be required. The laundry steam needs could be met from the steam raised from the incinerator exhaust. A 3-year payback represents an excellent long-term investment on a building which has a probable life of 20 years or more. The biggest perceived drawback of the scheme was the additional noise and vibration which would be caused by having the generators running continuously instead of in emergencies only.

An alternative to the diesel engine schemes was to install a 1.5 MW gas turbine (the smallest proven unit commercially available at the time). This could supply all the power needs of the site, and all the heat requirements except on the coldest winter days. The fuel would have been interruptible natural gas, but if the interruptible gas supply were not available, gas oil would have to be burnt at much higher cost. The other disadvantage is that if the gas turbines were shut down for maintenance, all the site heat and power needs would have to be supplied from boilers and imported electricity. Multiple dual-fuel engines are a more flexible option.

The payback for CHP schemes on this and other buildings is generally longer than for process plants, because the base heat load for space heating disappears in summer. Since electricity is generally cheaper in summer, it is usually best to shutdown the CHP plant and import electricity. Therefore, savings are only made over 6–9 months of the year.

9.6.6 Conclusions

Process integration techniques are not confined to process plants; they are of value in many other situations. On this site, it had been assumed that there was little scope for energy saving and that the only possibility would be heat exchange between exhaust air and fresh air streams. Process integration shows that, in fact, this is a relatively unimportant and uneconomic option. The Cascade analysis gave the following information:

- targets for energy use in winter and summer
- variation of utility use throughout the day and night
- identification of the incinerator heat recovery project
- identification of the change in operating times to smooth boiler demand
- evaluation of diesel engine and gas turbine CHP schemes.

For buildings in general, as with this site, the pinch will be at ambient temperature because of the space heating loads and other heating requirements (e.g. hot water production). Any substantial heat source at medium or high temperatures (such as the incinerator here) is potentially valuable for heat recovery, and CHP schemes will be highly promising, especially as longer payback times are generally acceptable than on process plants.

Table 9.11 summarises the cost data for the various projects identified (1990 prices). It can be seen that;

1. High-temperature heat recovery (from the incinerator) is much more cost-effective than low-temperature (from air exhausts, drain water, etc.)

Table 9.11 Economic analysis of projects for hospital site

Project description	*Savings (£ per annum)*	*Capital cost(£)*	*Payback time (years)*
Incinerator heat recovery			
Basic	14,000	15,000	1.1
With rescheduling	20,000	15,000	0.8
Rescheduling and waste import	24,000	15,000	0.6
Other heat recovery projects	33,000	165,000	5.0
CHP from diesel engines			
Standby generator conversion	87,000	150,000	1.7
New engine	100,000	430,000	4.3
Total for diesel CHP projects	187,000	580,000	3.1
CHP from gas turbine	230,000	900,000	3.9

2. Rescheduling the incinerator operations gives savings by eliminating start-up losses associated with running a boiler for "peak-lopping". These outweigh the additional labour costs. Additional benefits come from reduced boiler use and depreciation. Importing municipal waste can increase savings further.
3. The CHP systems give bigger savings than heat recovery but at longer paybacks. They can be viewed as a good long-term investment. The economics are made less favourable in buildings such as this because the summer heat load is much lower than the winter one.
4. Conversion of standby generators to CHP is highly promising. A significant proportion of these savings comes from eliminating distribution losses in the long steam mains.
5. The marginal payback on a new diesel engine is longer than that for standby generator conversion. The payback on similarly sized diesel-based and gas turbine-based CHP schemes was comparable; on this site the former was preferable for reasons of flexibility.

This case study highlights the value of heat recovery from incinerators, of standby generator conversion, and of using localised heating (including CHP) rather than a centralised system with long steam mains. It is noteworthy that in the last 20 years, use of gas engines to provide space heating and power simultaneously to buildings, in the way described above, has become increasingly common. Another good example is described in Section 5.6.2.

9.7 Conclusions

The five case studies above have shown that pinch analysis can be useful in a very wide range of situations. The techniques are not confined to large continuous processes (or even to processes at all). Savings do not just come from conventional heat recovery projects, but also from process improvements highlighted by the analysis. Payback times vary considerably, but projects can be selected both as short-term moneysavers or long-term investments – and remember that as energy is a direct cost, any savings made appear directly as a one-to-one increase in gross margin and profit, so a £1 energy saving can be worth £5 of extra sales with their associated costs. Close collaboration with plant managers and operators is an essential part of a study; they provide the initial data, give valuable insights into practicalities, and are the people who will actually run the final plant to achieve the projected savings. In many cases the insights gained by plant personnel are one of the most valuable, though intangible, benefits from the analysis.

The studies also give many practical examples of how to implement pinch analysis – particularly in the tricky areas such as stream data extraction which are more difficult to systematise. We hope that you, the engineer, will be stimulated by these examples to go out and try the techniques on your own plant, process or site. You might be surprised at some of the things you discover!

Exercises

E9.1 For the crude oil distillation unit (Section 9.2), develop a range of lower energy networks using the retrofit technique, starting with the base case and trying to eliminate pinch violators. What conclusions do you reach on this methodology in this case, compared with starting from an MER network?

E9.2 For the aromatics plant (Section 9.3),

2A. Network relaxation

By a formal process of loops and paths, evolve the network of Figures 9.22 and 9.23 with stream splits to the Phase III network of Figure 9.25 with no stream splits.

2B. Stream data modification

By studying the mixing junction between D1 bottoms and stream 7 to produce stream 4, find the temperature of the bottoms flow from D1, assuming specific heat capacity Cp is the same for all three streams. Modify the stream data to remove the mixing junction. What effect does this have on the energy targets?

2C. Zonal (structural) targeting

a. Perform zonal targeting for the two separate sections (streams 1–3 and 4–11), calculate the new targets and pinch temperatures, and examine the new composite curves and GCC. What are the implications for network design?
b. Construct an MER network for Zone 2 only (streams 4–11).
c. Relax this network to generate a range of alternative designs with varying levels of capital expenditure. Compare with the networks generated from the full-plant MER design and retrofit starting from the existing network.

References

Ahmad, S. and Linnhoff, B. (1989). Supertargeting: Different process structures for different economics. Transactions of ASME, *J Energy Res Technol*, 111(3): 131–136.

Amidpour, M. and Polley, G. T. (1997). Application of problem decomposition to process integration, *TransIChemE*, 75(Part A): 53–63, January 1997.

Clayton, R. W. (1988). Cost reductions on a bulk chemicals plant identified by a process integration study at Coalite Fuels and Chemicals Ltd. ETSU R&D Report RD/34/37, Energy Efficiency Office.

EEDS Profile 204 (1988). ETSU (Energy Technology Support Unit), Harwell, Energy Efficiency Office, UK Department of Energy.

Kemp, I. C. (1990). Applications of the time-dependent cascade analysis in process integration. *J Heat Recovery Sys CHP*, 10(4): 423–436.

Linnhoff, B. Townsend, D. W., Boland, D., Hewitt, G. F., Thomas, B. E. A., Guy, A. R. and Marsland, R. H. (1982). A User Guide on Process Integration for the Efficient Use of Energy, 1st edition. IChemE, Rugby.

10 Conclusions

Pinch analysis is now a mature subject. It has developed from the early work on targeting and heat exchanger network design to cover a wide range of aspects of process design, particularly those related to energy usage. Many new techniques have been developed and the methodology has become more complex; nevertheless, at its core are still the fundamental insights given by energy targets and the pinch concept.

Many points have been made and conclusions drawn throughout the book, but the key findings can be summed up as follows:

(a) Pinch analysis can be applied effectively to almost any plant, process or site, large or small.
(b) An accurate heat and mass balance is essential.
(c) Stream data extraction remains the most difficult task and in some situations there is no clear best choice between two or three different methods.
(d) The energy targets, the problem table and the identification of the pinch are probably the most valuable results from the analysis.
(e) A good approximation to the optimum ΔT_{min} can generally be obtained without using area and capital cost targeting methods: where the latter are used, only a rough estimate is often needed.
(f) Possible process changes should always be looked for at the beginning of targeting, and can save more money than heat recovery.
(g) Knowledge of the pinch is always useful, but there are situations where the pinch design method may not be the most appropriate way to design a heat exchanger network.
(h) A simple user-friendly Excel spreadsheet is available to perform the key targeting calculations.
(i) The process engineer is left in control; computer calculations are useful tools, not substitutes for the engineer's insights and expertise.

Because pinch analysis gives useful insights on almost any process, it could be said that every process engineer should calculate the pinch and targets on his plant, just as he would do a heat and mass balance, especially now that simple software is freely available.

Ultimately, process understanding is the key to safe, efficient and effective design and operation. If this book has helped you understand your plant, process or site better, it has succeeded in its task.

Happy analysing!

Notation

List of Symbols and Units*

A	Heat transfer area (m^2)
C_P	Specific heat capacity (kJ/kg K)
COP_p	Coefficient of performance for a heat pump (-)
COP_r	Coefficient of performance for a refrigerator (-)
CP	Heat capacity flowrate (kW/K)
H	Flow enthalpy (kW)
ΔH	Change in flow enthalpy (kW)
ΔH_{com}	Heat of combustion (kJ/kg)
h	Specific enthalpy (kJ/kg)
h	Film heat transfer coefficient of an individual stream (kW/m^2K)
k, K	Number of temperature intervals or segments (-)
L	Number of loops in a network (-)
m, $\dot{m}$	Mass flowrate (kg/s)
N	Number of process streams plus utilities, or process stream branches (-)
Q	Heat flow (kW)
Q_{Hmin}	Minimum feasible hot utility (kW)
Q_{Cmin}	Minimum feasible cold utility (kW)
q	Heat flow of an individual stream (kW)
S	Entropy (kJ/kg K)
s	Number of separate components (subsets) in a network (-)
S	Shifted temperature (°C or K)
S_S	Shifted supply temperature of process stream (°C or K)
S_T	Shifted target temperature of process stream (°C or K)
T	Temperature (°C or K)
T_S	Supply temperature of process stream (°C or K)
T_T	Target temperature of process stream (°C or K)
ΔT	Temperature difference (K)
ΔT_{cont}	ΔT_{min} contribution of an individual stream (K)
ΔT_{min}	Minimum allowed temperature difference (K)
$\Delta T_{threshold}$	Boundary value of ΔT_{min} between a threshold and pinched problem (K)
ΔT_{LM}	Log mean temperature difference (K)
t	Time (s)
U	Overall heat transfer coefficient (kW/m^2K)
u	Number of heat exchange units (i.e. heaters, coolers, exchangers) (-)

* SI units or common multiples (e.g. kW) are used in this list. Compatible units in other systems are listed in Section 8.4.5. Dimensionless units are denoted by a hyphen -.

u_{min}	Minimum number of units (-)
W	Shaft work for heat engine or heat pump (kW)
w	Work per unit mass flow (kJ/kg)
X	Heat load shifted around a loop or along a path (kW)
α	Heat flow across the pinch (kW)
η	Heat engine efficiency (-)
η_c	Reversible (Carnot) heat engine efficiency (-)
η_{mech}	Mechanical efficiency (-)

Subscripts

C, COLD	Relating to cold stream
H, HOT	Relating to hot stream
MER	At maximum energy recovery or minimum energy requirement
1, 2, ... A. B, ... i, n,	counters

Glossary of terms

Appropriate placement Positioning of utilities, heat engines, heat pumps or an extracted process (e.g. separation system) above or below the pinch and grand composite curve for best overall energy performance.
Background process The stream data for the remainder of the process after the extracted streams have been removed.
Balanced composite curves Composite curves including the hot and cold utility streams.
Balanced grand composite curves Grand composite curves including the hot and cold utility streams.
Balanced grid Network grid diagram including the hot and cold utility streams.
Cascade Set of heat flows through a heat recovery problem, in strict descending temperature order (as calculated in Problem Table analysis – see **Problem Table**).
Cascade analysis The method of batch process analysis based on breaking the process into time intervals and developing time-dependent heat cascades.
Cold stream Process stream requiring heating.
Composite curve Combined temperature-enthalpy plot of all hot or cold streams in a problem.
***CP*-Table** Tabulated values of stream heat capacity flowrates, immediately above or below the pinch.
Cycle time The total duration of a batch.
Cyclic matching Repeated matching of pairs of process streams.
Data extraction Definition of data for energy integration studies, from a given flowsheet.
Debottlenecking Increasing the production capacity of a plant by identifying and removing rate-limiting steps, such as slow processing stages or heavily occupied equipment items.
Direct heat exchange Heat exchanged between two streams in the same time interval of a batch process.
Direct contact heat transfer Heat exchanged by two streams which mix directly (e.g. steam injection).
Energy relaxation Process of reducing energy recovery in a heat exchanger network for the purpose of design simplification.
Extracted streams or **extracted process** A set of streams removed from the process stream data to test them for appropriate placement.
Feasible cascade Heat cascade in which net heat flow never becomes negative and is zero at the pinch.
Flowing stream A stream which receives or releases heat as it flows through a heat exchanger.
Gantt chart A representation of which streams exist in given time intervals of a batch process, also called a **time event chart**.

Grand composite curve (GCC) Plot of heat flow vs. temperature from a heat cascade (see **Cascade** and **Problem Table**).

Grid System of horizontal and vertical lines with nodes, for representing heat exchanger networks.

Heat cascade A table of the net heat flow from high to low temperatures divided up into temperature intervals.

Heat engine System converting high-grade heat to lower-grade heat and producing power.

Heat exchanger network (HEN) System of utility heaters and coolers and process interchangers.

Heat pump System upgrading heat from a lower to a higher temperature using power or high-grade heat.

Heat storage Heat recovery by taking heat out of one time interval in a batch or time-dependent process and supplying it to a later time interval.

Hot stream Process stream requiring cooling.

Individual heat cascades Heat cascades for a time interval considered in isolation from all other time intervals.

Infeasible cascade Heat cascade with zero hot utility and some negative values of net heat flow.

***In-situ* heating/cooling** A stream which is heated or cooled in a vessel over a period of time.

Intermediate condenser An additional condenser in a column working above the main condenser temperature.

Intermediate reboiler An additional evaporation stage in a column working above the main reboiler temperature.

Interval temperature Obsolete name for **shifted temperature**.

Loop System of connections in a heat exchanger network which form a closed pathway.

Maximum energy recovery (MER) Best possible energy recovery in a heat exchanger network for a given value of ΔT_{min}; also known as minimum energy requirement.

Maximum heat exchange (MHX) The maximum amount of heat which can be recovered by direct heat exchange in a batch process.

Maximum heat recovery (MHR) The maximum amount of heat which can be recovered for a batch process at given process conditions by direct heat exchange and heat storage added together.

MHR or MHX network A heat exchanger network achieving the MHR or MHX target.

Multiple utilities Utility or utility system whose temperature or temperature range falls within the temperature range of the process stream data.

Near-pinch Point in a heat cascade where net heat flow is very small but increases at temperatures on either side.

Network optimisation Evolution of a heat exchanger network to give most convenient heat exchanger sizes, allowing for existing area.

Network pinch Point in heat exchanger network where temperature driving force is lowest.

Overall heat cascade A time-dependent heat cascade for a batch process which includes the effects of heat storage.
Path System of connections in a heat exchanger network forming a continuous pathway between the utility heater and a utility cooler.
Pinch Point of zero heat flow in a cascade (alternatively, point of closest approach of composite curves in a "heating and cooling" problem).
Pinch design method Method of heat exchanger network design which exploits the constraints inherent at the pinch.
Pinch match Process interchanger which brings a stream to its pinch temperature (i.e. hot streams above the pinch, cold streams below).
Pinch region Range of temperatures over which cascade net heat flow is zero (or very low).
Pocket Region in the grand composite curve where neither external heating nor cooling is required.
Problem Table System of analysing process stream data for a heat recovery problem which exploits **temperature interval** sectioning of the problem, and predicts minimum utilities consumptions, pinch location, and cascade heat flows.
Process change Altering the stream data by changing the temperature and/or heat load of one or more streams.
Process sink profile Section of the grand composite curve above pinch temperature.
Process source profile Section of the grand composite curve below pinch temperature.
Profile Temperature-enthalpy plot of a stream or a composite stream.
Pumparound Liquid drawn from a distillation column which releases sensible heat and is returned to the column.
Rescheduling Altering the time period during which a stream exists.
Retrofit or **Revamp** Any change to an existing chemical process, but in this context, mostly changes for improvement in energy efficiency.
Shifted composite curves Plots of combined enthalpy of all hot and all cold streams against shifted temperature, touching at the pinch.
Shifted temperature Stream temperatures altered to include the effect of the required ΔT_{min}, usually by reducing hot stream temperatures by $\Delta T_{min}/2$ and increasing cold stream temperatures by $\Delta T_{min}/2$.
Site sink profile Plot of heat required by all processes on a site at given temperatures.
Site source profile Plot of heat released by all processes on a site at given temperatures.
Split grand composite curve Plot of the grand composite curve for the background process and the extracted streams on the same graph.
Stream splitting Division of a process stream into two or more parallel branches.
Subset Set of process streams or process streams, plus utilities, within a heat recovery problem which are in overall enthalpy balance.
Supply temperature Temperature at which a process stream enters a heat recovery problem.
Target A design performance limit, determined prior to design.

Target temperature Temperature at which a process stream leaves a heat recovery problem.

Temperature interval Section of a heat recovery problem between two temperatures which contains a fixed stream population.

Threshold problem Heat recovery problem that shows the characteristic of requiring either only hot or only cold utility, over a range of ΔT_{min} values from zero up to a threshold (or throughout).

Tick-off rule Heuristic of maximising the heat load on an interchanger by completely satisfying the heat load on one stream.

Time average model (TAM) Averaging heat flows by dividing the total heat load over the batch period by the total batch cycle time.

Time-dependent heat cascade A set of heat cascades for different time intervals, forming a matrix.

Time event chart A Gantt chart, plotting the time periods when different streams exist.

Time interval A period of time during which stream conditions do not change appreciably and for which a target can be obtained.

Time slice model (TSM) Division of a batch problem into time intervals and finding the targets for the individual cascades, with zero heat storage.

Top level analysis Study of a site's heat and power needs using existing utility consumption of plants, rather than targets.

***U.A.* analysis** Procedure of calculating *UA* values ($=Q/\Delta T_{LM}$) for matches in a heat exchanger network, for the purposes of preliminary costing and optimisation.

Utility System of process heating or process cooling.

Unit Process interchanger, heater or cooler.

ΔT_{min} Minimum temperature difference allowed in the process between hot and cold streams.

ΔT_{min} contribution (ΔT_{cont}) Temperature difference value assigned to individual process streams. Match-dependent ΔT_{min} values are given by the sum of the contributions in a match.

Further reading

Specific references have been included in each chapter. However, there are a number of general books and publications which are particularly useful.

ESDU International (1987–1990). ESDU (Engineering Sciences Data Unit) Data Items 87001, 89030 and 90017 on Process Integration. Available by subscription from ESDU International plc, London, UK.

Gundersen, T. (2000). A Process Integration Primer – IEA (International Energy Agency), Implementing agreement on process integration.

Karp, A., Smith, R. and Ahmad, S. (1990). Pinch Technology: A Primer. EPRI Report CU-6775, Electric Power Research Institute, Palo Alto, California.

Rossiter, A. P. (1995). *Waste Minimisation through Process Design*. McGraw-Hill, New York, USA. ISBN 007053957X.

Shenoy, U. V. (1995). *Heat Exchanger Network Synthesis; Process Optimisation by Energy and Resource Analysis*. Gulf Publishing Co, Houston, Texas, USA. ISBN 0881453196.

Smith, Robin (2005). *Chemical Process Design and Integration*. John Wiley and Sons, Chichester, UK. ISBN 0-471-48680-9 (hardback)/0-471-48681-7 (paperback). (Previous edition published by McGraw-Hill, New York, USA, 1995.)

Also:

UMIST Process Integration Research Consortium reports and publications. Available only by corporate subscription to members of the consortium.

Training courses in Process Integration, University of Manchester (previously UMIST, University of Manchester Institute of Science and Technology).

Appendix – using the spreadsheet software

The spreadsheet for pinch calculations may be downloaded from "http://books.elsevier.com/companions/0750682604".

Entering the data

When the spreadsheet is opened, it takes you directly to the INPUT tab, where the data is entered.

On the INPUT tab, enter the global ΔT_{min} chosen for the problem in cell F6, followed by the stream data. Note that the data must be in compatible units, as described in Section 8.4.5. The default spreadsheet display assumes that SI-based metric units are in use, but you can use your own units and alter the cell headings. You can also choose whether to enter the data as specific heat capacity and mass flowrate, heat capacity flowrate (*CP*) or heat flow.

Enter the stream data, line by line. As you begin each line, cells requiring input are highlighted in yellow. Streams can be numbered, named or both. If a stream changes heat capacity significantly over the temperature range, enter each segment separately (e.g. as 3a, 3b, 3c, etc.). Enter latent heat streams by assigning them a small temperature difference, say 0.1°C.

Results output

As data is entered, the spreadsheet automatically calculates the full stream data, which can be seen on the right-hand side of the first tab and on other "results" tabs. The program calculates whether a stream is hot or cold depending on the difference between the supply and target temperatures. All heat capacity flowrates (*CP*) should be positive, never negative.

The furthest left-hand tab, INDEX, gives a list of all the results tabs, with hyperlinks. The results tabs are as follows:

TARGETS – Problem table, energy targets, pinch temperature and type of problem (pinch, threshold, multiple pinch or pinch region)
CC – Hot and cold composite curves
SCC – Shifted composite curves
GCC – Grand composite curve
GRID – Network grid diagram, shifted temperatures
AS – Stream data plot, actual temperatures
SS – Stream data plot, shifted temperatures
AT – Interval tables (heat loads and temperatures), actual temperatures
ST – Interval tables (heat loads and temperatures), shifted temperatures
DTMIN – Variation of hot and cold utility targets and pinch temperature with ΔT_{min}.

The last of these calculations, for varying ΔT_{min} over a range, uses a macro and will not work if your spreadsheet security levels are set to High; therefore, on the spreadsheet menu go to Tools | Macro | Security and set to Medium (recommended) or Low. To obtain the plots, enter the maximum and minimum value of the required range and the program will then calculate over the range in 20 equal steps of ΔT_{min}.

Further program details may be found on the website.

Index

Lightning Source UK Ltd.
Milton Keynes UK
UKOW06n2040100816

280395UK00001B/34/P

9 780750 682602